PHYSICS OF SPACE: GROWTH POINTS AND PROBLEMS

PHYSICS OF SPACE: GROWTH POINTS AND PROBLEMS

Proceedings of the second 'Rencontres de l'Observatoire',
Observatoire de Paris, Meudon, France

Edited by

NICOLE MEYER-VERNET, MICHEL MONCUQUET
and FILIPPO PANTELLINI
Observatoire de Paris, Meudon, France

Reprinted from *Astrophysics and Space Science*
Volume 277, Nos. 1–2, 2001

SPRINGER SCIENCE+BUSINESS MEDIA, LLC

Library of Congress Cataloging-in-Publication Data

Physics of space : growth points and problems / edited by Nicole Meyer-Vernet, Michel Moncuquet, and Filippo Pantellini.
p.cm
ISBN 978-94-010-3813-3 ISBN 978-94-010-0904-1 (eBook)
DOI 10.1007/978-94-010-0904-1
1. Cosmic physics- -Congresses. I. Meyer-Vernet, Nicole. II. Moncuquet, Michel. III. Pantellini, Filippo.

QC801 .P48 2001
523.01- -dc21 00-066765

Cover illustration taken from 'Un autre monde',
by Grandville (edited by H. Fournier, Paris, 1844).

Printed on acid-free paper

TABLE OF CONTENTS

Jean-Louis Steinberg

PRÉFACE

Les deuxièmes "Rencontres de l'Observatoire", qui ont eu lieu à l'Observatoire de Paris à Meudon du 10 au 14 Janvier 2000, ont réuni autour du thème "Problèmes émergents en physique de l'espace" 120 physiciens et astrophysiciens venus d'une vingtaine de pays différents.

Nous avons voulu honorer à cette occasion Jean-Louis Steinberg pour ses contributions majeures à la recherche spatiale, à la radioastronomie et à la physique de l'espace. L'approche explicitement pluridisciplinaire de ce colloque, qui ne s'est pas laissé confiner dans les limites étroites de la physique spatiale ni dans celles imposées par certains programmes officiels, suit l'esprit de sa carrière scientifique: sortir des limites des sujets déjà étudiés ou sur le point de l'être, et appliquer les connaissances acquises pour explorer de nouveaux domaines.

Ce dernier quart de siècle a vu une croissance vertigineuse des performances spatiales. La technologie moderne ne permet pas encore de jongler avec les univers comme le prestidigitateur de Grandville (Grandville, Un autre monde, ed. H. Fournier, Paris, 1844); mais quelques décades ont suffi pour voir des instruments sophistiqués explorer les frontières du système solaire, et la communauté de la recherche spatiale a dépassé rapidement le sujet étroit de l'environnement soleil-terre pour s'intéresser à l'ensemble de l'héliosphère, où les mêmes processus physiques sont à l'œuvre. L'héliosphère n'est qu'un environnement cosmique parmi d'autres, mais c'est le seul accessible aux mesures *in situ* et il joue à ce titre un rôle fondamental comme laboratoire naturel de physique et d'astrophysique, élargissant ainsi le domaine de la physique de l'espace à l'ensemble des environnements cosmiques régis par des processus similaires.

Dans ce domaine élargi et malgré le temps limité disponible, cette conférence se proposait de dégager les idées qui ont modifié notre vision de la physique de l'espace, en insistant sur les ponts entre la physique des plasmas et les autres disciplines, et de mettre en évidence quelques sujets émergents et plusieurs problèmes fondamentaux encore non résolus. Parmi ceux-ci, la physique des plasmas sans collisions, notamment l'accélération des particules et la dissipation, joue un rôle important en astrophysique. C'est aussi le cas de la physique des vents et des jets: presque tous les objets cosmiques éjectent de la matière, alors que même le vent stellaire le plus proche – celui qui vient du soleil – reste incompris quatre décades après sa première détection *in situ*. La conférence a également abordé le domaine des plasmas poussiéreux qui sort à peine de l'enfance malgré leur ubiquité en astrophysique. L'étude des environnements planétaires et cométaires s'est développée considérablement ces dernières années sous l'impulsion des nouvelles données; pourtant nous sommes loin d'avoir compris le fonctionnement des magnétosphères – même la plus proche d'entre elles, et il était intéressant de confronter ces études avec le puzzle des magnétosphères de pulsars. Plusieurs nouvelles techniques de détection dans l'espace sont en train d'émerger et certaines sont évoquées ici. Fi-

Astrophysics and Space Science **277:** ix–xi, 2001.

nalement, les étapes parcourues ainsi que quelques futurs possibles ont été abordés par des pionniers de la discipline.

Le succès de cette conférence est largement dû à l'enthousiasme des participants et aux efforts des conférenciers pour faire partager leur domaine à leurs collègues de disciplines différentes. Il doit beaucoup aux suggestions des membres du Comité International et du Comité de Programme Scientifique pour le choix des thèmes et des conférenciers. Ces Rencontres ont bénéficié d'un soutien majeur financier et logistique de l'Observatoire de Paris dont nous remercions le Président Pierre Couturier. La conférence a également reçu le concours financier de l'Agence Spatiale Européenne, de la Station de Radioastronomie de Nançay, du Centre National d'Etudes Spatiales, des Universités de Paris 7 et Paris 11, du Centre National de la Recherche Scientifique et du Ministère de l'Education Nationale, de la Recherche et de la Technologie.

L'organisation matérielle a été prise en charge par l'Observatoire de Paris et le Département de Recherche Spatiale, et nous remercions Marie-Rose Rama-Rao secrétaire du colloque, Christiane Adam secrétaire du conseil scientifique, Jacqueline Thouvay administrateur du DESPA, ainsi que les membres locaux du Comité Scientifique, qui ont consacré beaucoup de temps et d'énergie à la préparation de ces Rencontres. Nous remercions également Agnès Fave, Roger Hulin, Baptiste Cecconi, Julien Queinnec et Jean-Jacques Poisot pour leurs contributions efficaces à l'organisation pratique, ainsi que Sylviane Chaintreuil, Christine Denis et de nombreux collègues de l'Observatoire qui ont facilité le déroulement de la conférence.

La qualité scientifique de ce volume doit beaucoup aux lecteurs qui ont accepté d'évaluer les contributions que nous leur avons soumises dans des délais généralement brefs, et nous remercions particulièrement Kaspar Arzner, Stuart Bale, Abdallah Barakat, Gérard Belmont, David Burgess, Michael Combi, Stanley Cowley, Dominique Delcourt, John Dorelli, R. Paul Drake, Dan Dubin, Joe Giacalone, Nat Gopalswamy, Roland Grappin, Ove Havnes, Phil Isenberg, Karine Issautier, Randy Jokipii, Michael Kaiser, Jan Kuijpers, Marty Lee, Emmanuel Lellouch, Bertrand Lembège, Robert MacDowall, Milan Maksimovic, Eckart Marsch, Fritz Neubauer, Eugene Parker, Christophe Sauty, Darrell Strobel, Adam Szabo, Marco Velli, Rudolf Von Steiger, Frank Verheest, Rudolf Wegmann, Philippe Zarka, Gaetano Zimbardo, ainsi que plusieurs autres collègues ayant évalué les articles.

Nous sommes particulièrement heureux de remercier Eugene Parker qui nous a fait l'honneur d'introduire ce volume.

Les éditeurs:
Nicole Meyer-Vernet
Michel Moncuquet
Filippo Pantellini

Comité International

Michel Blanc (France), Roger-Marie Bonnet (E.S.A.), Joseph Burns (U.S.A.), Stanley Cowley (U.K.), Vitaly Ginzburg (Russie), Joseph Lemaire (Belgique), Dominique Lequéau (France), Hiroshi Matsumoto (Japon), Marcia Neugebauer (U.S.A.), Eugene Parker (U.S.A.), René Pellat (France), Christopher Russell (U.S.A.), Evry Schatzman (France), David Southwood (E.S.A.), Robert Stone (U.S.A.), Vadim Tsytovich (Russie), James Van Allen (U.S.A.).

Comité de Programme Scientifique

Nicole Meyer-Vernet (France), Fran Bagenal (U.S.A.), Dominique Bockelée-Morvan (France), Jean-Louis Bougeret (France), David Burgess (U.K.), Tom Cravens (U.S.A.), Sang Hoang (France), Mihaly Horanyi (U.S.A.), Karine Issautier (France), Yolande Leblanc (France), Emmanuel Lellouch (France), Robert MacDowall (U.S.A.), André Mangeney (France), Eugene Mareev (Russie), Michel Moncuquet (France), Helmut Rucker (Autriche), Padma Shukla (Allemagne), Pier-Luigi Veltri (Italie).

FOREWORD

The second 'Rencontres de l'Observatoire', held at the Paris Observatory at Meudon on January 10–14, 2000, brought together more than 120 physicists and astrophysicists from nearly twenty countries around the subject 'Physics of space: growth points and problems'.

This conference was held in honour of Jean-Louis Steinberg, as a tribute for his major contributions to space research, radioastronomy and physics of space. The program tried to cover the physics of space in an interdisciplinary way, in the spirit of an important lesson we learned from Jean-Louis: to go farther than the subjects already studied or about to be so, and to apply the knowledge acquired to explore new territories.

The last quarter of the century has seen the vertiginous growth of space achievements. Even though modern scientists cannot play with universes as Granville's juggler ('An other world', ed. H. Fournier, Paris, 1844), they have been able within a few decades to explore much of the heliosphere with beautifully instrumented space probes. Even though the heliosphere is merely one particular cosmic environment, it is presently the only one accessible to *in situ* measurements and hence plays a unique role as a natural laboratory for physics and astrophysics. In this context space physics, which was initially confined to the study of the sun-earth environment, has not only engulfed the whole heliosphere but tends to include all the cosmic environments driven by similar processes.

In this spirit, the conference tried to highlight recent achievements which have changed our view of the physics of space and was articulated about a few growth points and fundamental problems, with emphasis on the bridges between space plasma physics and other disciplines. Among the fundamental problems which have been addressed, the physics of collisionless plasmas – in particular particle acceleration and dissipation, plays a major role in astrophysics. The same is true of the physics of winds and jets: most cosmic bodies eject matter into space, and yet the wind coming from the sun is still not understood forty years after its first *in situ* detection. A session was partly devoted to the physics of dusty plasmas, which is just emerging from infancy despite the ubiquity of these environments. The last decade has seen an explosion in the volume of data on planetary and cometary environments which has boosted the research in these domains; and yet we are far from understanding how magnetospheres are driven; even in the case of the earth there is no consensus on how the engine works, and it was interesting

Astrophysics and Space Science **277:** xiii–xv, 2001.

to confront this field with the puzzle of pulsars' magnetospheres. A number of novel space detection techniques are emerging and a special session was devoted to them. Finally, important landmarks of space physics history and possible futures have been discussed by pioneers of the field.

The success of this conference is largely due to the enthusiasm of participants and to the speakers' efforts to share their field with colleagues from different disciplines. It is also due to the organising committees and we are grateful to the members of the International Advisory Committee and to the other members of the Scientific Program Committee for their help in the choice of subjects and speakers. These 'Rencontres' could not have taken place without the financial and logistic support of the Observatoire de Paris and we thank its president Pierre Couturier. We also acknowledge the generous financial support received from the European Space Agency, the Radioastronomy Centre of Nancay, the 'Centre National d'Etudes Spatiales', the Universities of Paris 7 and Paris 11, the 'Centre National de la Recherche Scientifique' and the French 'Ministère de l'Education Nationale, de la Recherche et de la Technologie'.

The Observatoire de Paris and the Space Research Department (DESPA) took care of the practical organisation; we thank Marie-Rose Rama-Rao, conference secretary, Christiane Adam, scientific council secretary, Jacqueline Thouvay, DESPA administrator, as well as the local members of the Scientific Program Committee who spent much time and energy in the preparation of the conference. We also thank Agnès Fave, Roger Hulin, Baptiste Cecconi, Julien Queinnec et Jean-Jacques Poisot for their efficient contributions to the practical organisation, as well as numerous colleagues from the Observatory.

The scientific quality of these Proceedings owes much to the referees who accepted to evaluate the contributions submitted within a generally short delay, and we thank in particular Kaspar Arzner, Stuart Bale, Abdallah Barakat, Gérard Belmont, David Burgess, Michael Combi, Stanley Cowley, Dominique Delcourt, John Dorelli, R. Paul Drake, Dan Dubin, Joe Giacalone, Nat Gopalswamy, Roland Grappin, Ove Havnes, Phil Isenberg, Karine Issautier, Randy Jokipii, Michael Kaiser, Jan Kuijpers, Marty Lee, Emmanuel Lellouch, Bertrand Lembège, Robert MacDowall, Milan Maksimovic, Eckart Marsch, Fritz Neubauer, Eugene Parker, Christophe Sauty, Darrell Strobel, Adam Szabo, Marco Velli, Rudolf Von Steiger, Frank Verheest, Rudolf Wegmann, Philippe Zarka, Gaetano Zimbardo, as well as several other referees.

It is a pleasure to thank Eugene Parker who accepted to introduce this volume.

Nicole Meyer-Vernet
Michel Moncuquet
Filippo Pantellini
Guest editors

International Advisory Committee

Michel Blanc (France), Roger-Marie Bonnet (E.S.A.), Joseph Burns (U.S.A.), Stanley Cowley (U.K.), Vitaly Ginzburg (Russia), Joseph Lemaire (Belgium), Dominique Lequéau (France), Hiroshi Matsumoto (Japan), Marcia Neugebauer (U.S.A.), Eugene Parker (U.S.A.), René Pellat (France), Christopher Russell (U.S.A.), Evry Schatzman (France), David Southwood (E.S.A.), Robert Stone (U.S.A.), Vadim Tsytovich (Russia), James Van Allen (U.S.A.).

Scientific Program Committee

Nicole Meyer-Vernet (France), Fran Bagenal (U.S.A.), Dominique Bockelée-Morvan (France), Jean-Louis Bougeret (France), David Burgess (U.K.), Tom Cravens (U.S.A.), Sang Hoang (France), Mihaly Horanyi (U.S.A.), Karine Issautier (France), Yolande Leblanc (France), Emmanuel Lellouch (France), Robert MacDowall (U.S.A.), André Mangeney (France), Eugene Mareev (Russia), Michel Moncuquet (France), Helmut Rucker (Austria), Padma Shukla (Germany), Pier-Luigi Veltri (Italy).

A CRITICAL REVIEW OF SUN-SPACE PHYSICS

E.N. PARKER
Fermi Institute and Departments of Physics and of Astronomy, University of Chicago, Chicago, Illinois, USA

Abstract. The Sun and the heliosphere form a single dynamical system, driven by the convection in the Sun and the magnetic fields generated by that convection. The magnetic fields are the primary channel for producing the high temperatures and high velocities that extend outward to form the heliosphere. The essential point is that, while the general picture seems to be reasonable, several important steps in the process are not understood, and several concepts commonly employed in explanation are false physics. These scientific gaps should not be forgotten in the rush to pursue new and exotic discoveries.

1. Introduction

This review takes a critical look at present understanding of the internal dynamical properties of the active Sun and the outward extension creating the heliosphere. The Sun, the solar wind, and the heliosphere are considered to be a single dynamical system presenting a variety of dynamical phenomena as the energy progresses outward from the thermonuclear core, through the convective zone and the visible surface into the surrounding space and the heliosphere. We comment on 'the good, the bad, and the ugly' aspects of Sun-space physics

The last 50 years has provided a map of solar and heliospheric activity, and some of the physics of these diverse phenomena is understood. However, it must be realized that there are fundamental gaps in the physics, to such an extent that we cannot state at the present time why the Sun is obliged by the basic laws of physics to produce the heliosphere. The scientific procedure is to observe and describe, and then to understand in terms of the basic laws of physics, e.g. the equations of Newton, Maxwell, Lorentz, Boltzmann, etc. It must be appreciated that the large dimensions, large mass, and large temperatures on the Sun and the heliosphere provide phenomena that cannot be achieved in the terrestrial laboratory.

Hard work by many physicists over many years has established and corroborated the theory of the solar interior, which we shall call the *thermal Sun*. Through inversion of the observed helioseismic p-mode spectrum, the speed of sound can be inferred throughout the Sun to an accuracy of about one part in 500. Bahcall(1999) and coworkers (Bahcall and Pinsonneault, 1995; Bahcall and Francis, 1996; Bahcall *et al.,* 1996; Bahcall *et al.,* 1997) have shown that, with some fine tuning of the theoretical model, e.g. the gravitational settling of the heavier ions, the model

Astrophysics and Space Science **277:** 1–11, 2001.

differs nowhere by more than one part in 500 from the helioseismic model. Thus it appears that the elemental abundances throughout the Sun are the same as at the visible surface, modified only by the thermonuclear burning of hydrogen to form helium and by the gravitational settling.

The thermonuclear core is presently exploited as a neutrino factory to pursue the physics of neutrino oscillations. It should be recalled that the original hope of R. Davis (1964) and J. Bahcall (1964) was to test the theoretical model of the solar interior by observing the emitted neutrinos. As often happens, careful measurements turned up a surprise. Subsequently the Superkamiokande measurements of upward and downward propagating energetic neutrinos, produced by cosmic ray bombardment of Earth, show an asymmetry implying neutrino oscillations between states and, hence, a nonvanishing neutrino rest mass. The Sudbury Neutrino Observatory (in Canada), Superkamiokande (in Japan) and Borexino (in Italy) are coming into operation to study the energy spectrum of the emitted solar neutrinos, thereby identifying to some degree the nuclear reactions in the Sun producing the neutrinos at different energies. It will be a few years before the results are in, perhaps with implications for dark matter and cosmology, and perhaps with interesting extensions of the standard model of the fundamental particles.

2. Magnetic Activity

Turning to the magnetic fields and the resulting activity of the Sun, the basic impediment is that the essential dynamics is either hidden below the visible surface, obscured at scales of 100 km or less at the visible surface, or too transparent, e.g. the corona, to see clearly. Progress depends upon the continuing space program to probe the corona and the heliosphere, with such contemporary space missions as Yohkoh, Ulysses, Voyager, SOHO, ACE, TRACE, CLUSTER, SOLO, Solar-B etc.

To study the active surface of the Sun, the need is to resolve the basic structure at scales below 0.1″ (75 km) to follow the motions and interactions of the individual magnetic fibrils, their internal structure, and the behavior of the surrounding photospheric gases as the moving fibrils drive the activity in the atmosphere above.

The suprathermal activity of the Sun is magnetic in nature, and the whole system is created and driven by the hydrodynamic overturning in the convective zone of the Sun (extending from the visible surface down to a depth of 2.0×10^5 km). The theoretical difficulties begin with the convection. The simple mixing length representation of the convection, assigning a single characteristic length $\lambda(h)$ and velocity $v(h)$ to the turbulent convective cells at depth h, is inadequate for treating the nonuniform rotation, the meridional circulation, and the dynamo effects. Direct numerical simulation is confounded by the strong vertical stratification of the convective zone, equivalent to about 14 density scale heights and some 20 pressure scale heights. Numerical simulation is able to handle perhaps 5 scale heights at the present time, with suitable parametrizing of the turbulence, and the result does not provide the rotation profile inferred from helioseismology, nor does it provide a definitive result for the meridional circulation (~10 m/sec).

This is, then, the appropriate place to note Wigner's dictum that, 'The important problems in physics are rarely solved. They are either forgotten or declared to be uninteresting.' The basis for the dictum is partly human nature, and in recent years, partly bureaucratic pressure. However, there is no fundamental reason why we cannot rise above Wigner's dictum and go a lot farther with the physics of the Sun and the heliosphere.

3. Magnetic Field Generation

Consider the generation of the cyclic magnetic fields of the Sun, presumably created in the convective zone by a combination of nonuniform rotation, cyclonic convection, and turbulent diffusion of magnetic field. The difficulty with the physics of field generation is that it requires the equivalent of diffusion over the depth of the convective zone in an 11-year period. Note, then, that if η is the effective magnetic diffusion coefficient, the characteristic diffusion distance in a time t is $(4\eta t)^{1/2}$. Equating this to 2×10^{10} cm for t = 3×10^8 sec yields $\eta \sim 3 \times 10^{11}$ cm^2/sec. Turbulent mixing (say, λ = 300 km, v = 1 km/sec) of a scalar field provides an effective diffusion coefficient of the order of $0.1\,\lambda v = 3 \times 10^{11}$ cm^2/sec, pretty much what is needed to construct a quantitative numerical model of the solar dynamo.

Unfortunately the magnetic field is a vector field with internal tensions that do not submit to the extreme stretching and intensification indicated by the theory of turbulent diffusion of a scalar field, such as a puff of smoke. Observations of an activity complex at a fixed longitude on the rotating Sun show the emergence of 3×10^{23} Maxwells of magnetic field (Gaizauskas *et al.*, 1983), implying a mean azimuthal field B in the lower convective zone of the order 10^4 gauss, assuming a meridional cross sectional area of 2×10^9 cm by 2×10^{10} cm. This is comparable to the strength of the convective motions, i.e. $B^2/4\pi \sim \rho\langle v^2\rangle$, and it follows that the field is too strong to be significantly mixed (stretched, intensified, and tangled) by the convection. So we have no understanding of the diffusion that is necessary for the generation of magnetic field in the solar dynamo.

The next point is that we have no idea why the magnetic field generated by the dynamo is in the observed fibril form (widely separated clumps of 1500 gauss fields with diameters of 100 km). Numerical simulations of the buoyant rise of azimuthal magnetic field from the base of the convective zone show that the individual flux bundles must start with internal fields of $0.5\text{–}1.0 \times 10^5$ (Choudhuri, A.R. and P.A. Gilman, 1987; Choudhuri, 1993; D'Silva, S. and A.R. Choudhuri, 1993; Fan, Fisher and DeLuca, 1993; Fan, Fisher and McClymont, 1994; Schussler *et al.*, 1994; Caligari, F. Moreno-Insertis and Schussler, 1995). Unless the mean field is so strong as this, it implies that the field is in a fibril state, with the fibrils as intense as the necessary 10^5 gauss. One wonders if the conventional treatment of the solar dynamo in terms of the mean field is not approaching the problem from the wrong direction. Perhaps the individual fibril is the basic magnetic entity.

When we come to treat the dynamical activity of the magnetic fibrils emerging through the surface of the Sun, the degree of twisting of the fibril is a crucial quantity, about which we know nothing. The point is that the widely separated fibrils expand enormously in the rarefied atmosphere above the visible surface so that the twisting below the surface propagates up into the expanded region as a torsional Alfven wave, enhancing the potential for dynamical instability and microflaring.

Another question is the nature of the flux bundles that produce the ephemeral active regions, which do not show the same strong 11-year variability exhibited by the ordinary large active regions (Martin and Harvey, 1979; Golub and Vaiana, 1980)

A related question is the variation $\Delta L(t)$ of the luminosity of the Sun – the total irradiance – with the general level of magnetic activity, in the amount of one or two parts in 1000 through the two 11-year activity cycles since the measurements began in 1978. Foukal and Lean (1988) have shown that a major part, but not all, of the variation is associated with the increased faculae and plages, which more than offset the darkening caused by sunspots. It may be that the additional stirring caused by the upward rise of emerging flux bundles enhances the convective heat transport to the surface (Parker, 1995). However, the physics of the faculae and plages and the sunspot formation is not understood, so the whole thing is very mysterious.

4. Long Term Variation

It should be kept in mind that the general level of cyclic solar activity varies between wide extremes. The activity since 1720 has varied by a factor of two or three (defined in terms of sunspot numbers). At the present time (2000) the activity is near the upper bound of this so called 'normal' operation, having increased from near the lower bound around 1900. In addition to the variable normal operation, the Sun has spent 10 different centuries out of the last 70 centuries in a hypoactive state, e.g. the Maunder Minimum of the 17th century, when the activity remained at a very low level for 50–100 years (Eddy, 1973, 1983). The Sun has also spent 8 different centuries in a hyperactive state, e.g. the Medieval maximum of the 12th century, when atmospheric ^{14}C production was strongly suppressed by the substantially reduced cosmic ray intensity at Earth. We have no idea what the sunspot numbers were in that prolonged period of abnormally high activity.

The monitoring of the luminosity variations of other solar-type stars in association with their magnetic activity over the last fifteen years shows an approximate universal relation between irradiance and activity, with occasional abrupt excursions in irradiance of one part in 200 (Zhang *et al.*, 1994; Baliunas *et al.,* 1995;

Soon *et al.*, 1996). This relation suggests that the brightness of the Sun deviated both above and below the mean by about one part in 200 during the extended periods of hyperactivity and hypoactivity, respectively. Over the same periods the mean annual temperature in the northern temperate zone of Earth varied up and down by 1–2 C, with $\Delta T/T \sim \Delta L/L$. It should not go unnoticed that the doubling of magnetic activity, associated with the doubling of the total magnetic flux through the surface of the Sun (Lockwood, Stamper, and Wild, 1999) since 1900, suggests an increase in mean brightness of the order of 0.5–1 part in 1000, mostly between 1900 and 1950, during which time most of the warming of the terrestrial climate and surface sea water occurred. Whatever the practical implications, the hyper and hypo active periods raise fundamental questions about the convection and the dynamo operating in the Sun.

The next obvious question, then, is the change in the structure and dimensions of the heliosphere from 1900 to 1950 and 2000, and through the Maunder Minimum and the Medieval maximum. Experience over the last 30 years suggests that the solar wind was dominated by fast wind from extended coronal holes during the Maunder Minimum, while the equatorial solar wind probably exhibited persistent streams of slow dense wind during the Medieval Maximum. One wonders then if there were more collisions between fast and slow streams beyond the orbit of Earth during the Maximum.

5. Solar Corona and Heliosphere

Turning attention to the solar corona, the primary question is the source of the heat that creates it. The X-ray corona confined in the 100 gauss bipolar magnetic fields of active regions is presumed to be heated by nanoflares ($\leq 10^{24}$ ergs/flare) in the surfaces of tangential discontinuity arising from the continual slow quasi-static convective shuffling of the photospheric footpoints of the field by the granules and supergranules. The photospheric intermixing of the footpoints interlaces the lines of the bipolar field, so that their relaxation to equilibrium automatically forms surfaces of tangential discontinuity (in the ideal case of vanishing electrical resistivity). The presence of a small resistivity initiates rapid reconnection, which dissipates the magnetic free energy of the interlaced field and continues as long as the footpoint motions maintain the interwoven field line topology (Parker, 1988, 1994). This concept of coronal heating needs to be checked by observations at high spatial and temporal resolution, of course.

It has been suggested by Martin (1988) and by Porter and Moore (1988) that the coronal holes, whose expansion provides the fast solar wind, are heated by the microflares in the supergranule boundaries. Observations by SOHO and TRACE find such a large number of microflares that the idea appears to be correct. Thus, given the observed magnetic fields and flares, we can begin to see why the Sun has both an X-ray corona and coronal holes expanding to provide the fast solar wind.

The slow wind streams are of more obscure origin, evidently coming from the corona around the periphery of the active regions, and perhaps, or perhaps not, related to the diffuse X-ray corona.

The strong excursions of the interplanetary field from the mean spiral form represent Alfven waves of large amplitude, mostly propagating outward in the wind, but with some propagating inward as well. Their origin is not clear, and whether they come partly from a much stronger wave field near the Sun remains unknown (cf. Burlaga, 1968, 1969; Burlaga and Ness, 1968, 1969; Belcher and Davis, 1971).

The interaction of fast and slow solar wind streams beyond the orbit of Mars, where their boundaries are more nearly azimuthal, has proved to be very interesting, providing a saw toothed profile of the wind velocity and density, with forward and backward shocks, particle acceleration, etc. (Burlaga, 1997).

The region of the termination shock and particularly the subsonic shocked solar wind beyond, forming the downstream wake of the heliosphere in the local interstellar wind (Parker, 1963), lie beyond the range of present spacecraft, although there is still hope that Voyager may reach the termination shock before it is finished. The anomalous cosmic ray component is believed to be created by the acceleration at the terminal shock of ions that originally fell into the heliosphere as interstellar neutral atoms and became ionized somewhere in the vicinity of Jupiter (Garcia-Munoz, Mason and Simpson, 1973; Fisk, Koslovsky and Ramaty, 1974; Pesses, Jokipii and Eichler, 1981; Jokipii, 1986.

6. Impact of the Solar Wind on Magnetospheres

Attention has been given to the interaction of the solar wind with the atmosphere of the nonmagnetic planet Venus, and the bare surface of the Moon, and the surfaces of asteroids and dust grains. Indeed, the embedding of solar wind ions in the surfaces of moon rocks has provided a valuable record of the long term properties of the solar wind.

The impact of the solar wind against the magnetospheres of Earth, Jupiter, Saturn, Uranus, and Neptune drives the magnetospheric activity and space weather around each planet. The rapid spinning of the gigantic magnetospheres of Jupiter and Saturn adds a new energy source to the impact effects. The Jupiter relativistic electron beacon (Simpson, 1974; Shenette, Conlon and Simpson, 1974) is a particularly exotic example. The role of Io in the activity of the Jovian magnetosphere is a whole field of study in itself.

Planetary magnetospheric activity begins with the size and shape of the magnetosphere under the impact of the solar wind, including the upstream bow shock and the fluctuations with changing plasma and magnetic conditions in the wind. The sunward reconnection with the interplanetary magnetic field and the associated tailward transport of flux bundles is an important ingredient in the geomagnetic substorm and may be presumed to occur in other planetary magnetospheres as well. Magnetospheric convection and the return flow of magnetic flux to the sunward side arises for any planet with a nonconducting layer in its atmosphere. Magnetospheres are inflated by the build up of trapped particles, as in the main phase of the geomagnetic storm, relaxing back to normal as the trapped energetic particles charge exchange with the ambient atmospheric atoms and ions (Dessler and Parker, 1959, 1968). When one looks into the small-scale plasma physics of the structure of the magnetopause, there turns out to be no equilibrium state for certain oblique incidence of the solar wind particles (Parker, 1967; DeKeyser and Roth, 1998), suggesting rapid erosion of the magnetopause by the wind. Then when equilibrium of the magnetopause exists, there are instabilities (Parker, 1958a; Eviatar and Wolf, 1968), whose dynamical effects are also not clear. The transport of flux bundles from the sunward magnetopause into the geotail may be presumed to create a magnetic field in the geotail with wandering field lines, resulting in the spontaneous creation of surfaces of tangential discontinuity as the field relaxes toward equilibrium. This may be the origin of the surfaces of discontinuity, or current sheets, responsible for the aurora (Parker, 1994).

The physics of planetary magnetospheres has come a long way from Akasofu's recognition of the substorm phenomenon and the identification of flux transfer events, etc (Akasofu, 1966, 1981; Akasofu and Chapman, 1972; Baumjohann and Haerendel, 1990). The substorm is describable now in terms of basic physical principles (Zhu, 1994, 1995).

7. The Role of Physics

Coming back to the statement of purpose at the beginning of this review, that the ultimate goal is to understand why the Sun and the heliosphere and the planetary magnetospheres are obliged by the basic laws of physics to behave as they do, we encounter what might be termed the 'ugly' side to the scientific effort, which we cannot afford to overlook. That is the unfortunate tendency to replace the principles of Newton and Maxwell with declared principles that give an illusion of clever scientific explanation, when in fact the declared principles are contradicted by Newton and Maxwell.

There is not space enough here to summarize the many popular misconceptions (see, for instance Parker, 1996, 1997), so we concentrate on the electric circuit analog so widely employed in 'explaining' magnetospheric dynamics. The popular idea is that the electric current **j** flowing in a deformed magnetic field **B** ($\mathbf{j} = c\nabla\times\mathbf{B}/4\pi$) satisfies the same electric circuit equation, e.g. $RI + L\,dI/dt = V$, as the electric current I in a fixed electric circuit in the laboratory. In fact the currents in a plasma are created by $\nabla\times\mathbf{B}$ as the field is deformed by the moving plasma and flow passively under entirely different conditions from the current in the fixed electric circuit.

To show the profound differences between the behavior of the current in a large-scale plasma and field and the current described in the electric circuit analog, consider the popular 'current interruption' phenomenon, reputed to produce enormous potential differences when the current is blocked by the sudden loss of electrical conductivity (Alfven and Carlquist, 1967). The simple case of a long straight twisted magnetic flux bundle with its central axis along the z-axis suffices, using the cylindrical polar coordinates (r, φ, z). Denote the magnetic field by $B_r = 0$, $B_\varphi = b(r)$, $B_z = B(r)$, and suppose the fluid to be incompressible and without electrical resistivity. The current density is $j_z = (c/4\pi r)d[rb(r)]/dr$. The electric circuit analog focuses on the net current I flowing in one direction along the central core of the flux tube, with the return current in the outer regions. The inductance L per unit length is conveniently defined as $2W/I^2$, where W is the magnetic energy per unit length of the magnetic field b(r),

$$W = 2\pi \int_0^\infty \frac{drrb(r)^2}{8\pi}$$

At time $t = 0$ an insulating sheet is introduced at $z = 0$, blocking the passage of j_z across $z = 0$. The electric circuit analog predicts an enormous electric potential difference across the insulating sheet as a consequence of the inductance of the electric circuit. A twisted flux tube of length Λ has a total inductance ΛL. The potential difference V(t) is then

$$V(t) = \Lambda L \frac{dI}{dt}$$

where dI/dt is essentially a delta function.

Suppose, however, that we approach the problem from the point of view of Newton and Maxwell. Noting that the fluid has no resistivity and so cannot support an electric field in its own frame of reference, it follows that there is the electric field $\mathbf{E} = -\mathbf{v}\times\mathbf{B}/c$ in the laboratory frame, so that Faraday's induction equation reduces to the familiar MHD form,

$$\frac{\partial \mathbf{B}}{\partial t} = \nabla \times (\mathbf{v} \times \mathbf{B})$$

It should be noted that this is the basic induction equation for the large-scale variations of **B** in the presence of any fluid medium, whether collision dominated or collision free and whether fully or partially ionized, that cannot support a significant electric field in its own moving frame of reference (Parker, 1957, 1996). The Hall effect, arising when neutral atoms are present, is small in the large-scale limit. The dissipative terms, neglected in the large-scale limit, depend in detail on the nature of the medium, of course.

The momentum equation for the large-scale bulk motion of the plasma or fluid medium is

$$\rho \frac{dv_i}{dt} = +\frac{\partial}{\partial x_j}(M_{ij} - p_{ij})$$

which is nothing more than the statement of conservation of the total momentum flux $\rho v_i v_j + p_{ij}$ of the bulk velocity v_i and the thermal momentum flux p_{ij} under the influence of the Maxwell stress tensor M_{ij}.

For the twisted flux tube these dynamical equations reduce to

$$\frac{\partial B_z}{\partial t} = 0, \quad \frac{\partial B_\varphi}{\partial t} = B\frac{\partial v_\varphi}{\partial z}, \quad \rho\frac{\partial v_\varphi}{\partial t} = \frac{B}{4\pi}\frac{\partial B_\varphi}{\partial z}$$

with the fluid pressure given by

$$p(r, z, t) + \frac{B_z^2(r)}{8\pi} + \frac{B_\varphi^2(r, z, t)}{8\pi} = \text{constant}$$

The boundary condition for $t > 0$ is $j_z(r,0,t) = 0$, so that $B_\varphi(r,0,t) = 0$. The solution for $z > 0$ is

$$v_\varphi(r, z, t) = \frac{b(r)}{(4\pi\rho)^{1/2}}\{1 - S[z - C(r)t]\}$$

$$B_\varphi(r, z, t) = b(r)S[z - C(r)t]$$

where $C(r) = B(r)/(4\pi\rho)^{1/2}$ represents the Alfvén speed in $B_z(r)$ and $S(X)$ is the unit step function, equal to +1 when $X > 0$ and zero when $X < 0$. For $z < 0$,

$$v_\varphi(r, z, t) = -\frac{b(r)}{(4\pi\rho)^{1/2}}S[z + C(r)t]$$

$$B_\varphi(r, z, t) = b(r)\{1 - S[z + C(r)t]\}$$

The solution represents torsional Alfvén waves propagating away from $z = 0$ in both directions. The passage of the wave front, representing a kink in the magnetic field where B_φ drops to zero, converts the magnetic energy of b(r) into rotational kinetic energy, leaving the longitudinal magnetic field $B_z = B(r)$ unchanged. There is no inductive effect producing a large potential difference across the insulating sheet, because the current flow I is not interrupted. Instead, the current along the axis of the twisted flux tube is diverted radially at the moving wave fronts $z = \pm C(r)t$ to connect with the return current in the outer regions of the twisted tube. There is a radial electric field $E_r = -v_\varphi B(r)/c$ in the laboratory frame of reference, which has opposite signs on each side of $z = 0$ and provides a potential difference

$$V(r) = 2\int_0^r ds v_\varphi(r)B(r)/c$$

across $z = 0$. Note that this limited potential difference has nothing to do with the inductance ΛL of the twisted flux tube.

The essential point is that the electric current flows in the moving frame of reference of the fluid or plasma, in which there is no electric field, and therefore no powerful inductive effects. The current, having no force or inertia, freely switches its connectivity in response to the changes in $\nabla\times\mathbf{B}$ created by $\mathbf{v}$ and η. Since there is no vector $\mathbf{B}$ whose curl fails to have vanishing divergence, electric charge is always conserved, and the principal current flows are not interrupted in any conducting part of the fluid. This is in contrast with the electric circuit analog wherein the current is constrained to a path fixed in the laboratory frame of reference and hence subject to blockage and the associated enormous inductive effects.

8. Summary

In summary, we are challenged by many fundamental questions concerning the physics of the Sun and heliosphere, beginning with the convection and the magnetohydrodynamic solar dynamo, the associated fibril magnetic fields, the formation and structure of sunspots from the fibril fields, and coronal heating and the associated coronal expansion producing the solar wind and the heliosphere. Particle acceleration in flares and shock fronts, both near the Sun, in and around planetary magnetopheres, and out in the heliosphere continues to be a topic of central interest. The magnetospheres of Earth, Jupiter, and Saturn *et al.*, the physics of Io, and the regions around Venus, Mars, and asteroids each provide a unique world for scientific inquiry. So there are profound and exciting challenges facing us, all the way from the convective zone of the Sun to the downstream interstellar wake. The physics is the ultimate goal.

References

Akasofu, S.I.: 1966, *Space Sci. Rev.* **6**, 21.
Akasofu, S.I.: 1981, *Space Sci. Rev.* **28**, 121.
Akasofu, S.I. and Chapman, S.: 1972, *Solar Terrestrial Physics*, Clarendon Press, Oxford, 8.7.2–8.9.
Alfven, H. and Carlquist, P.: 1967, *Solar Phys.* **1**, 220.
Bahcall, J.N.: 1964, *Phys. Rev. Lett.* **12**, 300.
Bahcall, J.N.: 1999, *Current Science* **77**, 1487.
Bahcall, J.N., Calaprice, F., McDonald, A.M. and Totsuka, Y.: 1996, *Physics Today*, July, pp. 30.
Bahcall, J.N. and Francis, H.: 1996, *Physics World*, September, p. 41.
Bahcall, J.N. and Pinsonneault, M.H.: 1995, *Rev, Mod. Phys.* **67**, 781.
Bahcall, J.N., Pinsonneault, M.H., Basu, S. and Christensen-Dalsgaard, J.: 1997, *Phys. Rev. Lett* **78**, 171.
Baliunas, S.L. *et al.*: 1995, *Astrophys. J.* **438**, 269.
Baumjohann, W. and Haerendel, G.H.: 1990, in: C.T. Russell, E.R. Priest and L.C. Lee (eds.), *Physics of Magnetic Flux Ropes*, Geophysical Monograph **58**, American Geophysical Union, Washington, D.C., pp. 627–635.
Belcher, J.W. and Davis, L.: 1971, *J. Geophys. Res.* **76**, 3534.
Burlaga, L.F.: 1968, *Solar Phys.* **4**, 67.
Burlaga, L.F.: 1969, *Solar Phys.* **7**, 57.
Burlaga, L.F. and Ness, N.F.: 1968, *Canadian J. Phys.* **46**, 59.
Burlaga, L.F. and Ness, N.F.: 1969, *Solar Phys.* **9**, 467.

Burlaga, L.F., *Interplanetary Magnetohydrodynamics*, Oxford University Press, New York.

Caligari, P., Moreno-Insertis, F. and Schussler, M.: 1995, *Astrophys. J.* **441**, 896.

Choudhuri, A.R.: 1993, *Solar Phys.* **123**, 217.

Choudhuri, A.R. and Gilman, P.A.: 1987, *Astrophys. J.* **316**, 788.

Davis, R.: 1964, *Phys. Rev. Lett.* **12**, 303.

DeKeyser, J. and Roth, M. 1998, *J. Geophys. Res.* **103**, 6653.

Dessler, A.J. and Parker, E.N.: 1959, *J. Geophys. Res.* **64**, 2239.

Dessler, A.J. and Parker, E.N.: 1968, *J. Geophys. Res.* **73**, 3091.

D'Silva, S. and Choudhuri, A.R.: 1993, *Astron Astrophys.* **272**, 621.

Eddy, J.A.: 1973, *Climate Change* **1**, 173.

Eddy, J.A.: 1983, in: B.M. McCormac (ed.), *Weather and Climate Response*, Associated University Press, Boulder, Colorado, p. 1.

Eviatar, A. and Wolf, R.A.: 1968, *J. Geophys. Res.* **73**, 5562.

Fan, Y., Fisher, G.H. and DeLuca, E.E.: 1993, *Astrophys. J.* **405**, 390.

Fan, Y., Fisher, G.H. and McClymont, A.N.: 1994, *Astrophys. J.* **436**, 907.

Fisk, L.A., Koslovsky, B. and Ramaty, R.: 1974, *Astrophys. J. Lett.* **190**, L35.

Foujal, P. and Lean, J. 1988, *Astrophys. J.* **328**, 347.

Gaizauskas, V., Harvey, K.L., Harvey, J.W. and Zwaan, C.: 1983, *Astrophys. J.* **265**, 1056.

Garcia-Munoz, M., Mason, G.M. and Simpson, J.A.: 1973, *Astrophys. J. Lett.* **182**, L81.

Golub, L. and Vaiana, G.S.: 1980, *Astrophys. J. Lett.* **235**, L119.

Jokipii, J.R.: 1986, *J. Geophys. Res.* **91**, 2929.

Lockwood, M., Stamper, R. and Wild, M.N.: 1999, *Nature* **399**, 437.

Martin, S.: 1988, *Solar Phys.* **117**, 243.

Martin, S. and Harvey, K.L.: 1979, *Solar Phys.* **59**, 105.

Parker, E.N.: 1957, *Phys. Rev.* **107**, 924.

Parker, E.N.: 1958, *Phys. Fluids* **1**, 171.

Parker, E.N.: 1963, *Interplanetary Dynamical Processes*, Interscience. Div., John Wiley and Sons, New York.

Parker, E.N.: 1967, *J. Geophys. Res.* **72**, 2315, 4365.

Parker, E.N.: 1988, *Astrophys. J.* **330**, 474.

Parker, E.N.: 1994, *Spontaneous Current Sheets in Magnetic Fields*, Oxford University Press, New York.

Parker, E.N.: 1995, *Astrophys. J.* **440**, 415.

Parker, E.N.: 1996, *J. Geophys. Res.* **101**, 10587.

Parker, E.N.: 1997, *Astrophys. J.* **471**, 489.

Pesses, M.E., Jokipii, J.R. and Eichler, D.: 1981, *Astrophys. J. Lett.* **246**, L85.

Porter, J.G. and Moore, R.L.: 1988, in: *Proc. 9th Sacramento Peak Summer Symp*, 1987, National Solar Observatory, Sunspot New Mexico, p. 30.

Schussler, M., Caligari, P., Ferriz-mas, A. and Moreno-Insertis, F.: 1994, *Astron. Astrophys.* **281**, L69.

Shenette, O.L., Conlon, T.F. and Simpson, J.A.: 1974, *J. Geophys. Res.* **79**, 3551.

Simpson, J.A. *et al.*: 1974, *Science* **183**, 306.

Soon, W.H., Posmentier, E.S. and Baliunas, S.L.: 1996, *Astrophys. J.* **472**, 891.

Zhang, Q., Soon, W.H., Baliunas, S.L., Lockwood, G.W., Skiff, B.A. and Radick, R.R.: 1994, *Astrophys. J. Lett.* **427**, L111.

Zhu, X.: 1994, *J. Geophys. Res.* **99**, 11327, 11343.

Zhu, X.: 1995, *J. Geophys. Res.* **100**, 1847.

I

FUNDAMENTAL PROCESSES

PROCESSUS FONDAMENTAUX

ACCELERATION AND TRANSPORT OF ENERGETIC CHARGED PARTICLES IN SPACE

J.R. JOKIPII
The University of Arizona, Tucson, Arizona 85721, U.S.A.

Abstract. Energetic charged particles or cosmic rays are found in space wherever the ambient matter density is small enough to allow them to exist. Their observed kinetic energies vary from just above the local thermal energies to in excess of 10^{20} eV. The energy spectrum is quite smooth, suggesting that a common mechanism is responsible for particles of all energies. Their arrival directions are distributed evenly in solid angle, with the observed anisotropies being quite small even at energies $\approx 10^{18}$ eV. Various scenarios for their acceleration are discussed, including 2nd-order Fermi acceleration, shocks and cosmic-ray viscosity. It is concluded that the most likely mechanism for the acceleration of most energetic particles to high energies is diffusive shock acceleration. This mechanism has the benefit of producing a power-law energy spectrum with a spectral index which is very insensitive to parameters and which is close in magnitude to that observed in a variety of contexts. It is also reasonably fast and efficient. Anomalous cosmic rays in the heliosphere are discussed as one example of the success of the shock-acceleration picture.

1. Introduction

Cosmic rays or energetic charged particles are found in nature wherever the ambient density is low enough that collisional losses are negligible. The energy spectrum observed at Earth during quiet times (undisturbed by solar activity) is illustrated in Figure 1. The spectrum extends over a very wide range of kinetic energies from just above thermal energies to more than 10^{20} eV. The anisotropy (relative to the local fluid frame) is small – being less than or of order 10^{-2} at $\approx 10^{18}$ eV and smaller at lower energies. The spectrum of energetic particles extends smoothly down to thermal energies, (as is illustrated in Figure 2), so the low-energy limit of cosmic rays is simply a matter of definition.

As is seen in Figure 1, the energy spectrum is a remarkably smooth power law above some 10^9 eV, with only a minor change in slope occurring between 10^{15} and 10^{16} eV. The structure in the spectrum at 10^9 eV and below is due to the fact that we observe from within the heliosphere, which profoundly alters the lower-energy spectrum. It is believed that the interstellar spectrum extends smoothly down to energies significantly below 10^8 eV, but this is not certain. The anomalous cosmic-ray oxygen, shown in Figure 1, is a consequence of the interaction of the heliosphere with the interstellar medium.

Astrophysics and Space Science **277:** 15–26, 2001.

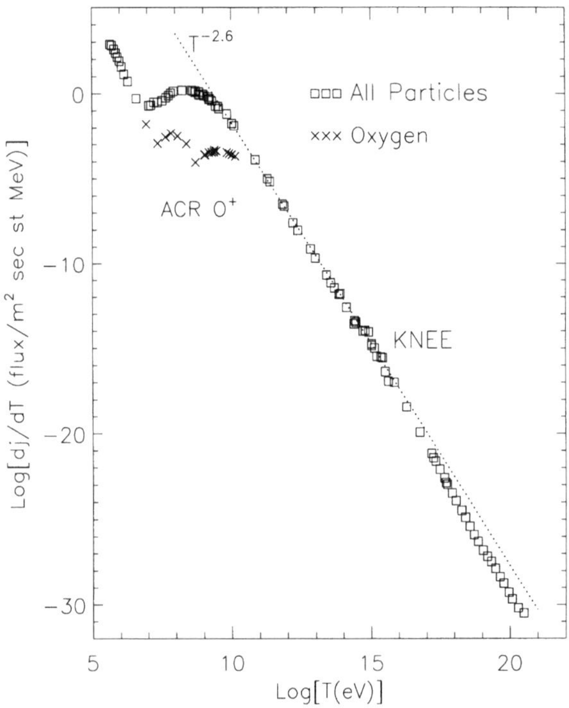

Figure 1. Observed energy spectrum of cosmic rays during quiet times. Shown are the all-particle spectrum and superposed, for illustration, the Oxygen spectrum, including the anomalous component at $10^7 - 10^8$ eV. The turnup below 10^6 eV is thought to be solar in origin.

Observed photons, emitted from distant galaxies and regions of our galaxy due to the interaction of cosmic rays with ambient matter or electromagnetic fields, show that cosmic rays exist in many places. Moreover, the inferred energy spectra of the cosmic rays in these remote sources are also power laws with slopes not much different from that observed at Earth. This fact, together with the smoothness of the observed spectrum at Earth over a large range in energy, suggests a common acceleration mechanism for the bulk of the cosmic rays.

The observed composition of cosmic rays tells us that they are typically confined to our galaxy for several thousand rectilinear transit times, probably by the fluctuations in the interstellar magnetic field. It is remarkable that the observed power spectrum of interstellar electron density fluctuations is also a smooth power law over a wide range of spatial scales. Since the electron density fluctuations are likely related to the magnetic fluctions which confine the cosmic rays, the smooth density power spectrum is almost certainly related to the smooth cosmic-ray spectrum. As is pointed out below, the interaction of cosmic rays with magnetic irregularities is primarily determined by irregularities with a scale approximately

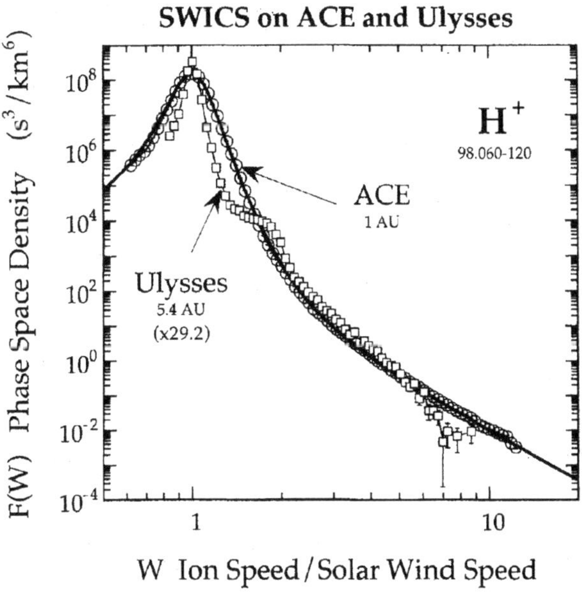

Figure 2. Energy spectrum of charged particles at low energies, showing the high-energy tail to the thermal distribution which is believed to continue and merge smoothly with the more-energetic charged particles (Gloeckler, private communication, 2000).

equal to the cosmic-ray gyro radius in the interstellar magnetic field, which is about 3×10^{-6} Gauss.

2. Acceleration of Charged Particles

In the absence of collisions with other particles, the only means for accelerating particles in a collisionless plasma is the ambient electric field $\mathbf{E}(\mathbf{r}, t)$. The increment in particle energy ΔT in the time interval Δt is given by

$$\Delta T = q \int_{t}^{t+\Delta t} \mathbf{w} \cdot \mathbf{E}(\mathbf{r}, t) dt, \tag{1}$$

where the integral must be taken along the particle trajectory. In what follows, I take the energetic particles to be rare enough that they do not affect the background plasma or electromagnetic field.

Clearly, we must know the particle motion to determine the energy change. This fact leads to the general requirement that acceleration must be coupled to the spatial transport, and both $\mathbf{E}(\mathbf{r}, t)$ and $\mathbf{B}(\mathbf{r}, t)$ must be considered together.

In this paper I will mainly consider quite energetic particles, for which the most important spatial plasma scales are larger than the energetic-particle gyro-radii.

At these scales the plasmas are generally hydromagnetic, and the electric field is determined from the fluid velocity $\mathbf{U}$ and magnetic field by the relation

$$\mathbf{E} = -\frac{\mathbf{U} \times \mathbf{B}}{c} \tag{2}$$

where c is the speed of light. The physical problem of acceleration, then, consists of solving Equations 1 and 2 for a given situation.

It is easy to go astray. One incorrect picture, apparently simple enough to be attractive to several authors, is essentially the following: Consider a slab of plasma of scale L which moves with uniform velocity $\mathbf{U}$ across a uniform magnetic field $\mathbf{B}$. Because of the induced electric field given in Equation (2), there is an electric potential drop across the slab of magnitude $\Delta\phi = eUBL/c$, where e is the unit of electrical charge. It is often simply assumed that the charged particle can gain the potential energy. This is significant, being some 10^{16} eV across the galactic disk and a few hundred MeV across the heliosphere. Unfortunately this mechanism doesn't work for a typical cosmic-ray particle, whose gyro-radius is a small fraction of the scale L. Since $\mathbf{E}$ is normal to $\mathbf{B}$, the particle simply gyrates and drifts normal to the electric field, and cannot move more than twice its gyro-radius in the direction of $\mathbf{E}$. Something more is required for the particles to gain significant energy.

A robust and widely applicable transport theory, applicable to particles whose speed w is significantly larger than U has been developed and applied extensively over the past four decades. It may be considered as an expansion in powers of the small ratio U/w. Unfortunately, a general theory for lower-energy particles for which U/w is not small is not available – for such particles one must use numerical simulations. The available theory utilizes the fact that the plasmas in space are turbulent, with broadband fluctuations over scales including those comparable with the particle gyro-radii. The particle motions are described statistically – the turbulent magnetic fluctuations 'scatter' the particles in angle, driving them to near-isotropy in the fluid frame (or, more generally, the frame of the scatterer) with a time scale τ_{scat}. This scattering occurs much more rapidly than energy change.

To zeroth order in U/w the resulting motion may be approximated as diffusive, and the phase space distribution function $f(p, \mathbf{r}, t)$ of the particles having a momentum magnitude p at position $\mathbf{r}$ and time t satisfies a diffusion equation

$$\frac{\partial f}{\partial t} = \frac{\partial}{\partial x_i}\left[\kappa_{ij}\frac{\partial f}{\partial x_j}\right] + Q, \tag{3}$$

where κ_{ij} is the diffusion tensor, determined by the spectrum of the magnetic fluctuations, and Q represents any source. Magnetic fluctuations at scales of the order of the energetic-particle gyroradius are the most important in determining κ_{ij}, although the field-line mixing, which affects the cross-field transport depends on very long length scales.

The flow velocity $\mathbf{U}$ does not appear in this order, and because $\mathbf{E} = 0$ in this limit, neither convection nor energy change appear. Equation (3) gives a crude, but sometimes acceptable description of the behavior of very high energy particles occasionally emitted by the Sun. However, it is too simple for most purposes.

To the next (first) order in U/w we obtain a much more useful equation, which can be written

$$\frac{\partial f}{\partial t} = \frac{\partial}{\partial x_i}\left[\kappa_{ij}\frac{\partial f}{\partial x_j}\right] - U_i\frac{\partial f}{\partial x_i} + \frac{1}{3}\frac{\partial U_i}{\partial x_i}\frac{\partial f}{\partial \ell n(p)} + Q \tag{4}$$

In this equation we have, in addition to the diffusion, convection and acceleration/deceleration caused by the electric field. Note that the electric field does not appear explicitly – it is nonetheless contained in the terms containing the flow velocity **U**. Equation (4) was first written down by Parker (1965). It is the basis of most current work on cosmic-ray transport and acceleration, and is often called the Parker equation. The equation is a good approximation for energetic particles ($U/w \ll 1$) if the scattering by magnetic irregularities is faster than the macroscopic time scales, so the distribution is nearly isotropic. It applies at shocks (discontinuities in **U**) and the entire theory of diffusive shock acceleration (Krymsky, 1976; Axford, *et al.*, 1977; Bell, 1978; Blandford and Ostriker, 1978) may be obtained from the equation by simply putting a step-function flow velocity **U** into it.

From this we see that it applies equally to parallel and perpendicular shocks (Jokipii, 1982, 1987). If the divergence of the flow velocity **U** is zero, there is no energy change to this order in U.

If we carry out the expansion to second order in U/w, we obtain new effects involving the viscosity of the cosmic rays and the acceleration of the flow (Berezhko and Krymsky, 1981; Earl *et al.*, 1988; Williams and Jokipii, 1993). These give additional terms representing acceleration of the cosmic rays caused by the fluid velocity shear, other gradients in velocity, and fluid acceleration. They can give acceleration in a divergence-free flow, which is absent in the first-order equation.

The above approximations neglect the possible random motions of the scattering magnetic irregularities relative to the flow velocity. If these are included (e.g., forward and backward moving Alfvén waves), further acceleration is introduced. This is the well-known 2nd-order Fermi acceleration, introduced by Fermi (1949), long before the present transport equation was developed. The effect of 2nd-order Fermi acceleration on the distribution function f may be written

$$\left(\frac{\partial f}{\partial t}\right)_{2nd-order Fermi} = \frac{1}{p^2}\frac{\partial}{\partial p}\left\{p^2 D_{pp}\frac{\partial f}{\partial p}\right\} \tag{5}$$

and represents diffusion in momentum, with momentum diffusion coefficient, D_{pp}. This term, which is generally much slower than shock acceleration, may simply be added to the right side of Equation (4) and is often invoked where shocks are believed not to be present. Since the Alfvén speed and fluid speed are often comparable, the viscous acceleration and 2nd-order Fermi acceleration are formally of the same nature, although the Fermi term is the one most-often discussed. In many cases the viscous terms can be neglected because of the effects of the magnetic field in suppressing transport normal to the field.

3. Discussion of Specific Acceleration Mechanisms

The basic transport equation discussed above has been used in a variety of contexts to discuss the acceleration of energetic particles. The most successful and widely applicable is that of standard diffusive shock acceleration, which results from the application of the first-order Equation (4) to a planar shock wave. As we will see, this has the singular property of producing a power law momentum spectrum which is not sensitive to the parameters, and which is close to that observed. For a general review, see Drury (1983).

3.1. Diffusive shock acceleration

Consider a steady, plane shock propagating in a uniform medium. Define the x-direction as the direction of propagation and let particles be introduced uniformly and steadily at the shock, at an injection momentum p_0. Work in the shock-normal coordinate system, with the shock at the fixed position $x = x_{sh}$. The shock ratio r is defined as the ratio of upstream to downstream flow speed U_1/U_2. It is readily found that the steady solution to the Parker equation in this case is given by

$$f(p) = Ap^{-\frac{3r}{(r-1)}} H(p - p_0) F(x, p) \tag{6}$$

where $H(p)$ is the Heaviside step function and the function $F(x, p)$ is independent of p at the shock, is indepent of x behind the shock, and decreases exponentially upstream as $\exp(-U_1(x_{sh} - x)/\kappa_{xx}(p)$. Note that in the limit of a strong shock, where $r \to 4$ the momentum dependence becomes $f(p) \propto p^{-4}$, which corresponds to an energy spectrum $dj/dT = p^2 f \propto p^{-2}$ which, if steepened somewhat due to energy-dependent transport and loss from the galaxy is not far from the observed spectrum at relativistic energies (e.g., Figure 1).

This power law spectrum is independent of shock speed, diffusion coefficients and other parameters. Since shocks in astrophysics tend to be strong, this may be the desired 'universal' acceleration spectrum. Because of the idealizations involved and because of the energy-depend transport, which changes the spectrum somewhat, the observed spectrum will deviate some from the shock generated spectrum. In addition, there will be a high-energy cutof and, in some cases, the accelerated particles will modify the shock.

The acceleration by the shock is not instantaneous, of course. Solving the time-dependent version of Equation (4), with the injection at momentum p_0 turned on at a time t_0, reveals that the spectrum above p_0 is still the universal power law given in Equation (6), but with a high-momentum cutoff, p_c which increases at a rate

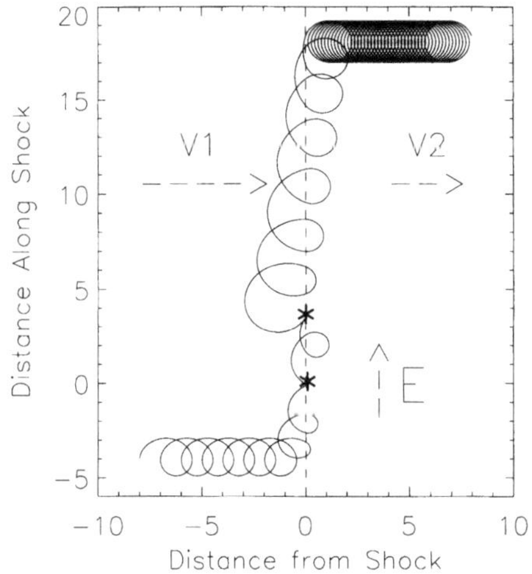

Figure 3. Illustration of mechanism of diffusive acceleration at a typical shock, which is propagating normal to the magnetic field. Shown is a particle trajectory gyrating in the magnetic field and being scattered by magnetic fluctuations. In this case the energy gain comes mainly from drifting in the convection electric field. If the shock were parallel, the particles would gain energy by scattering back and forth across the shock. In spite of the difference in the acceleration mechanism, the resulting spectra are as given in the text.

$dp_c/dt \approx 4U_1^2 p_c/\kappa_{xx}$. Hence, quasi-perpendicular shocks will in general accelerate particles faster than will quasi-parallel shocks, since for quasi-perpendicular shocks $\kappa_{xx} = \kappa_\perp$ which is generally significantly smaller than $\kappa_\parallel$, which is the relevant diffusion coefficient for quasi-parallel shocks. In fact, the physics of acceleration at quasi-perpendicular shocks is quite different than that at quasi-parallel shocks, as illustrated in Figure 3.

Diffusive shock acceleration has so many attractive aspects – it is quite fast (especially at quasi-perpendicular shocks), it naturally produces a power-law energy spectrum which is quite close to that observed in many places, and the shocks which can do the acceleration are quite common – that it is regarded by many as possibly the only important acceleration mechanism. Certainly, it is highly likely that most galactic cosmic rays are accelerated at supernova shock waves by this mechanism.

The above discussion is in the test-particle limit, where the effects of the cosmic rays on the flow are neglected. This is likely to be a poor approximation in some cases. In this case, simple picture presented here becomes only the first approximation.

3.2. 2ND-ORDER FERMI ACCELERATION

2nd-Order Fermi acceleration, contained in Equation (5), has an extensive history. It is quite often the mechanism of choice for possible diffuse *re*-acceleration of cosmic rays in the interstellar medium and in solar flares. Nonetheless, it has at least one considerable disadvantage. It does not produce a power law spectrum with the desired value in a robust way. The shape of the spectrum depends sensitively on the transport parameters in the acceleration region. Nonetheless, it remains popular.

4. The Example of Anomalous Cosmic Rays

The anomalous cosmic rays are an ionic component of the cosmic-ray flux which is observed, in the heliosphere, to be significant at energies ranging from 20 MeV to more than a GeV. They consist of a number of ion species ranging from Hydrogen to beyond Neon, but anomalous carbon is significantly under-abundant. The highest kinetic energy observed is more than 1.6 GeV. Observations from detectors on board heliospheric spacecraft show that their intensity continues to increase out to the maximum distance reached by current spacecraft, indicating that they are accelerated in the outer solar system. They were first seen in the early 1970's, and were initially called 'anomalous' cosmic rays (herein ACR) because of their unusual composition and because their spectrum was unlike that of the already well-known galactic or solar cosmic rays. This name has stuck, even though we now know much more about the nature and origin of these particles – arguably more than we know about galactic cosmic rays.

In 1974, Fisk, Kozlovsky and Ramaty suggested that the ACR were the result of the acceleration, in the heliosphere, of recently-ionized interstellar neutral atoms. The neutral atoms stream into the solar system from the interstellar medium and are then ionized by solar ultraviolet radiation, or by charge exchange with the solar wind. This hypothesis explained very nicely the unusual composition of the ACR, since only initially-neutral interstellar particles can enter the heliosphere to participate in the process. Carbon, which is almost completely ionized in the interstellar medium and hence is mostly prevented by the interplanetary magnetic field from entering the solar system, is notably depleted in the ACR. Recent observations have established that the bulk of the ACR are indeed singly charged, as expected in this picture.

In 1981, Pesses, Jokipii and Eichler pointed out that many features of the anomalous component could be explained if the major part of the acceleration of the newly-ionized particles occurs at the termination shock of the solar wind, by the mechanism of diffusive shock acceleration. Subsequent sophisticated numerical simulations carried out since the mid-1980's (Jokipii, 1986; Steenberg and Moraal, 1999) clearly showed that energetic, singly charged particles could be accelerated very efficiently at the termination shock. The essential observed features of the

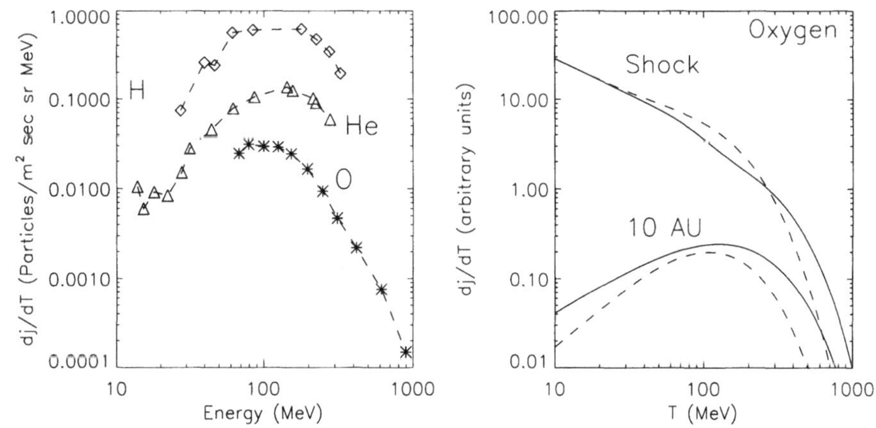

Figure 4. The observed and calculated spectra of ACR. The left panel shows a plot of the observed spectrum of ACR at 21 AU. The right presents plots of the computed energy spectra of anomalous Oxygen near the heliospheric equator, for two radii, and for the two signs of the heliospheric magnetic field. The solid lines correspond to the case where the northern heliospheric magnetic field is directed outward $A > 0$, corresponding to the 1996 sunspot minimum. The dashed lines are the same for $A < 0$, corresponding to the 1986 sunspot minimum.

spectrum, time variation, and spatial gradients could be explained very naturally in terms of this picture. This picture of the creation of freshly-ionized interstellar neutrals and subsequent acceleration to high energies at the termination shock of the solar wind is now the standard paradigm. A typical observed ACR spectrum (with the solar and galactic contributions removed) is shown in Figure 4.

The general configuration of the inner heliosphere is well understood, and extrapolation to the termination shock probably does not introduce significant uncertainties. The configuration beyond the shock is uncertain, and we adopt a simple configuration which contains the basic features expected.

The solar wind velocity in the model calculation is taken to be radial out to a spherical termination shock at a radius R_{ts} at which it drops by a factor of r_{sh} (which is 4 for a strong shock), and then decreases as $1/r^2$ (small Mach number flow) out to an outer boundary R_b, where the energetic particle are presumed to escape. Typically, R_{ts} is taken to be some 70–100 AU and R_b some 30%–50% larger.

During the years around each solar minimum, the interplanetary magnetic field is organized into two hemispheres separated by a thin, nearly equatorial current sheet, across which the field reverses direction. In each hemisphere the field is approximately a classical Archimedean spiral, with the sense of the field being outward in one hemisphere and inward in the other. The field direction alternates with each 11-year sunspot cycle, so that during the 1996 sunspot minimum, the northern field was directed outward from the sun (conventionally denoted as $A >$ 0), but in 1986 the northern field pointed inward ($A < 0$). This field is assumed

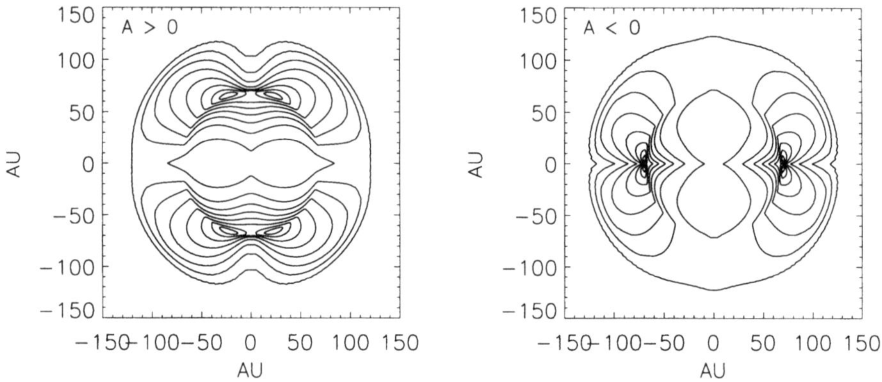

Figure 5. Equal intensity contours computed for 8 MeV/nucleon anomalous oxygen, as a function of position in a meridional plane, as computed for the model parameters discussed in the text. The left panel is for $A > 0$ and the right is for $A < 0$.

to continue beyond the termination shock, with the spiral angle reflecting the local solar wind speed V_w.

There is now evidence that the **polar** magnetic field differs considerably from this spiral (Jokipii and Kóta, 1989; Jokipii *et al.*, 1995), so the polar field is modified in our simulations. The magnetic-field structure for the years near sunspot maximum is not simple, so the following discussion is most-relevant during the several years around sunspot minimum.

The basic numerical model (including two spatial dimensions, heliocentric radius and polar angle) follows the acceleration of low-energy, singly-charged particles injected into the solar wind *a la* Fisk, Koslovsky and Ramaty, and their subsequent acceleration at the termination shock. The injection energy used is typically of the order of 100 keV, because of the limited dynamical range available in numerical solutions. The method of solution is then to follow these particles in time as they are accelerated at the shock and propagate throughout the heliosphere, until the distribution reaches a steady steady state. The characteristic time to approach a steady state is found to be 2–3 years or less, except for the highest-energy particles.

The theoretical results from a simulation, in which the injection of low-energy particles was spatially uniform over the shock, are shown in Figures 4 and 5. Here nominal values for the parameters were used. $V_w = 400$ km sec^{-1}, $\kappa_\parallel = 1.5 \times 10^{22} P^{.5} \beta$ cm^2 sec^{-1}, $\kappa_\perp = .1\ \kappa_\parallel$, $R_{ts} = 70$ AU, and $R_b = 130$ AU (here P is the particle rigidity in GV, and β is the ratio of the particle speed to the speed of light). A modified polar field (Jokipii and Kóta, 1989) was used. Note that the relative normalization (amplitude) of the two spectra are not significant and may be changed by small variation of the parameters. The computed energy spectra of anomalous protons, helium, neon, etc are in good agreement with the observed spectra.

Recent work (Mewaldt *et al.*, 1996; Jokipii, 1996, and Barghouty *et al.*, 2000) has included the possibility of further stripping of the anomalous cosmic rays as they propagate in the heliosphere. They find that above an energy of about 300 MeV, the singly-charged anomalous cosmic rays change to being multiply charged. The observations, for a variety of species, can be readily accounted for in the present picture, extended to include multiply-charged species. There is no space to discuss this in detail here.

5. Summary and Conclusions

The ubiquitous presence of energetic particles in collisionless astrophysical plasmas, and the fact that in many places the energy spectra are power laws with similar indices, suggests that there is a common acceleration for most of the observed cosmic rays. This mechanism appears to be that of diffusive shock acceleration, where the particles are maintained at near isotropy by magnetic scattering, so the diffusive approximation is valid. This mechanism operates at any shock, including perpendicular and parallel shocks, as long as the scattering is rapid enough. The example of anomalous cosmic rays and the solar wind termination shock suggests that the mechanism can explain the observations in some detail.

Acknowledgements

This work was supported, in part by NASA under grants NAG5-7793 and NAG5-6620, and by the NSF under Grant ATM 9616547. I thank G. Gloeckler for providing Figure 2.

References

Axford, W.I., Lear, E. and Skadron, G.: 1978, *Proc. 15th Int. Cosmic Ray Conf.*, Plovdiv, Bulgaria, 11, 132.
Barghouty, N. *et al.*: 2000, in: *Acceleration and Transport of Energetic Particles Observed in the Heliosphere*, Proceedings of ACE2000, Indian Wells, CA, in press.
Bell, A.R.: 1978, *Mon. Not. R. Astron. Soc.* **182**, 147.
Berezhko, E.G. and Krymsky, G.F.: 1981, *Soviet Astron. Lett.* **7**, 352.
Blandford, R. and Ostriker, G.: 1978, *Astrophys. J.* **221**, L29.
Drury, L.O.'C.: 1983, *Rep. Prog. Phys.* **46**, 973.
Earl, J., Jokipii, J.R. and Morfill, G.: 1988, *Astrophys. J.* **331**, L91.
Fermi, E.: 1949, *Phys. Rev.* **75**, 1169.
Fisk, L.A., Kozlovsky, B. and Ramaty, R.: 1974, *Astrophys. J. Lett.* **190**, L35.
Jokipii, J.R.: 1982, *Astrophys. J.*.
Jokipii, J.R.: 1986, *J. Geophys. Res.* **91**, 2929.

Jokipii, J.R.: 1987, *Astrophys. J.* **313**, 842.
Jokipii, J.R.: 1992, *Astrophys. J.* **393**, L41.
Jokipii, J.R.: 1996, *Astrophys. J.* **466**, L47.
Jokipii, J.R. and Kóta, J.: 1989, The polar heliospheric magnetic field, *Geophys. Res. Lett.* **16**, 1.
Jokipii, J.R. *et al.*: 1995, *Geophys. Res. Lett.* **22**, 3385.
Krymsky, R.F.: 1977, **Dokl. Akad. Nauk SSSR 234**, 1306.
Mewaldt, R. *et al.*: 1996, *Astrophys. J.* **466**, L47.
Parker, E.N.: 1965, *Planet. Space Sci.* **13**, 9.
Pesses, M.E., Jokipii, J.R. and Eichler, D.: 1981, *Astrophys. J.* **246**, L85.
Steenberg, C.D. and Moraal, H.: 1999, *J. Geophys. Res* **104**, 24879.
Williams, L.L. and Jokipii, J.R.: 1993, *Astrophys. J.* **417**, 725.

VARIOUS REGIMES OF COSMIC RAY DIFFUSION IN TURBULENT MAGNETIC FIELDS

GUY LAVAL
CPHT, UMR 7644 CNRS-Ecole Polytechnique, 91128 Palaiseau, France

Abstract. Charged particle transport in magnetized plasmas is one of the most important issues of plasma physics with a specific emphasis for magnetic confinement fusion, space physics and astrophysics. In Tokamak research, the understanding and the assessment of confinement performances have been hindered by the turbulence which induces complicated way of transporting the plasma energy to the machine walls. In cosmic ray research, turbulence plays also a major role for understanding transport and acceleration across galaxies and the heliosphere. In Tokamaks and cosmic ray physics, the transport induced by turbulent electric and magnetic fluctuations has been investigated by using theoretical works, numerical simulations and measurements. Between the two domains, cross fertilization examples are frequent. In the last decades, the diffusion induced by stationary magnetic fluctuations in a collisional plasma has been extensively studied as an exciting theoretical challenge as well as for fulfilling experimental needs. The results revealed a large variety of regimes with rather unexpected behaviors. This tutorial paper gives an elementary description of the identified mechanisms and tries to assess their interest for cosmic ray transport understanding.

1. Introduction

It has been very early recognized that turbulent magnetic fields govern the propagation, the acceleration and the confinement of cosmic rays in the galaxy and the heliosphere (Parker, 1958). Diffusive transport equations, including drifts and convection (Parker, 1965), have been set up and their solutions have been used to determine the diffusion tensors from ground base and satellite measurements. Later, the diffusion coefficients were related to the power spectrum of magnetic fluctuations by using quasi-linear theory (Jokipi, 1972).

The cosmic ray diffusion is a complicated process as the fluctuation spectrum extends over many scales. On the one hand, when charged particles interact with static magnetic disturbances, the adiabatic motion in the mean field is destroyed by the part of the spectrum with wavelength close to the particle gyration radius. Consequently, the particles diffuse along and across the average field lines with diffusion coefficients $\chi_{//}$ and $\chi_{\perp}$. On the other hand, the long wavelength perturbations preserve the adiabatic motion but they generate field line wandering which induces particle spreading across the mean field direction and which has to be considered as an additional transverse diffusion process (Jokipi, 1966a). These diffusion mechanisms are coupled and it may not be sufficient to add the diffusion coefficient to describe the overall transport (Barge *et al.*, 1984).

Astrophysics and Space Science **277:** 27–34, 2001.

Interestingly enough the same question has been raised in laboratory plasmas to take into account the coupling between collisional diffusion and the braiding of field lines resulting from instabilities or external non-axially symmetric windings of tokamaks (Rechester *et al.*, 1978). In these plasmas, $\chi_{//}$ and $\chi_{\perp}$ are given by $\chi_{//} = V_{th}^2/\nu$ and $\chi_{\perp} = a_L^2\ \nu$, where V_{th} is the rms particle velocity, ν the collision frequency and a_L the average particle gyration radius. There is a clear-cut separation between the physical processes leading to field line wandering and particle diffusion. This simplification has allowed a thorough detailed examination of the heat and particle transport. It showed that, depending on parameters, the coupling could be very important and leads to various overall transverse diffusion regimes.

The purpose of this paper is to put forward the main results which have been obtained for laboratory plasmas, with emphasis on the transition between the collisional and the collisionless regime. Simple proofs will be given by using elementary models. Then the applicability to cosmic rays will be discussed.

2. The Collisional Model

A simple collisional model is sufficient to introduce the various regimes and obtain the corresponding diffusion rates. We assume that particles stick to lines of forces and move freely along them with a constant velocity V between two successive collisions. Collisions occur with a frequency ν. A collision entails two perturbations of the particle motion: firstly, the velocity along the field line is reversed or unchanged with equal probabilities and, secondly, the guiding center of the particle jumps, with equal probabilities, to any other field line, provided that its displacement remains smaller than the average gyration radius a_L. With such a model, collisions induce diffusive particle transport along and across field lines with the diffusion coefficients $\chi_{//} = V^2/\nu$ and $\chi_{\perp} = a_L^2\ \nu$, as already mentioned. The mean squared parallel and transverse displacements of the particle guiding center are written:

$$(\Delta r_{//})^2 = 2\chi_{//}t \quad (\Delta r_{\perp})^2 = 2\chi_{\perp}t \tag{1}$$

The mean magnetic field $\mathbf{B}_0$ is assumed to be uniform and we have $\mathbf{B} = \mathbf{B}_0 + \mathbf{B}_1$, where $\mathbf{B}_1$ is a small fluctuating stationary component which depends on the spatial coordinates with $\mathbf{B}_1.\mathbf{B}_0 \approx 0$. If we are allowed to apply the quasi-linear theory, the field lines diffuse across the $\mathbf{B}_0$ average field lines with a mean square displacement $(\Delta r_{\perp})^2 = 2D_{FL}s$, where s is the field line length and $D_{FL} \approx (B_1/B_0)^2 L_{//}$, $L_{//}$ being the B_1 correlation length in the $\mathbf{B}_0$ direction. If we assume that the B_1 spectrum is isotropic in the transverse plane, namely a function of $| k_{\perp} |$ and $k_{//}$, the validity of quasi-linear theory implies $B_1/B_0 << 1/kL_{//}$, k^{-1} being the B_1 correlation length in the transverse direction. It means that the correlated displacement of the field line remains smaller than the typical B_1 transverse wavelength.

If collisions are negligible so that particles stream freely along field lines, the field line diffusion will induce particle transverse diffusion with a diffusion coefficient $D_C = VD_{FL}$ which has to be added to $\chi_\perp$. On the contrary, if $\chi_\perp$ is negligible but parallel diffusion cannot be ignored because the particle mean free path $\lambda = V/\nu$ is much shorter than the typical field line length, we can write $s^2 = 2\chi_{//}t$, $(\Delta r_\perp)^2 = 2D_{FL}s \approx 2D_{FL}(2\chi_{//}t)^{1/2}$. This regime is well known in astrophysics (Getmantsev, 1963) and is called compound diffusion. For such parameters, we find that the transverse particle transport induced by field line diffusion is sub-diffusive since $D = \lim_{t\to\infty} (\Delta r_\perp)^2/t = 0$. We conclude that the field line diffusion may induce particle transverse diffusion with a coefficient ranging from D_C to zero. In the following sections, different regimes will be shown to exist according to the size of the parameters D_{FL}, $L_\perp$, a_L, ν, V.

In order to obtain the diffusion coefficients in these different regimes, we use the general result of random walk which allows to write:

$$D = <\Delta r_\perp^2> /2\tau \tag{2}$$

where τ is the decorrelation time of the random walk, $<\Delta r_\perp>^2$ being the mean square value of the particle transverse displacement during τ. The decorrelation occurs when the particle guiding center is moving on a field line which is decorrelated from the initial one. If we assume that the average particle displacement l_d along field lines during τ is larger than $L_{//}$, we have $<\Delta r_\perp>^2 = 2D_{FL}$ and Eq.(2) becomes:

$$D = D_{FL} l_d/\tau \tag{3}$$

3. The Stochastic Instability

As recalled in part II, the fluctuating $\mathbf{B_1}$ field component generates field line diffusion provided that the quasi-linear conditions are fulfilled, which implies a short enough parallel correlation length of magnetic fluctuations. In such a case, the field lines wandering comes with another important phenomenon which is called the stochastic instability and can be described as follows. If two points M and M' lie on the same perturbed field line Γ_0, s is defined as the length of the field line linking M and M'. If, now, Γ_1 is a field line which is very close to Γ_0 in the vicinity of M, so that its shortest distance ϵ to M is such that $k\epsilon << 1$, its shortest distance ϵ' to M', on the average, will be given by $\epsilon' = \epsilon \exp(hs)$, with $h \approx k^2 D_{FL}$, provided that $k\epsilon' < 1$, that quasi-linear theory is valid and for an isotropic B_1 spectrum in the transverse plane (Isichenko, 1991). This exponential divergence of neighboring field lines is a sign of the chaos induced by the perturbation $\mathbf{B}_1$. The exponential amplification with a spatial rate h remains valid as long as the field lines Γ_0 and Γ_1 remain correlated.

4. The Parallel Diffusion Regime

We consider first the case where l_d is much larger than the mean free path $\lambda = V/\nu$. We can approximate the longitudinal particle motion by diffusion and set $l_d = (2\chi_{//}\tau)^{1/2}$, and we have now:

$$D = D_{FL}(2\chi_{//})^{1/2}/\tau^{1/2} \tag{4}$$

First, we assume $hl_d << 1$, so that the stochastic instability can be neglected during the decorrelation process. The most simple way for computing τ is to assume that decorrelation results from transverse collisional diffusion which gives $\tau = \tau_{KP} = (2\chi_\perp k^2)^{-1}$. It entails:

$$D = D_{KP} = 2D_{FL}k(\chi_{//}\chi_\perp)^{1/2} \tag{5}$$

The diffusion coefficient D_{KP} given by Eq.(5) has been obtained first by Kadomtsev and Pogutse (Kadomtsev *et al.*, 1978). When computing τ, we have neglected the stochastic instability of neighboring field lines which can amplify the particle spreading across the $\mathbf{B}_0$ direction and consequently reduce τ. The diffusion coefficient D_{KP} is valid provided that the the amplification factor is negligible over l_{dKP}, namely $hl_{dKP} = h(\chi_{//}\tau_{KP})^{1/2} << 1$ which reads $kD_{FL} << (\chi_\perp/\chi_{//})^{1/2}$. Then, in such a case (Isichenko, 1991), D_{KP} is much smaller than and the field lines wandering does not contribute to the transport across the $\mathbf{B}_0$ field lines.

Whenever the B_1spectrum is strongly anisotropic in the transverse plane, D_{KP} may lead to enhanced diffusion. As an example, if $\mathbf{B}_0 = B_0\mathbf{e}_z$,$\mathbf{B}_1 = B_{1x}\mathbf{e}_x + B_{1y}\mathbf{e}_y$, where B_{1x} is a function of y,z and B_{1y} is a function of x,z, with $|B_{1x}| >> |B_{1y}$, the transverse diffusion coefficient is anisotropic but can be much larger than $\chi_\perp$ in the $\mathbf{e}_x$ direction because, in such a case, $h \approx (B_{1x}B_{1y}/B_0{}^2)k_xk_yL_{//}$ is much smaller than $k_y{}^2D_{FLxx}$ since $D_{FLxx} = (B_{1x}{}^2/B_0{}^2)L_{//}$ so that the condition $h\ l_d << 1$ does not entail $D_{KP} << \chi_\perp$.

Going back to the isotropic case, we have now to take into account the amplification of the decorrelation by the stochastic instability of neighboring field lines. Let δr be the distance between two neighboring field lines. If collisional transverse diffusion is added to the stochastic instability, $<\delta r^2>$ is obtained from:

$$d < \delta r^2 > /ds = 2h < \delta r^2 > +2s\chi_\perp/\chi_{//} \tag{6}$$

which gives $<\delta r^2> = k^{-2}$ for $s = l_{dRR} = h^{-1}Ln(1/kd)$ where d is defined by $hd = (\chi_\perp/2\chi_{//})^{1/2}$. We notice that $kd << 1$ is a mere consequence of the inequality $h\ l_{dKP} > 1$ which is assumed to be fulfilled in this regime. By inserting $l_d = l_{dRR}$ in Eq.(4), we find $\tau^{1/2} = (2\chi_{//})^{-1/2}\ h^{-1}Ln(1/kd)$ and the effective transverse diffusion coefficient for particles D writes:

$$D = D_{RR} = 2D_{FL}h\chi_{//}/Ln(1/kd) \tag{7}$$

This diffusion coefficient has been obtained by Rechester and Rosenbluth (Rechester *et al.*, 1978) by using similar heuristic arguments. It cannot be recovered from perturbation expansions like quasi-linear theory. Numerical simulations have confirmed its validity in special cases and it has been obtained from rigorous calculations (Laval, 1993) when the stochastic wandering of field lines can be described with a saw tooth map model. The diffusion coefficient D_{RR} displays a spectacular enhancement of the effective transverse diffusion coefficient with respect to D_{KP} for kd<<1 which implies generally $\chi_\perp << \chi_{//}$ for an isotropic spectrum.

In Eq.(7), it is seen that D_{RR} may well exceed the collisionless diffusion coefficient $D_C = VD_{FL}$ if $\chi_{//}$ is taken large enough. This is unphysical, showing that the diffusion approximation for the parallel motion is no longer valid for long parallel mean free path.

5. The Transition to the Long Mean Free Path Regime

We now assume a long mean free path so that $h\lambda > 1$. For short mean free path, it was possible to use the diffusion approximation for the parallel motion. This approximation yields only rms displacements as a function of time. If $h\lambda < 1$, the fluctuations will not change the order of magnitude for l_d and τ but in the long mean free path regime where $h\lambda > 1$, we have to estimate the displacement along field lines in a more precise way since the decorrelation process can take place within one or a few collision times.

Let us first assume $\lambda >> L_{//}$. In such a case, in the meantime between collisions, the particle follows a given field line Γ_0 and its path is a random walk with a step frequency $V/L_{//}$ and a step amplitude $L_{//}B_1/B_0$. The field line wandering induces a transverse displacement which is given by

$$(\Delta r_\perp)^2 = 2VL_{//}(B_1/B_0)^2 t = 2D_c t. \tag{8}$$

If a collision occurs at time t_1 and if the particle velocity is reversed by this collision, we have for $t > t_1$

$$(\Delta r_\perp)^2 = 2D_c t_1 + 2D_c(t_1 - t) \tag{9}$$

since the particle moves back on the same field line, while Eq.(8) still holds for $t > t_1$, if the collision does not reverses the particle velocity.

Moreover, after a collision at $t = t_1$, the particle may be found on any neighboring field line at a distance smaller than a_L with equal probability. This transverse displacement is amplified by the stochastic instability and reaches k^{-1} for a parallel displacement $l_d = l_{dc} = h^{-1}\mathrm{Ln}(1/ka_L)$.

If the mean free path λ is larger than l_{dc}, and if the velocity is reversed during the collision, the particle travels backwards and its rms distance δ to Γ_0 increases exponentially. Then, the transverse displacement is again given approximately by Eq.(9) until time $t = t_1 + l_{dc}/V$, when $\delta = k^{-1}$. For $t > t_1 + l_{dc}/V$, the particle moves on a field line which is decorrelated from Γ_0 and Eq.(9) is no more valid. For t> t_1+ l_{dc}/V , the transverse displacement is growing again so that

$$(\Delta r_\perp)^2 = 2D_c(t - l_{dc}/V). \tag{10}$$

For νt>>1, it gives $(\Delta r_\perp)^2$ = $2D_c(1- l_{dc}\nu/2V)t$, where the probability for a velocity reversal has been taken into account. Consequently, in this very long mean free path regime, we find:

$$D = D_c(1 - l_{dc}/2\lambda) \tag{11}$$

and the diffusion coefficient is bounded from above by D_c.

If now hλ >1 with λ < l_{dc}, the model is more complicated. If at time t=0, a particle is on the field line Γ_0 at s=s_0 and if a collision occurs, the particle will be on adecorrelated field line when it reaches s = $s_0 \pm l_{dc}$. Several collisions take place before decorrelation and here we have to take them into account to describe the parallel motion. Let τ_1 be the particle escape time from the segment $| s - s_0 | < l_{dc}$. At t = τ_1, the average $\Delta r_\perp$ is given by:

$$(\Delta r_\perp)^2 = 2D_c l_{dc}/V. \tag{12}$$

It must be noticed that the transverse collisional displacements but the first one have been neglected. This approximation is is valid for hλ >1 where only a few collisions are needed to move on a decorrelated field line. After n such decorrelation processes, the transverse particle displacement will be $(\Delta r_\perp)^2$ = 2 $D_c n l_{dc}$/V at time t=τ_1+τ_2+...τ_n. Consequently, the effective diffusion coefficient D is written:

$$D = D_c l_{dc}/(V < \tau >) \tag{13}$$

where $< \tau >$ is the average escape time from the the interval $| s - s_0 | < l_{dc}$. The average decorrelation time $<\tau>$ may be easily obtained for this simple collisional model with the result (Laval, 1993):

$$< \tau > = \frac{l_{dc}}{V_{th}} \left(\frac{\nu l_{dc}}{2V_{th}} + 1 \right) \tag{14}$$

When inserted into Eq.(13), it gives:

$$D = D_c \left(\frac{\nu l_{dc}}{2V_{th}} + 1 \right)^{-1} \tag{15}$$

This result has been compared with exact computations (Laval, 1993) and it has been found that there was a very good agreement in a broad range of parameters for hλ >1. For l_{dc}<<λ, we recover Eq.(11) but for l_{dc}>>λ, the Rechester-Rosenbluth diffusion coefficient is obtained provided that a_L is substituted to d. This discrepancy results from the approximation which lead to Eq.(12).

These results may be summarized by defining three regimes for an isotropic magnetic fluctuation spectrum. First, if $kD_{FL} < (\chi_\perp/2\chi_{//})^{1/2}$, field line wandering does not contribute to transverse particle transport. If $k(\chi_\perp/2\chi_{//})^{1/2} < k^2 D_{FL} < V/\chi_{//}$, the validity of the quasi-linear theory for field line diffusion requires $(\chi_\perp/2\chi_{//})^{1/2} < 1$ and the transverse particle diffusion coefficient is then given by $D = D_{RR} = (kD_{FL})^2 \chi_{//}/\mathrm{Ln}(2k^2 D_{FL}{}^2 \chi_{//}/\chi_\perp)$ which may lead to strongly enhanced transverse particle transport. Finally, for $k^2 D_{FL} > V/\chi_{//}$, the transverse diffusion coefficient is approximated by $D = D_{FL}\ V \left(\frac{|\ Ln(ka_L)\ |\ V_{th}}{k^2 D_{FL} \chi_{//}} + 1 \right)^{-1}$ and reduces to $D = D_C$ for $\chi_{//} >> |\ Ln(ka_L)\ |\ V_{th}/k^2 D_{FL}$, which is the collisionless limit.

6. The Cosmic Ray Transport in the Heliosphere

Parallel transport of cosmic rays has been studied for decades as it is of primary importance for understanding solar flare events. The parallel diffusion coefficient $\chi_{//}$ is usually related to 'slab' magnetic perturbations with wave vectors parallel to the averaged field $\mathbf{B}_0$. These perturbations generate pitch angle scattering and induce a diffusive parallel motion. The mean free path is expected to increase with rigidity (Giacalone, 1998) and with $B_0{}^{-1}$. This mean free path is well determined in the range 0.08 –0.3 AU for rigidity R between 1 and 1000MV and its value is lying between 0.05 and 0.1AU.

Cross field diffusion is not so well known. Its importance has grown with the interest for the outer heliosphere and to understand the galactic cosmic ray confinement as well as their penetration into the heliosphere. The importance of field line wandering has been pointed out very soon (Jokipi, 1966b). The role of field line separation has attracted attention (Jokipi, 1973) as well as the transverse diffusion suppression by the parallel diffusion. Numerical direct simulations (Giacalone *et al.*, 1994) of particle motions in 3D random magnetic turbulence have shown that transverse diffusion is restored. However the consistency with quasi-linear theory is questionable and the rigidity dependence is not well understood.

It would be an enticing program to try applying the previous collisional theories to transverse cosmic ray transport. The interpretation of observations is quite controversial (Giacalone *et al.*, 1994) and a unified theory would be welcome. The relationship between $\chi_\perp$ and $\chi_{//}$ is almost the same as for collisional diffusion (Craven, 1997), so that the transposition would be easy. It could also include the effect of drifts and of time dependent fields. Drifts may be taken into account by setting $\chi_\perp = V_d k^{-1}$ where V_d is the transverse drift velocity: the non-stationary perturbations give rise to an equivalent transverse diffusivity $\chi_\perp = \omega k^{-2}$, where ω is a characteristic frequency (Isichenko, 1991). The theory could be confronted with extended numerical simulations and observations by comparing the observed rigidity dependence with the rather complicated one which is predicted by the theory.

However, the conditions, in space, have many features which make necessary to reexamine the theory. In the interplanetary medium, $\chi_\perp/\chi_{//}$ is found between

0.02 and 0.08 whereas in Tokamaks it may be as small as 10^{-7}, and, consequently, the asymptotic regimes found in magnetically confined plasmas will be much more difficult to display in space. In the heliosphere, the magnetic fluctuation spectrum is broad and continuous over many scale lengths. It is not obvious to determine which scale will generate $\chi_{\perp}$ and $\chi_{//}$ and which one is taken into account to describe field line wandering and stochastization. These difficulties will be aggravated by the lack of a systematic theory which could contain all cases whatever the type of fluctuation is considered. Any progress in the direction of a general theory, similar to quasi-linear theory and including an arbitrary spectrum of fluctuations, would be welcome.

In this tutorial paper, the quasi-linear description of field line wandering has always been considered as valid. Consequently, it ruled out any considerations about anomalous transport regimes like percolative diffusion (Isichenko, 1991) and Lévy flights (Zimbardo *et al.*, 1995). This exciting field has been ignored in a concern to keep this paper at a suitable level of difficulty. However these aspects may be important as recently recognized in the transport of the solar wind plasma across the magnetopause (Galeev *et al.*, 1986). They need also to be thoroughly investigated and the progress of computing facilities is of major importance in this respect.

References

Barge, P., Millet, J. and Pellat, R.: 1984, Importance of transverse diffusion for cosmic ray propagation, *Astrophys. J.* **284**, 817.

Craven, T.E.: 1997, in: *Physics of Solar System Plasmas,* Cambridge Universty Press, p. 257.

Galeev, A.A., Kusnetsova, M.M. and Zelenyi, L.M.: 1986, *Space Sci. Rev.* **44**, 1.

Getmantsev, G.G.: 1963, *Soviet Astron.* **6**, 477.

Giacalone, J.: 1998, *Space Sci. Rev.* **83**, 351.

Isichenko, M.B.: 1991, *Plasma Phys. Controlled Fusion* **33**, 809.

Jokipi, J.R.: 1966a, Cosmic19-ray propagation. 1. Charged particles in a random magnetic field, *Astrophys. J.* **146**, 480.

Jokipi, J.R.: 1966b, *Ap. J.* **146**, 480.

Jokipi, J.R.: 1972, Fokker-Planck equation for charged-particle transport in random fields, *Astrophys. J.* **172**, 319.

Jokipi, J.R.: 1973, *Astrophys. J.* **183**, 1029.

Kadomtsev, B.B. and Pogutse, O.P.: 1978, in: *Plasma Physics and Controlled Nuclear Fusion Research Proc. 7-th Int. Conf.*, Innsbruck: 1978, **I**, IAEA, Vienna: 1979, p. 649.

Laval, G.: 1993, *Phys. of Fluids* **B5**, 711.

Parker, E.N.: 1958, Origin and dynamics of cosmic rays, *Phys. Rev.* **109**, 1328.

Parker, E.N.: 1965, The passage of energetic particles through interplanetary space, *Planet. Space Sci.* **13**, 9.

Rechester, A.B. and Rosenbluth, M.N.: 1978, *Phys. Rev. Lett.* **40**, 38.

Zimbardo, G., Veltri, P., Basile G. and Principato, S.: 1995, *Phys. Plasmas* **2**, 2653 .

NON GAUSSIAN AND NON LOCAL TRANSPORT IN THE EARTH'S DISTANT MAGNETOTAIL

A. GRECO, G. ZIMBARDO and P. VELTRI
Physics Department, University of Calabria and Istituto Nazionale di Fisica della Materia, Unita' di Cosenza, I-87036 Arcavacata di Rende, Italy

A.L. TAKTAKISHVILI
Abastumani Astrophysical Observatory, Tiblisi, Georgia

L.M. ZELENYI
Space Research Institute, Moscow, Russia

Abstract. We study the ion dynamics in a magnetic field reversal with a constant electric field and with a model of three dimensional magnetic turbulence. By computing the mean square displacements in the plane of the current sheet we find superdiffusive and superballistic transport regimes. Since velocity increases with the length of the free path, we have accelerated Lévy flights. The possibility to generate power law velocity distribution functions is pointed out, as well as the long memory effects and non local properties of ion transport.

1. Introduction

Particle transport in a magnetic field reversal in the presence of turbulence represents an important issue both because this could control the particle loss in fusion machines and because magnetic field reversals, where most of the energy dissipation takes place, are an almost ubiquitous structure of astrophysical plasmas. We study the effect of a constant electric field and of magnetic fluctuations, which model the observed magnetic turbulence (Hoshino *et al.*, 1994), on particle transport in a magnetic field reversal. To this end, we use the magnetic field model of the Earth's distant magnetotail developed by Veltri *et al.*, 1998, and integrate ion trajectories as a function of the value of the cross tail electric field E_y. We find that ion motion is non Gaussian, that the mean square displacement grows as $\langle \Delta x^2 \rangle \sim t^\gamma$, with $\gamma \neq 1$, exhibiting Lévy flights and *superballistic* transport, $\gamma > 2$, in the plane of the magnetic field reversal. In order to understand if this superballistic transport is a transient effect we integrate ion trajectories for a very long time. This allows to characterize the ion dynamics in an asymptotic regime, even though the achieved distances are sometimes much larger than the magnetic tail size.

Astrophysics and Space Science **277:** 35–38, 2001.

2. Numerical Model and Statistical Interpretation

We present here the magnetic field model which represents the distant magnetotail. This consists of a field reversal which, in the GSM coordinate system, reads as $\mathbf{B}_0(z) = B_{0x}(z)\mathbf{e}_x = B_0 \tanh(z/\lambda)\mathbf{e}_x$, where λ is the current sheet half thickness, and of a three-dimensional spectrum of frozen fluctuations $\delta\mathbf{B}(\mathbf{r}) = \sum_{\mathbf{k},\sigma} \delta B_\sigma(\mathbf{k})$ $\mathbf{e}_\sigma(\mathbf{k}) \exp[i(\mathbf{k}\cdot\mathbf{r} + \phi_\mathbf{k}^\sigma)]$, where $\delta B_\sigma(\mathbf{k})$ is the amplitude of the mode with wave vector $\mathbf{k}$ and polarization σ, ($\sigma = 1, 2$), $\mathbf{e}_\sigma(\mathbf{k})$ are the polarization unit vectors, while $\phi_\mathbf{k}^\sigma$ are random phases. The Fourier amplitude is given by a power law spectrum, where the spectral index α is fixed as 3/2; this agrees with the power law spectrum observed for the magnetic field turbulence (Hoshino *et al.*, 1994) (see Veltri *et al.*, 1998 for more details). In addition, a constant electric field in the y direction is present, which corresponds to the dawn-dusk electric field found in the magnetotail. The equation of motion is $m_i d\mathbf{v}/dt = e(\mathbf{E} + \mathbf{v}\times\mathbf{B})$, where m_i, e, $\mathbf{v}$ are the ion mass, charge and velocity, and where $\mathbf{B} = \mathbf{B}_0(z) + \delta\mathbf{B}(\mathbf{r})$ and $\mathbf{E} = \mathcal{E} E_0 \mathbf{e}_y$, $E_0 = \text{const}$. Also, time is normalized to the inverse of the ion gyrofrequency $\Omega = eB_0/m_i$. As shown by Veltri *et al.*, 1998, a magnetic fluctuation level $\delta B/B_0 = 0.3$ is necessary to have a consistent description of the Earth's distant magnetotail.

To study the transport, we define $\Delta x = x(t) - \langle x(t)\rangle$ and $\Delta y = y(t) - \langle y(t)\rangle$, and integrate ion trajectories for a very long time, $t = 6000\,\Omega^{-1}$. We find that the mean square displacements $\langle\Delta x^2\rangle$ and $\langle\Delta y^2\rangle$ are well represented by the non Gaussian relations $\langle\Delta x^2\rangle \propto t^{\gamma_x}$ and $\langle\Delta y^2\rangle \propto t^{\gamma_y}$. The values of the anomalous diffusion exponents γ_x and γ_y obtained by a least square fit are reported in Table I of Zimbardo *et al.*, 2000. In general, values larger than 1 are always found, indicating the occurrence of Lévy, rather than Brownian, random walk. Even more, values up to 4 are obtained, corresponding to a transport process faster than the ballistic one ($\gamma = 2$). This is the result of the combined action of the electric field and of magnetic fluctuations: as the ions are bouncing between magnetic walls in the neutral sheet, they are accelerated by the steady electric field, producing a super-ballistic transport. The presence in xy plane of magnetic turbulence randomizes the ion speed, increasing the spread in particle positions, so that the value of γ is larger than 2.

Our findings on the growth of $\langle\Delta x^2\rangle$ and $\langle\Delta y^2\rangle$ can be interpreted with the Klafter-Blumen-Shlesinger (KBS) model of continuous time random walk (Klafter *et al.*, 1987), which allows for values of γ larger than 2, too. In this model, the probability density $\Psi(\ell, \tau)$ of doing a jump of length ℓ and of duration τ (with ℓ and τ expressed in units of the magnetic tail size L and Ω^{-1}, respectively) is given by $\Psi(\ell, \tau) = A\ell^{-\mu}\delta(\ell - c\tau^\nu)$, with $\ell \to \infty$, where A is a normalization constant, c is a proportionality constant, μ is the exponent which characterizes the power law distribution of jump lengths typical of Lévy flights, and ν is an exponent which relates the jump length to jump duration. As shown by Zimbardo *et al.*, 2000, the relation $\ell = c\tau$ holds for $\mathcal{E} \gtrsim 0.2$ and with $\nu \sim 2$ for our simulation, and the values

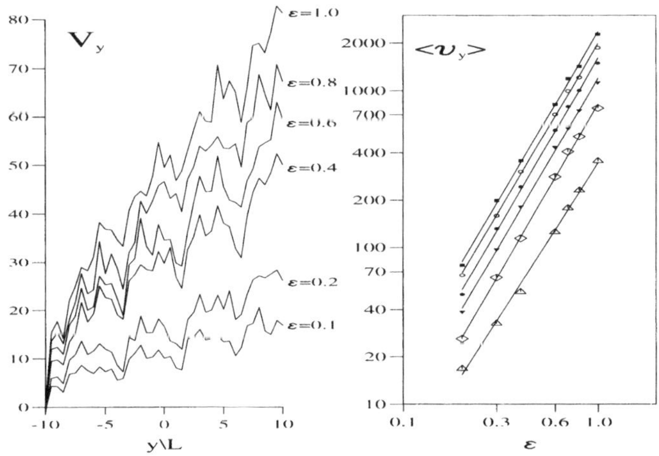

Figure 1. Bulk velocity V_y versus y for six values of electric field (right panel). Ensemble averaged single particle velocity $\langle v_y \rangle$ versus the electric field for six values of the integration time, represented by six different symbols (left panel).

of γ found from the simulation are in agreement with those predicted by the KBS model. The finding that ν_x and ν_y are larger than one means that longer jumps are carried out with higher velocity. This fact can be interpreted as the evidence of acceleration of ions during the Lévy flights. The fact that Lévy motion is not at constant velocity implies that the velocity distribution function $f(v)$ takes the form of a power law for large velocities, with $f(v) \propto v^{(3-\mu\nu-\nu)/(\nu-1)}$.

From a statistical point of view, however, if a physical observation is limited to a time span T (which corresponds to a scale length $L = cT^{\nu}$), the velocity distribution $f(v)$ ought to be truncated at $v* = cT^{\nu-1}$. Computing analytically, from the above velocity distribution function, the ensemble averaged single particle velocity $\langle v_y \rangle$ in y–direction, i.e., according to the definition $\langle v_y \rangle = \int_0^{v*} vf(v|T)dv$, versus the observation time T and versus $\mathcal{E}$, we find that this velocity satisfies the relation $\langle v_y \rangle \propto \mathcal{E}^2 T^{\nu-1}$. It appears that $\langle v_y \rangle$ has a quadratic dependence on $\mathcal{E}$ for a fixed integration time and, since $\nu \sim 1.9$ for every value of $\mathcal{E}$ (except for $\mathcal{E} = 0$), a linear one on T. These analytical results are confirmed by the computation of $\langle v_y \rangle$ from the equations of motion, as a function of the time t, as shown in Figure 1 (see also Greco *et al.*, 2000). Further, we have computed numerically the bulk velocity V_y (this differs from $\langle v_y \rangle$, which is a Lagrangian velocity, whereas V_y is an Eulerian velocity, computed at a fixed position) as a function of position y along the tail. As we can see from Figure 1, V_y does not reach an upper limit in the current sheet. We argue that this is due to the presence of Lèvy flights, which are a way to allow the particle memory not to be lost. In other words, long flights can bring information throughout the system and we can always find a jump longer than the system length. During the jump the velocity increases, and so does V_y. This implies that superballistic transport is not a transient regime, but rather it is intrinsic in the system configuration and this means that transport is non local.

3. Conclusion

In conclusion, we have shown that superballistic transport regimes are found in the plane of current sheet in a magnetic field reversal with magnetic turbulence and a steady electric field. Also, the anomalous diffusion exponents γ can be interpreted in terms of KBS model and power law distribution functions can be generated. The presence of accelerated Lèvy flights and the consequent non saturation of V_y mean that Ohm's law does not hold and we can only define a non-local, system-dependent conductivity.

Acknowledgements

This work was supported by the Italian MURST, by the Italian CNR, contracts no. 98.00129.CT02 and 98.00148.CT02, by ASI, contract no. ARS 98–82, and by INTAS 97–1612 grant.

References

Greco, A., Veltri, P., Zimbardo, G., Taktakishvili, A.L. and Zelenyi, L.M.: 2000, *Nonlin. Proc. Geophys.* **7**, 159.
Hoshino, M., Nishida, A., Yamamoto, M.T. and Kokubun, S.: 1994, *Geophys. Res. Lett.* **21**, 2935.
Klafter, J., Blumen, A. and Shlesinger, M.F.: 1987, *Phys. Rev. A* **35**, 3081.
Veltri, P., Zimbardo, G., Taktakishvili, A.L. and Zelenyi, L.M.: 1998, *J. Geophys. Res.* **103**, 14,897.
Zimbardo, G., Greco, A. and Veltri, P.: 2000, *Phys. Plasmas* **7**(4), 1071.

ACCELERATION AND TRANSPORT OF PARTICLES IN COLLISIONLESS PLASMAS: WAKES DUE TO THE INTERACTION WITH MOVING BODIES

MAXIM G. PONOMARJOV

State Academy of Aviation Technology, Promyshlennayastr, 1 (Box 22), 152300 Tutaev, Yaroslavl Region, Russian Federation

Abstract. A method is developed that allows the numerical and analytical description of the effects of ambient magnetic field on the time-dependent 3D structures of space plasma flows due to bodies in motion through a plasma. Some of these effects have been observed in space and ionosphere as stratified, flute and yacht sail like structures of plasma disturbances, jets, wakes and clouds. The method can be used for the simulations of Solar Wind flow taking into account the magnetic field effects and the interactions with the Interstellar Medium. These problems are of practical interest in fluid mechanics, space sciences, astrophysics, in turbulence theory. They also have some fundamental interest in their own right, as they enable one to concentrate on the effects of the ambient electric and magnetic fields.

1. Introduction

Plasma disturbances due to moving bodies have been intensively studied during the last 50 years. Reviews on the subject have been published by Alpert *et al.* (1963), (1964), (1965); Gurevich *et al.* (1980); Samir *et al.* (1989), Senbetu and Henley (1989). Usually, the following basic simplifications have been made in the definition of the existing theoretical models for bodies moving through a rarefied plasma: (1) In the case of a small Debye length compared to the body radius, the quasi-neutral condition of equal local density for ions and electrons is assumed to be valid; (2) If the object velocity is much larger than the average thermal velocity of the ions, it is common practice to neglect the ion thermal motion and to replace the unknown ion distribution function in front of the object by a beam moving in a straight line with the constant object velocity. It has been suggested that plasma expansion processes may have a direct relation to the problem of the flow of a rarefied plasma past a rapidly moving object.

Different type of interactions of the ambient particles with the object surface are considered as absorption, direct and diffuse reflection. The different types are described by the probability $w_k^i(\vec{u}, \vec{u}_1, \vec{r}, t)$, i.e. the probability that the k-particle, impinging on the object surface S at the point $\vec{r}$ with the velocity $\vec{u}_1$ will produce

Astrophysics and Space Science **277:** 39–44, 2001.

the i-particle to be emitted from the S at the $\vec{r}$ with the velocity $\vec{u}$ at time t. This probability is defined by the boundary condition at the body surface:

$$(\vec{n} \cdot \vec{u}) F_i(\vec{r}, \vec{u}, t) = - \int_{\vec{n} \cdot \vec{u}_1 < 0} \sum_k w_k^i(\vec{u}, \vec{u}_1, \vec{r}, t)(\vec{n} \cdot \vec{u}_1) F_k^f(\vec{r}, \vec{u}_1, t) d\vec{u}_1 \tag{1}$$

For the absorption, direct and diffuse reflection the probability has been first introduced by Alpert *et al.* (1963), (1964), (1965) while the cases of direct and diffuse reflection have been considered in detail by Ponomarjov (1995), (1996), (1999). Usually, at large distances from the body the plasma is considered to be undisturbed and in a state of thermodynamic equilibrium:

$$F_i|_{|\vec{r}| \to +\infty} \to N \left(\frac{\beta_i}{\pi} \right)^{3/2} \exp[-\beta_i \vec{v}^2] \tag{2}$$

To avoid considering the boundary conditions Equations (1), (2), Alpert *et al.* (1964) introduced a 'special term' in Boltzmann equation, which allowed to consider the effects of ambient particle interactions with the body surface. This term placed in the right hand of the time-independent Boltzmann equation and contains the implicit function that describes the body surface entirely and the gradient of this function. This pioneering approach covers a lot of interesting problems, some of which have not been considered in detail so far.

The present paper shortly describes the image method, which allows avoiding the treatment of the implicit function and its gradient. These circumstances allow considering surfaces with undefined normal and gradient. Examples of such surfaces are: sharp antennas, point-like objects, small two-sided plates. Besides, the image method covers the time-dependent problems too. It should be noted that for smooth bodies the image method gives results, which are in good agreement with results presented by Alpert *et al.* (1963), (1964), (1965); Gurevich *et al.* (1980).

2. Basic Method

According to the image method (Ponomarjov, 1995, 1996, 1999), the disturbances on the i-species in the plasma due to a point-like object or a small two-sided plate are simulated using the Boltzmann equation (3) with image and additional sources. This equation is for the perturbed distribution function of the i-species $F_i^{imw}(\vec{r}, \vec{v}, t)$. This function is so defined that $F_i^{imw}(\vec{r}, \vec{v}, t) d\vec{r} d\vec{v}$ is the number of perturbed particles which, at time t, have positions lying within the volume element $d\vec{r} = dxdydz$ about $\vec{r}$ and velocities lying within the velocity-space element $d\vec{v} = dv_x dv_y dv_z$ about $\vec{v}$.

$$\begin{aligned} &\frac{\partial F_i^{imw}}{\partial t} + \vec{v} \frac{\partial F_i^{imw}}{\partial \vec{r}} + \frac{q_i}{m_i} \left(\frac{1}{c} [\vec{v}, \vec{B}] + \vec{E} \right) \frac{\partial F_i^{imw}}{\partial \vec{v}} \\ &= Q_i^{im} + Q_i^{w} + \sum_j C_{ij}(F_i, F_j) \end{aligned} \tag{3}$$

where boundary conditions both near the small body surface and far from the small body are transformed into the image and additional sources:

$$Q_i^{im} + Q_i^{w} = Q_i^{im+} + Q_i^{im-} + Q_i^{w+} + Q_i^{w-} \tag{4}$$

and

$$Q_i^{im+} + Q_i^{w+} = \delta\{\vec{r} - \vec{r}_S\} \times$$

$$\times \begin{cases} \int_{\vec{n}\cdot\vec{u}_1<0} \sum_k w_k^i(\vec{u}, \vec{u}_1, \vec{r}, t)(\vec{n}\cdot\vec{u}_1) F_k^f(\vec{r}, \vec{u}_1, t) d\vec{u}_1 & \text{if } (\vec{n}\cdot\vec{u}) > 0 \\ -(\vec{n}\cdot\vec{u}) F_i^f(\vec{r}, \vec{u}, t) & \text{if } (\vec{n}\cdot\vec{u}) < 0 \end{cases} \tag{5}$$

$$Q_i^{im-} + Q_i^{w-} = \delta\{\vec{r} - \vec{r}_S\} \times$$

$$\times \begin{cases} \int_{\vec{n}\cdot\vec{u}_1<0} \sum_k w_k^i(\vec{u}, \vec{u}_1, \vec{r}, t)(\vec{n}\cdot\vec{u}_1) F_k^f(\vec{r}, \vec{u}_1, t) d\vec{u}_1 & \text{if } (\vec{n}\cdot\vec{u}) > 0 \\ (\vec{n}\cdot\vec{u}) F_i^f(\vec{r}, \vec{u}, t) & \text{if } (\vec{n}\cdot\vec{u}) < 0 \end{cases} \tag{6}$$

while $\sum_j C_{ij}(F_i, F_j)$ is the collision integral, which describes collisions between particles of the i-species and particles of all species. This term simulates the i-species distribution function evolution due to interparticle interactions.

In principle the collision integral can be approximated by the appropriate image and additional sources too. However below we consider collisionless flows only and assume that the effects of interparticle collisions are not significant compared to the effects of interactions with the body and the ambient magnetic field. The image point-like source $Q_i^{im} = Q_i^{im+} + Q_i^{im-}$ is the source of negative intensity, so that it simulates the absorption of i-species particles by the small plate. Q_i^{im+} and Q_i^{im-} are intensities of absorption by the front and reverse sides of the plate, respectively. The plate sides should be selected according to a given outward directed normal vector $\vec{n}$. The additional point-like source $Q_i^{w} = Q_i^{w+} + Q_i^{w-}$ simulates the reflection of i-species particles by the plate. Q_i^{w+} and Q_i^{w-} are the intensities of the reflection on the front and reverse sides of the plate, respectively. For 'non small' 2D and 3D bodies with a boundary surface S we introduce the following imaginary and additional sources:

$$Q_i^{imS} + Q_i^{wS} = \oint_S (Q_i^{im+} + Q_i^{w+}) dS \tag{7}$$

where the integration extends over the entire surface S, in order to obtain the distributed (on the surface S) image and additional sources $Q_i^{imS} + Q_i^{wS}$.

The following step is to find the distribution function of perturbed particles as solution of the Boltzmann equation (3) with the image and additional sources. The general solution has been obtained by Ponomarjov and Gunko (1995) who used the method of characteristics in the case of uniform magnetic and electric fields. The distribution function was obtained in quadratures. Concentrations of perturbed particles, velocities, pressures, heat flows and other perturbed quantities can be derived by computing the appropriate moment of the distribution function. For example, the perturbed charged particle concentration due to the reflection on the surface element dS can be obtained as the first moment of the distribution function:

$$\delta n = n^{w+} + n^{im-} = \int F^{imw}(\vec{r}, \vec{v}, t) dv_x dv_y dv_z \tag{8}$$

where the integral of the function over the velocity space can be reduced to elementary functions but there are quadratures, which consist of integrals over time-delays. Usually, these integrals are not reducible to elementary functions. However, for some particular cases they allow asymptotical approximations.

In order to illustrate how the asymptotic expressions can be obtained, the cross-sections of the oscillating iso-density contours are plotted for different times in Figure 1. In the figure a diffuse reflecting disk moves horizontally along to the magnetic field from left to right. The disk velocity V is $\sqrt{10}$ times greater than the mean thermal velocity of the ions (i.e. $\beta V^2 = 10$). The brighter the contours the higher the density. These contours show the dynamics of the developing stratifications and flute structures of the perturbed charged particle distribution due to rapidly moving small diffuse reflecting body through the ambient magnetic field. Indeed, one can see the long unstable periodical flute structures on the ion density behind the disk. The spatial period of these structure along the OX-axis corresponds to the distance covered by the disk during one ion gyro-period in the external magnetic field. These structures are very similar to the structures observed behind perfectly absorbing and directly reflecting plates (Ponomarjov, 1995, 1996). Both structure and wake move with the disk with the higher density region (pulsating at the ion gyrofrequency) clearly visible ahead of the disk. It should be noted that the extent of this region is significantly shorter than in the case of the directly reflecting plate (Ponomarjov, 1996).

3. Conclusions and Proposals for Future Work

We discuss the image method for a multi species plasma and different types of interactions with the surface of a moving body. For absorption, direct and diffuse reflections of particles with the surfaces the Boltzmann equation with image term and additional sources is solved analytically and 3D distributions of plasma disturbances are obtained in the ambient magnetic field in quadratures with asymptotic approximations (see also Ponomarjov, 1995, 1996, 1999). The next step may be the simulation the wakes in transitional regimes (from collisionless to MHD) and the generalization of the image method for these regimes. In addition there are a number of possible applications of the method to the simulation of 2 and 3 dimensional bodies in steady and unsteady flows. Finally, even the solar wind may be

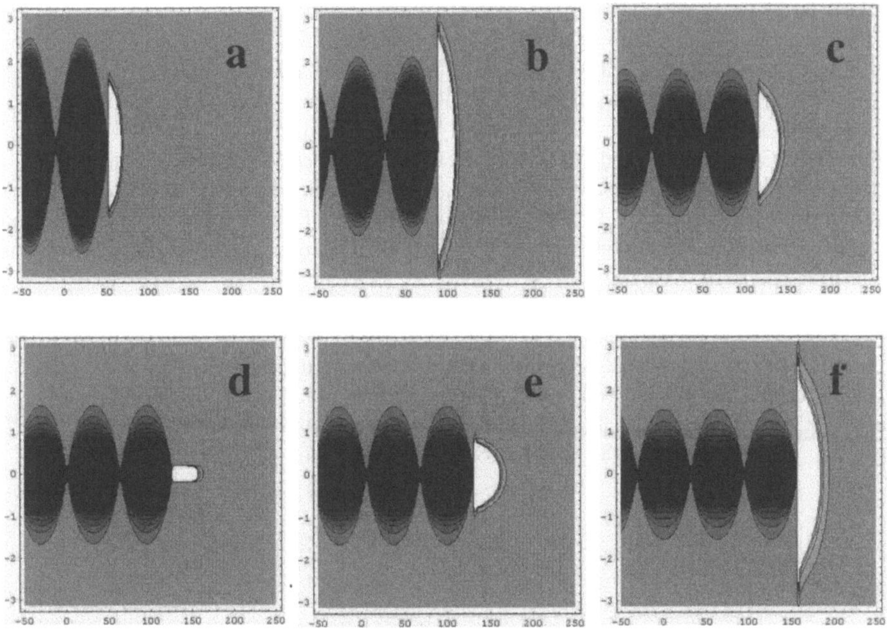

Figure 1. Contour plot of the perturbed ion density (in the plane $Z = 0$) due to the moving small disk at the following times: a) $t = \pi/48 + 5\pi/3$; b) $t = \pi/48 + 17\pi/6$; c) $t = \pi/48 + 11\pi/3$; d) $t = \pi/48 + 4\pi$; e) $t = \pi/48 + 26\pi/6$; f) $t = \pi/48 + 5\pi$. Dark (bright) regions correspond to low (high) densities. The disk surface diffusely reflects all incident ions. The disk moves with the velocity $\vec{V}$ along the external magnetic field $\vec{B}$. The magnetic field, the disk normal and the velocity are directed along the OX-axis (horizontal direction from left to right). The disk center is located at $Y = 0$ immediately behind the bright (white) region.

simulated using the above method. The Sun can be introduced into the Boltzmann equation as a distributed source. The interaction of the supersonic wind with the interstellar medium can be included via the collisions of the wind electrons with the heavy ions of the Interstellar Medium. This basic assumption allows to solve the Boltzmann equation analytically and to derive the Solar Wind distribution in quadratures. The densities and velocities of the Wind will be calculated as appropriate moments of the obtained distribution. In different specific cases their asymptotic approximations will be derived.

The basic goal of this paper is to interest institutions and specialists in related fields in the described method given the possibility of its adaptation and development for related topics. Thus, proposals for collaborations are welcome.

References

Alpert, Ya.L., Gurevich, A.V. and Pitaevskii, L.P.: 1963, *Space Sci. Rev.* **2**, 680.

Alpert, Ya.L., Gurevich, A.V. and Pitaevskii, L.P.: 1965, *Space Physics with Artificial Satellites*, Consultant Bureau, N.-Y.

Alpert, Ya.L., Gurevich, A.V. and Pitaevskii, L.P.: 1965, *Space Sci. Rev.* **6**, 419.

Baranov, V.B., Krasnobaev, K.V. and Kulikovskiy, A.G.: 1971, *Soviet Phys. Dokl.* **15**, 9.

Gurevich, A.V. and Pitaevskii, L.P.: 1980, Nonlinear Dynamics of Rarefied Plasmas and Ionospheric Aerodynamics, in: M.A. Leontovich (ed.), *Voprosy Teorii Plasmy* **10**, Atomizdat, (in Russian).

Ponomarjov, M.G.: 1996, *Phys. Rev. E* **54**, No. 1, 5591–5598.

Ponomarjov, M.G.: 1997, *Astron. Nachr.* **318**, 187–192.

Ponomarjov, M.G.: 1999, *Astrophys. Space Sci.*, Spec. issue: Proceedings of International Conference 'Progress in Cosmic Gas Dynamics', Moscow, Russia (in press).

Ponomarjov, M.G. and Gunko, Yu.F.: 1995, *Planet. Space Sci.* **43**, No. 10/11, 1409–1427.

Samir, U., Comfort, R.H., Singh, N., Hwang, K.S. and Stone, N.H.: 1989, *Planet. Space Sci.* **37**, 873.

Senbetu, L. and Henley, J.R.: 1989, *J. Geophys. Res. (A)* **94**, 5441.

OBSERVATIONAL EVIDENCE OF DISSIPATIVE SMALL SCALE PROCESSES: GEOTAIL SPACECRAFT OBSERVATION AND SIMULATION OF ELECTROSTATIC SOLITARY WAVES

Y. OMURA, H. KOJIMA, T. UMEDA and H. MATSUMOTO
Radio Science Center for Space and Atmosphere, Kyoto University

Abstract. In recent spacecraft observations, coherent microscale structures such as electrostatic solitary waves are observed in various regions of the magnetosphere. The Geotail spacecraft observation has shown that these solitary waves are associated with high energy non-thermal electrons flowing along the magnetic field. The solitary structures are generated as a result of a long time evolution of coherent nonlinear trapping of electrons as found in bump-on-tail, bi-stream and Buneman instabilities. It is noted that these solitary waves can be generated at distant regions far away from the spacecraft locations, because these trapped electrons, or electron holes, are drifting much faster than the local thermal plasmas. Some of the solitary waves are accompanied by perpendicular electric fields indicating that two- or three-dimensional potential structures are passing by the spacecraft. Depending on the local plasma parameters, these multi-dimensional solitary structures couple with perpendicular modes such as electrostatic whistler modes and lower-hybrid modes. In a long time evolution, these perpendicular modes are dissipated via self-organization of small solitary potentials, leading to formation of one-dimensional potential troughs as observed in the deep magnetotail. The above dissipative small-scale processes are reproduced in particle simulations, and they can be used for diagnostics of electron dynamics from spacecraft observation of multi-dimensional solitary waves in various regions of the magnetosphere.

1. Historical Development

In the year of 1957, when the Space Age was inaugurated by launch of the first satellite Sputnik 1, a paper entitled as 'Exact nonlinear plasma oscillations' (Bernstein, Greene and Kruskal, 1957) was published in *Physical Review*. It was shown that equilibrium solutions to the time-independent Vlasov-Poisson equations can be constructed if the velocity distribution functions of electrons and ions are expressed as functions of the Hamiltonian, i.e., the sum of kinetic energy and electrostatic potential energy. The physical model considered in the paper is one-dimensional with variation along the static magnetic field. It is also discussed that there exist potential structures such as bumps and dents that trap some electrons and ions. The equilibrium solution in such a collisionless plasma has generally been called as a BGK mode after the three authors of the paper. However, there was no recognition that BGK modes can really exist in space plasmas.

Ten years later, it was demonstrated by a computer simulation that there appears a solitary BGK mode after nonlinear evolution of an instability induced by

Astrophysics and Space Science **277:** 45–57, 2001.

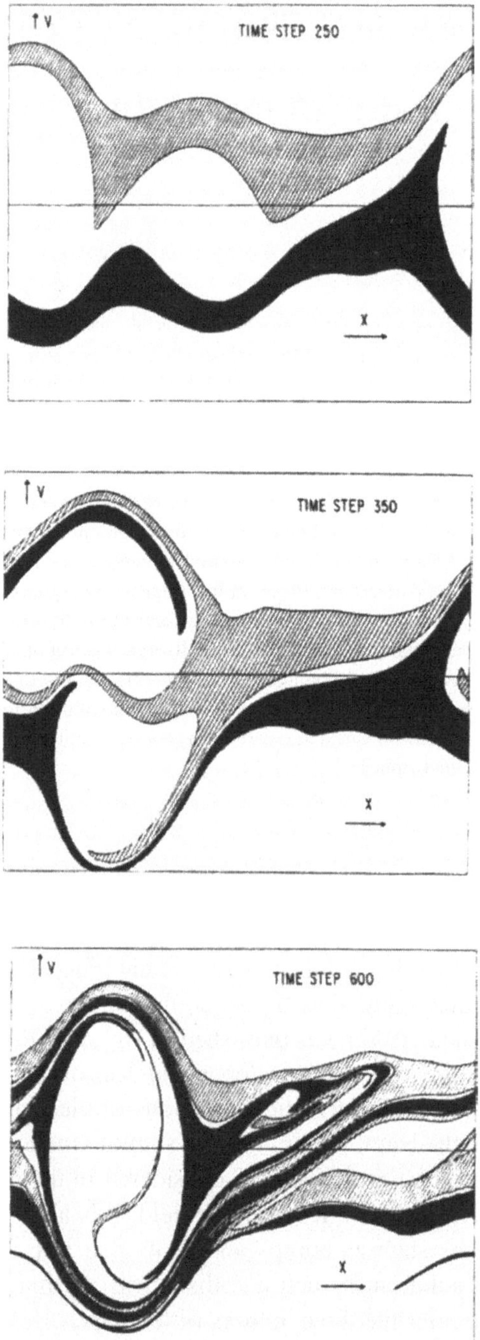

Figure 1. Vlasov Simulation of two-stream instability with a water-bag model (after Berk and Roberts, 1967).

two counter streaming electron beams (Berk and Roberts, 1967). The simulation was performed by solving the one-dimensional Vlasov-Poisson equations for two electron beams represented by simplified step-like velocity distribution functions, which is called 'water-bag' model. In spite of the simplified distribution functions, an essential physics was retained regarding the nonlinear process of formation a large BGK potential after coalescence of smaller potentials that trap the electrons as shown in Figure 1. The simulation was an excellent numerical study achieved by the limited computational capability of 1960's.

In 1979, an experiment was conducted in a laboratory plasma where a large amplitude soliton was generated by a pulsed electric field locally imposed by a pair of electrodes (Saeki *et al.*, 1979). After the passage of fast moving solitons, there arise small potentials that coalesce with each other to form solitary positive potentials. Through comparison with a Vlasov simulation of the bump-on-tail instability of electrons, they identified the positive potential as the one that traps electrons, and they named it as an electron hole because it appears as a hole in the velocity-position phase space.

While such progresses of physics on nonlinear processes were made in theoretical and numerical studies and laboratory experiments, Gurnett *et al.* (1976) reported a space observation of a new type of plasma wave emissions by the IMP-8 satellite in the plasma sheet boundary layer of the Earth's magnetotail. The emission was purely electrostatic, and it had a broadband spectrum from the low frequency to the electron plasma frequency. The emissions were named as 'broadband electrostatic noise (BEN)'. Since the plasma wave instrument could only measure frequency spectra, it was believed that the emission was superposition of waves unstable over a wide range of wave frequency. Some of the linear dispersion relations for strongly unstable plasmas consisting of electron and/or ion beams show a beam mode unstable over a wide frequency range. However, simulation studies, conducted during 1980's (e.g., Schriver *et al.*, 1987) showed that such a strong instability diffuses the energetic electrons and ions, and it saturates as a normal mode in a quasi-equilibrium thermal plasma. Thus such efforts to explain BEN by the linear theory were not successful.

The real nature of the BEN emissions was revealed by a plasma wave instrument on board the Geotail spacecraft in the plasma sheet boundary layer. The instrument had a capability of recording waveforms of plasma waves as well as their spectra (Matsumoto *et al.*, 1994). During the event of BEN emissions as identified by the frequency spectra obtained by the multi-channel analyzer (MCA) and the sweep frequency analyzer (SFA), the waveform capture (WFC) recorded a series of coherent solitary waves (electric field) with polarizations turning positive to negative as well as negative to positive. It was revealed that the BEN emissions are anything but broadband noise. These solitary waves are electrostatic in most cases, because the polarizations of the bipolar pulses are mostly along the static magnetic field without any transverse magnetic field components. The pulses are occasionally accompanied by monopolar variation of electric field perpendicular

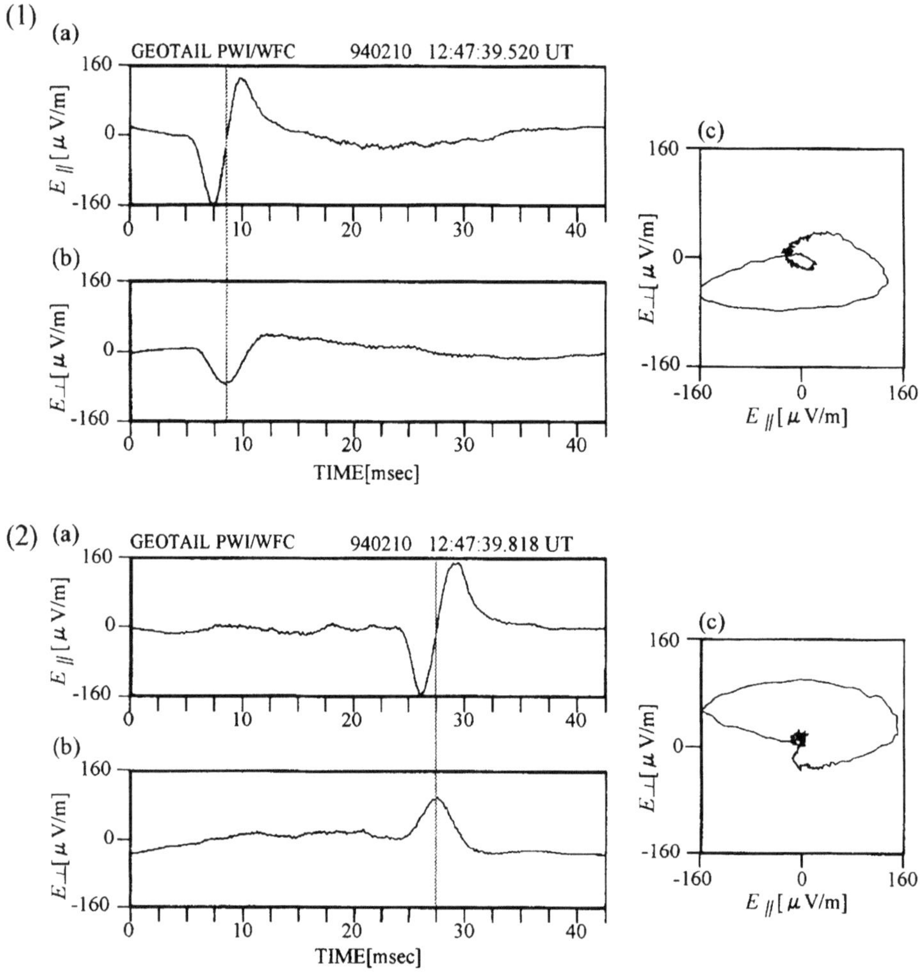

Figure 2. Geotail observation of electrostatic solitary waves.

to the ambient magnetic field as shown in Figure 2. Since the time variation of these solitary waves are of the order of a few milliseconds which is the same order of time as the electron plasma oscillation period in the magnetotail, the solitary pulses correspond to potential structures composed by dynamics of electrons, i.e., positive solitary potentials.

As we have described above, these positive potentials were reported previously as electron holes found in the Vlasov simulations and the laboratory experiment. Since we also had experiences to have observed electron holes in test runs of particle simulation codes for electron two-stream instabilities, the solitary waveforms observed by the Geotail spacecraft were somewhat familiar to us. From close similarity between the observed waveform and those found in the simulations, we readily assumed positive potentials moving along the magnetic field with a high

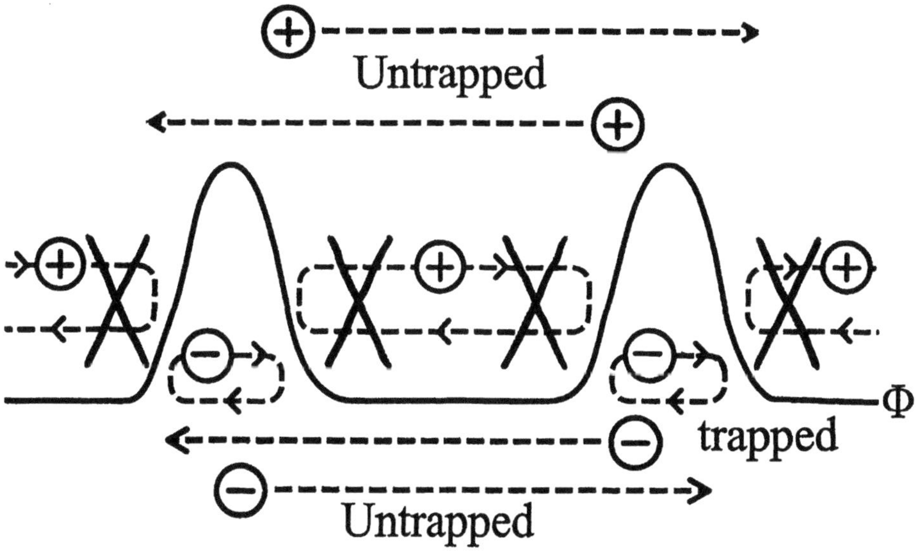

Figure 3. Schematic illustration of electron holes and ion dynamics.

velocity as an explanation of the bipolar waveforms of the electric field. However, it was not clear how such a two-stream instability that generates the electron holes is taking place in the plasma sheet boundary layer of the magnetotail. Since correlation with the ion dynamics was also indicated in the observation of BEN (Kojima *et al.*, 1994), we clarified the effect of the ion thermal velocity and the density ratios of two electron beams (Omura *et al.*, 1994).

2. Generation Mechanism of Electron Holes

One of the most important findings in a series of simulation runs with different parameters is a recognition that electron holes are BGK modes formed only by electron dynamics. The BGK equilibrium of the electron holes should not be disturbed by ion dynamics. Ions can disturb the equilibrium, when they are trapped by the potential wells formed by two positive potentials corresponding to the adjacent electron holes as schematically illustrated in Figure 3. Let us think of the frame of reference moving with the electron holes. If the drift kinetic energy is smaller than the potential energy of the well, the ions are trapped to destroy the potential well by their large inertia effect. The necessary condition for electron holes not to be disturbed by the ions is that the ions are moving with a kinetic energy larger than the depth of the potential well. In other words, the electrons forming electron holes should be moving at a high relative velocity with respect to the ions. In a simulation run with two initially counter streaming electron beams with drift velocities +V and -V and with equal densities, we find formation and coalescence of trapped particle vortices moving with velocities close to zero. If we assume a system where the total current parallel to the static magnetic field is zero, we find ions are at rest being trapped by the potential wells between the vortices which temporarily appear as electron holes. Because of the large momentum of ions, the trapped electrons are strongly disturbed, and they are split into two groups, forward and backward moving smaller electron holes (Omura *et al.*, 1996). If the temperature of ions are cold enough to allow a stable propagation of ion acoustic waves, the electron vortices decay into the ion acoustic modes instead of forming electron holes. As a necessary condition for large amplitude electron holes to be generated, the average velocity of the two electron beams should have a high velocity in the frame of ions. Generally speaking, we can find electron holes in regions where a field-aligned current exists along with a population of warm thermal ions. There is no requirement that the two electron beams have equal densities. As an extreme case, even when the density of one of the electron beams moving with ions is zero, i.e., when there is only a single electron beam moving against the thermal ions, large amplitude electron holes can be generated after a strong instability which is called Buneman instability.

The large amplitude of electron holes mentioned above corresponds to a potential depth that can trap electrons with kinetic energy of the velocity shear between two counter-streaming electron beams or electron and ion beams in the Buneman instability. The amplitude of electrostatic solitary waves (electron holes) observed by the Geotail spacecraft, however, is of the order of 0.1 mV/m, which is much smaller than those found in the two-stream or Buneman instability. With the electric field of 0.1 mV/m, the spatial scale must be 100 km to form even a relatively small potential of 10 V. The duration time of the observed solitary pulses is of 1–10 ms. If the spatial scale is 100 km, the pulses pass by the spacecraft at a velocity of 10 000–100 000 km s^{-1}. Since the 10 V potential is too small to trap such high energy electrons with kinetic energy of a few keV, the observation in the magnetotail contradicts with the generation mechanism via two-stream instability or Buneman instability.

It is not necessary, however, for electron holes to be large enough to trap the whole population of the two-streaming beam electrons as found in the two-stream instability. If there arises a coherent motion that results in a gradient of the velocity distribution function in the phase space, it is possible to construct a BGK equilibrium with a corresponding density gradient in space. A typical example of such coherent motions is found in a bump-on-tail instability induced by a weak electron beam placed on a thermal tail of a warm Maxwellian velocity distribution function. Since the electron beam forms a positive gradient that give rise to a positive growth rate, a coherent Langmuir wave with phase velocity slightly smaller than the drift velocity of the electron beam grows to form a continuous sinusoidal potential structure. The beam electrons are trapped in each potential well of this structure, and they are diffused in the velocity phase space through nonlinear trapping motion, forming a stable velocity distribution function. As we found in the nonlinear evolution of the two-stream instability, the adjacent potentials with trapped electrons coalesce with each other, gradually forming an isolated electron holes in a long time evolution of the bump-on-tail instability. The depth of the potential is determined by the saturation level of the instability. Especially for a relatively weak beam forming a small bump on the high energy tail of the velocity distribution function, small amplitude electron holes can be generated.

The Geotail spacecraft observed electrostatic solitary waves (ESW) in the plasma sheet boundary layer (Matsumoto *et al.*, 1994; Kojima *et al.*, 1999). We found a good correlation between the solitary waveform and the enhancement of high energy electron flux in the high energy tail of the velocity distribution function as shown in Figure 4 (Omura *et al.*, 1999). This shows that ESW are carried by the high energy electron flux drifting with a velocity from a few 1000 km s^{-1} to a few 10 000 km s^{-1}. The source region of ESW is located much far away from the point of observation.

Simulation studies of electron holes have been performed in a system of a periodic boundary, where electron beam and background thermal plasma interact uniformly in space. A more realistic model is a localized source of the electron beam with open boundary conditions. As shown in Figure 5, we continuously inject an electron beam forming a bump-on-tail distribution function from the left boundary. Along the passage of the electron beam, there occurs an instability forming coherent potential structures moving with the electron beams. The beam electrons are trapped in a series of potentials, and there arises an attraction force between electron holes (Krasovsky *et al.*, 1999). They coalesce with adjacent potentials forming isolated electron holes. Since the velocity distribution at the front end of the continuous electron beam is stabilized with the phase trapping, the electron beam can propagate through the background thermal plasma without disturbing them. We performed a long run with a very long one-dimensional open system as shown in Figure 6. We found that the electron holes as identified by enhancement of electric field energy and beam thermal velocity can propagate over a very long distance through the plasma. It is possible to observe electron holes at a distant place far away from the source region where the instability is taking place.

3. Multi-dimensional Electron Holes

Some of the ESW observed by the Geotail spacecraft and other recent spacecraft are accompanied by a perpendicular electric component. The stability of potential structures with such perpendicular components have not been studied theoretically yet except for a numerical simulation by Goldman *et al.* (1999), since the BGK theory is essentially for one-dimensional models. In a two-dimensional simulation study of the bump-on-tail instability, we first find excitation of two-dimensional potential structures in the initial stage of the linear instability. As the time goes on, the two-dimensional structures are aligned in the perpendicular direction with perpendicular electric field diminishing slowly. After the diffusion process in the perpendicular direction, there appear one-dimensional potential structures which

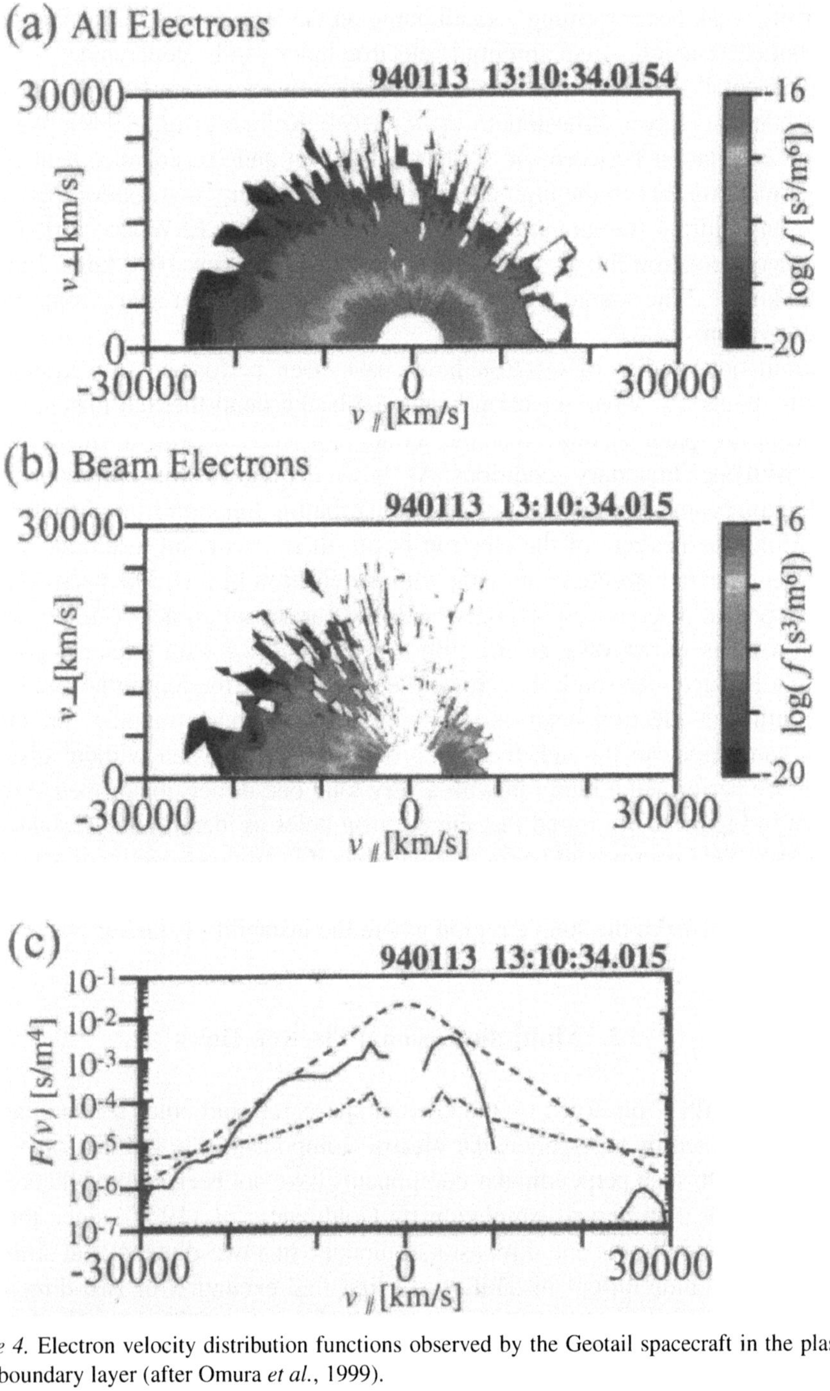

Figure 4. Electron velocity distribution functions observed by the Geotail spacecraft in the plasma sheet boundary layer (after Omura *et al.*, 1999).

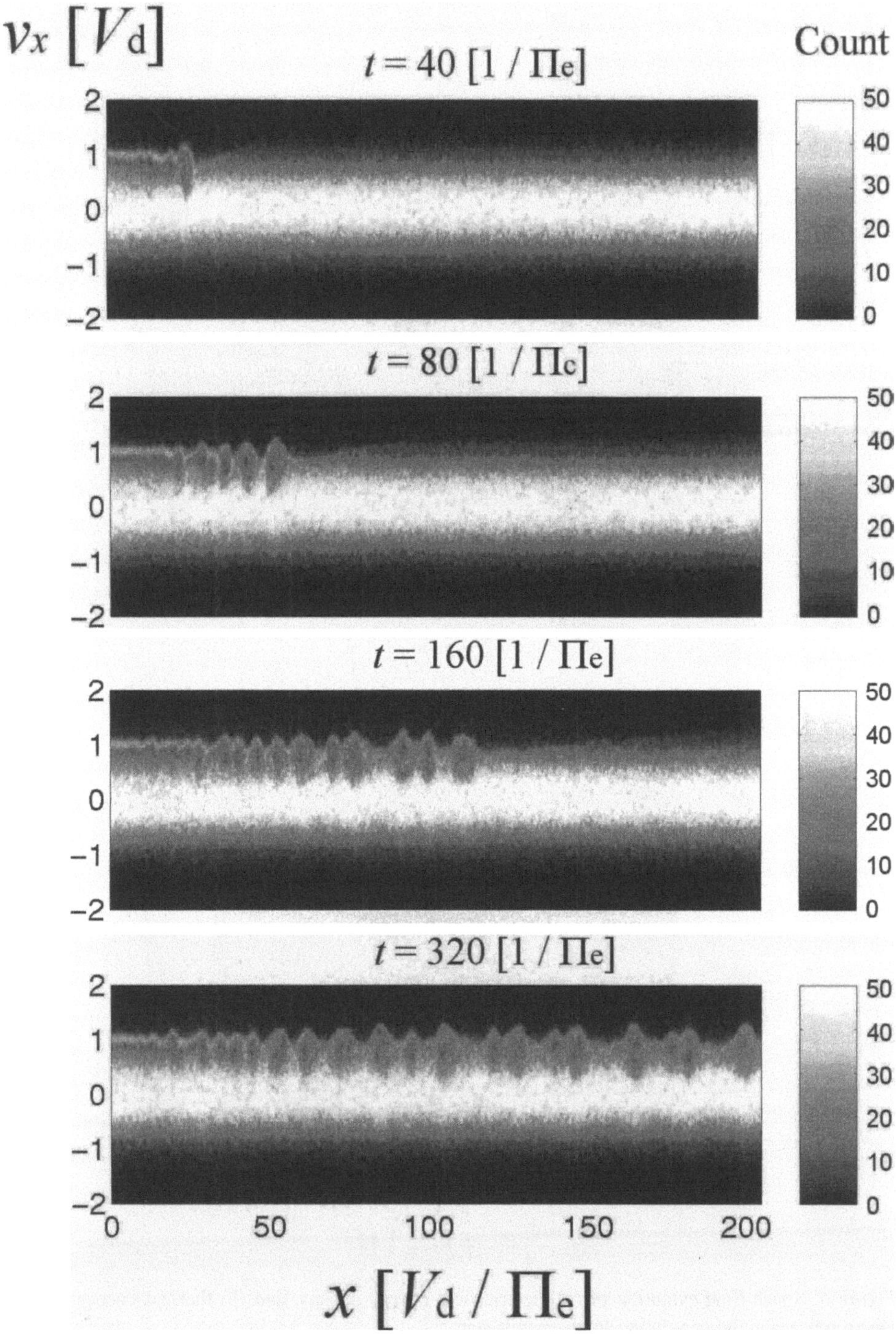

Figure 5. Phase space plots of electron bump-on-tail instability in an open system. The density of the electron beam is 5% of the total density. Vd (beam)/Vth(background) = 2.

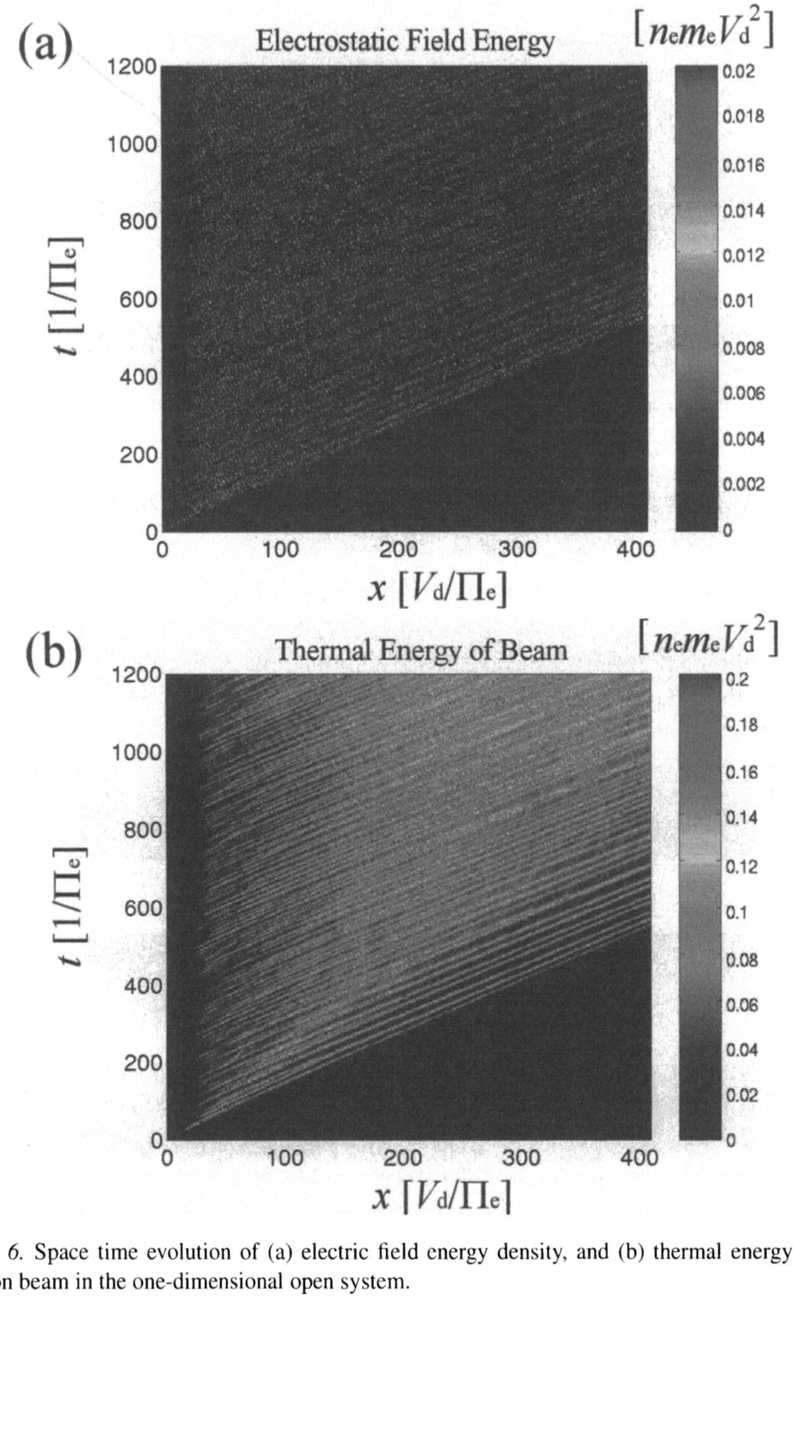

Figure 6. Space time evolution of (a) electric field energy density, and (b) thermal energy of the electron beam in the one-dimensional open system.

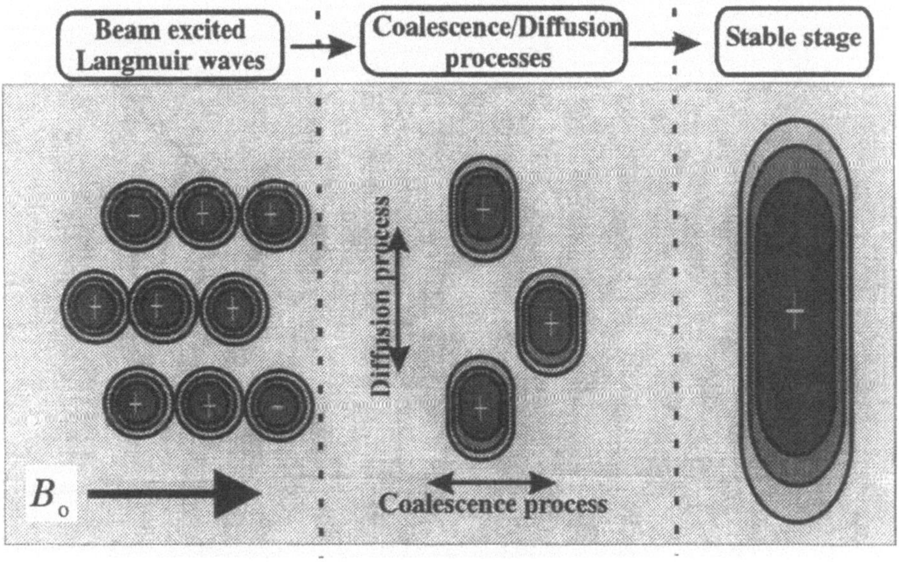

Figure 7. Schematic illustration of evolution of potential structures in a two-dimensional system.

are essentially the BGK modes (Miyake *et al.*, 1998). If we assume the two-dimensional structures represent a transient process in the long time evolution of the system as schematically illustrated in Figure 7, they indicate that the source region of the instability, where we can find an unstable electron beam, is closer for the two-dimensional potentials than that for the one-dimensional potentials. The structures of the potentials can be inferred from the polarization of observed waveforms as schematically illustrated by Figure 8. A preliminary study based on a statistical analysis of ESW data observed by the Geotail spacecraft have been conducted in order to identify the acceleration region of electrons that are associated with the magnetic reconnection process in the magnetotail (Omura *et al.*, 1999).

4. Conclusion

Solitary waves have been found in various regions of the magnetosphere other than the plasma sheet boundary layer via waveform observations by recent spacecraft such as the Wind (Bale *et al.*, 1998) Polar (Franz *et al.*, 1998), and Fast (Ergun *et al.*, 1998). Simulation studies show these solitary waves are electron holes resulting from dissipative small scale phenomena such as electron beam instabilities along the magnetic field. The characteristics of these electrons holes such as amplitude, duration time, and multi-dimensionality are different depending on variety of plasma parameters in the magnetosphere. More theoretical and numerical studies of multi-dimensional electron holes, which cannot be described by the simple BGK theory, are necessary for extraction of information concerning the micro-

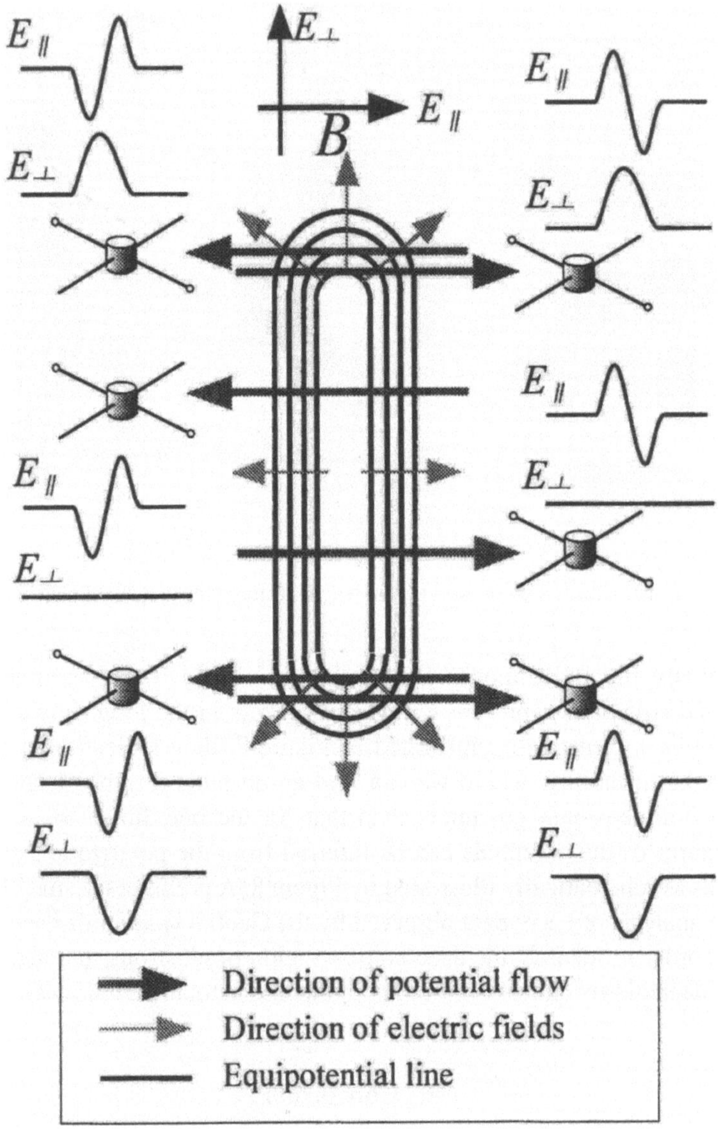

Figure 8. Relation between observed waveforms and a two-dimensional potential structure.

scopic electron dynamics that are difficult to be observed directly by spacecraft instruments.

Acknowledgement

The computer simulation was done on the KDK computer system at Radio Science Center for Space and Atmosphere, Kyoto University.

References

Bale, S.D., Kellogg, P.J., Larson, D.E., Lin, R.P., Goetz, K. and Lepping, R.P.: 1998, Bipolar electrostatic structures in the shock transition region: Evidence of electron phase space holes, *Geophys. Res. Lett.* **25** (15), 2929–2932.

Berk, H.L. and Roberts, K.V.: 1967, Nonlinear Study of Vlasov's Equation for a Special Class of Distribution Functions, *Phys. Fluids* **10**, 1595–1597.

Bernstein, I.B., Greene, J.M. and Kruskal, M.D.: 1957, Exact nonlinear plasma oscillations, *Phys. Rev.* **108**, 546–550.

Ergun, R.E., Carlson, C.W., McFadden, J.P., Mozer, F.S., Delory, G.T., Peria, W., Chaston, C.C., Temerin, M., Roth, I., Muschietti, L., Elphic, R., Strangeway, R., Pfaff, R., Cattell, C.A., Klumpar, D., Shelley, E., Peterson, W., Moebius, E. and Kistler, L.: 1998, FAST satellite observations of large-amplitude solitary structures, *Geophys. Res. Lett.* **25** (12), 2041–2044.

Franz, J.R., Kintner, P.M. and Pickett, J.S.: 1998, POLAR observations of coherent electric field structures, *Geophys. Res. Lett.* **25** (8), 1277–1280.

Goldman, M.V., Oppenheim, M.M. and Newman, D.L.: 1999, Nonlinear two-stream instabilities as an explanation for auroral bipolar wave structures, *Geophys. Res. Lett.* **26** (13), 1821–1824.

Gurnett, D.A., Frank, L.A. and Lepping, R.P.: 1976, Plasma waves in the distant magnetotail, *J. Geophys. Res.* **81**, 6059–6071.

Kojima, H., Matsumoto, H., Miyatake, T., Nagano, I., Fujita, A., Frank, L.A., Mukai, T., Patersonk, W.R., Saito, Y., Machida, S. and Anderson, R.R.: 1994, Relation between electrostatic solitary waves and hot plasma flow in the plasma sheet boundary layer: GEOTAIL Observations, *Geophys. Res. Lett.* **21**, 2,919–2,922.

Kojima, H., Omura, Y., Matsumoto, H., Miyaguti, K. and Mukai, T.: 1999, Characteristics of Electrostatic Solitary Waves Observed in the Plasma Sheet Boundary: Statistical Analyses, *Nonlinear Processes in Geophysics* **6**, 179–186.

Krasovsky, V.L., Matsumoto, H. and Omura, Y.: 1999, Interaction of Small Phase Density Holes *Phys. Scripta* **60**, 438–451.

Matsumoto, H., Nagano, I., Anderson, R.R., Kojima, H., Hashimoto, K., Tsutsui, M., Okada, T., Kimura, I., Omura, Y. and Okada, M.: 1994, Plasma Wave Observations with GEOTAIL Spacecraft, *Journal of Geomagnetism and Geoelectricity* **46**, 59–95, 1994.

Matsumoto, H., Kojima, H., Miyatake, S., Omura, Y., Okada, M., Nagano, I. and Tsutsui, M.: 1994, *Geophys. Res. Lett.* **21**, 2,915–2,918.

Miyake, T., Omura, Y., Matsumoto, H. and Kojima, H.: 1998, *J. Geophys. Res.* **103**, 11,841–11,850.

Omura, Y., Kojima, H. and Matsumoto, H.: 1994, Computer Simulation of Electrostatic Solitary Waves: A Nonlinear Model of Broadband Electrostatic Noise, *Geophys. Res. Lett.* **21**, 2,923–2,926.

Omura, Y., Matsumoto, H., Miyake, T. and Kojima, H.: 1996, Electron beam instabilities as generation mechanism of electrostatic solitary waves in the magnetotail, *J. Geophys. Res.* **101**, 2,685–2,697.

Omura, Y., Kojima, H., Miki, N., Mukai, T., Matsumoto, H. and Anderson, R.: 1999, Electrostatic solitary waves carried by diffused electron beams observed by GEOTAIL spacecraft, *J. Geophys. Res.* **104**, 14,627–14,637.

Saeki, K., Michelsen, P., Pecseli, H.L. and Juul Rasmussen, J.: 1979, *Phys. Rev. Lett.* **43**, 501–504.

Schriver, D. and Ashour-Abdalla, M.: 1987, Generation of high-frequency broadband electrostatic noise: The role of cold electrons, *J. Geophys. Res.* **92**, 5807–5819.

INSTABILITIES AND FORMATION OF COHERENT STRUCTURES

F. MOTTEZ
Centre d'études en Environnements Terrestre et Planétaires (CETP-CNRS)
E-mail: fabrice.mottez@cetp.ipsl.fr

Abstract. This tutorial paper is devoted to theoretical aspects of the coherent electrostatic structures that have been encountered in various space plasmas. These structures, called solitary waves, electrostatic shocks or double layers have been observed in the solar wind, the Earth bow shock, the auroral zones and the magnetotail. Most of these structures can be interpreted in terms of electron or ion phase space holes. Their 1D structure, their emergence from plasma instabilities, their mutual interactions, their bidimensional stability, and their ability to create large scale electric fields are discussed in this paper.

1. Introduction

Small scale coherent electrostatic structures have a limited extent in the direction parallel to the ambient magnetic field (usually of the order of a few Debye lengths), they can be seen on electric antennas, provided the required time resolution, but they have no magnetic counterpart. The fluctuation of electric potential associated to these structures is usually comprised between a fraction of Volt, to a hundred Volts. They generally move in the plasma; some of them seem to be stable, at least over a fraction of a second.

The discovery of small scale coherent electrostatic structures started in the high altitude auroral zone of the Earth with the S3-3 satellite (Temerin *et al.*, 1982), they were confirmed by the other probes devoted to the study of that region (Bostrom *et al.*, 1988; Ergun *et al.*, 1998). They focused attention because their presence seems to be connected to the auroral acceleration processes, to which a lot of work has been devoted. Since then, small scale coherent electrostatic structures were found in the mid altitude auroral zone (Dovner *et al.*, 1994), in the solar wind (Mangeney *et al.*, 1999), at the Earth bow schock (Bale *et al.*, 1998) and in the tail of the magnetosphere (Matsumoto *et al.*, 1994). Most of these observations are reviewed in Salem *et al.* (1999).

The origin of these structures can be found, although not systematically, in the nonlinear evolution of beam plasma instabilities. A lot of work based on theory and simulation has been done on the electrostatic coherent structures. The aim of this article is neither to present a complete review of these works nor to give a precise account of new original results, but to emphasize a few points that seem important for a good understanding of the physics of these structures. These points especially

Astrophysics and Space Science **277:** 59–70, 2001.

concern their one dimensional equilibrium, the processes of their creation through beam instabilities, the interaction between several coherent structures, the question of their bidimensional stability, and their role in acceleration processes.

Three (original) figures displayed in this paper come from a simulation that has already been described in details in a former paper (Mottez *et al.*, 1997). The aim of these figures is to illustrate some general concepts that are presented in this paper, it is not to report about recent scientific results.

2. The Main Features of an Electrostatic Coherent Structure

The onedimensional theory of solitary electrostatic waves has been developed by Bernstein, Greene and Kruskal (1957) long before the discovery of coherent structures in space plasmas. It is a nonlinear kinetic theory formulated in the onedimensional Vlasov-Poisson framework of equations.

The idea is to set 'arbitrarily' an electrostatic profile $\Phi(z)$. The particles of the plasma, according to the difference between their kinetic energy K and the electrostatic potential $q\Phi(z)$, will adopt different behaviors: trapped inside the structure, reflected (outside), free, and sometimes accelerated. A lot of possible distribution functions can be the solutions of the Vlasov-Poisson equations associated to $\Phi(z)$. Quite often, the distribution functions $f(z, v)$ of all the populations (free, reflected ...) can be fixed, 'arbitrarily', except one, for instance, $f_t(z, v)$ for the trapped particles. Then, the Vlasov-Poisson equation is solved for $f_t(z, v)$. If the solution is positive (a necessary condition for a distribution function), it is a model of a onedimensional electrostatic coherent structure, usually called a BGK wave.

Among the possible BGK waves, some involve a density of trapped particles lower than the density of free and reflected particles: they are called phase space holes (Schamel, 1982). This name does not necessarily means that the total density (i.e. the distribution $f(x, v)$ integrated over the velocities) is much lower, because the lack of particles in the 'trapping area' of the phase space can be counterbalanced by the density of passing particles. Most of the coherent electrostatic structures encountered in space plasmas are interpreted in terms of phase space holes, ion phase space holes (or shortly 'ion holes') or electron holes.

The small scale coherent structures encountered in space have an electrostatic potential energy that is usually smaller than the thermal energy of the plasma (Salem *et al.*, 1999) (apart from FAST measurements [Ergun *et al.*, 1998]). Therefore, their description must take into account both trapped and free particles: they require a kinetic approach. The electron hole structures are associated to a positive potential perturbation, as shown in Figure 1, they involve trapped and free electrons, and reflected and free ions. The ion holes are associated to a negative potential perturbation, they involve trapped and free ions, and reflected and free electrons. Some structures are associated to a net potential drop, they are called double layers.

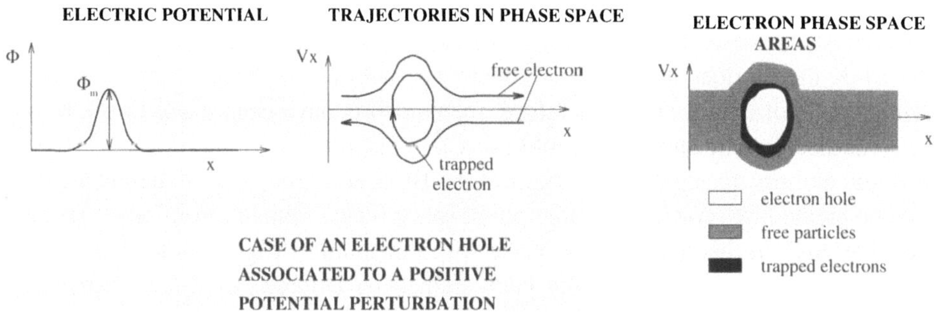

Figure 1. Structure of an electron phase space hole.

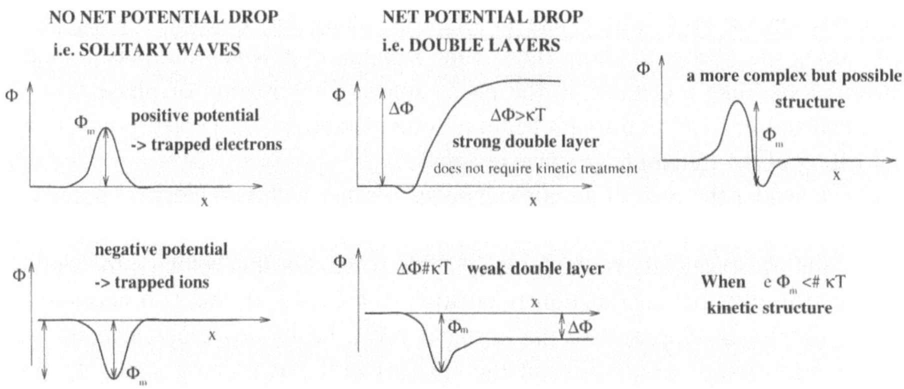

Figure 2. Some kinds of coherent structures.

The coherent electrostatic structures mentioned in this paper are referred as phase space holes. This is the point of view of someone who watches the phase space density. If someone is more interested in the motion of the particles in the phase space, the sames structures will be refered as phase space vortices. But, the phase space is not directly accessible from observational data. Therefore the same objects will be also called solitary waves, electrostatic shocks or double layers, according to the point of view of someone who watch the electric field measured onboard spacecraft. Let us notice that the word solitary wave does not imply that these structures must be isolated. They sometimes appear as isolated structures, and sometimes in large amounts and close to each other.

Most potential structures observed in the high altitude auroral zone ($\sim$ 10000 km) onboard Viking (Bostrom *et al.*, 1988) seem to be associated to ion phase space holes, whether they carry a net potential drop (double layers) or not (solitary waves). Large amplitude structures observed at lower altitude ($\sim$ 1000 km), by FAST are interpreted as electron phase space holes (Ergun *et al.*, 1998). The structures observed in the magnetotail (Matsumoto *et al.*, 1994) and bow schock (Bale *et al.*, 1998) are interpreted as electron phase space holes.

3. How to Create Phase Space Holes with Instabilities

The BGK theory does not say how coherent structures are created, they just describe a state of equilibrium. Therefore, many efforts have been made to see how coherent electrostatic structures could grow in a plasma.

Some authors questioned the origin of the BGK solution: could coherent structures be created by growing unstable phase space holes, initiated from phase space perturbations? In this approach, a phase space instability 'digs' a hole in the z, v phase space. This idea has initially been studied by Dupree (1982) and Tetreault (1991); it is not in the scope of this paper.

An another possibility is the creation of phase space holes through beam-plasma instabilities. The phase space of a plasma with a beam can be interpreted as a plasma distribution with a hole that is localized along the velocity v axis, and infinite along the z axis: the hole (that is the separation between the plasma and the beam) looks like a channel in the phase space. The creation of phase space holes (localized in z and v) through beam plasma instabilities just corresponds to a partial filling of the initial phase space channel.

Let us consider the case of an homogeneous plasma with two electron populations with different average velocities, i.e. an electron phase space with a channel. Let v_ϕ be the phase velocity of the most unstable wave. For this velocity, the slope of the electron distribution function is positive $\partial_v f(v_\phi) > 0$: this is a necessary condition for the development of the unstable wave. From a topological point of view, this corresponds to a border of the channel in the x, v phase space. If the spectrum of unstable waves has a narrow range of phase velocities, particles with a velocity close to the phase velocity oscillate with an average velocity equal to v_ϕ. In the theoretical frame of particle trapping by a monochromatic wave (case of an infinitely narrow spectrum of unstable waves) (O'Neil, 1965; Al'thsul and Karpman, 1966) the criterion for a particle to be trapped is

$$\Delta v = \|v - v_\phi\| < \frac{2qE}{mk} \tag{1}$$

were E is the amplitude of the wave, and k its wave number. The trapped particle oscillates in the potential well with a frequency

$$\omega_t = \sqrt{\frac{qEk}{m}}. \tag{2}$$

the trapping frequency ω_t is much higher than the wave frequency ω. The length Δz of the potential well is the wavelength

$$\Delta z = \frac{2\pi}{k}. \tag{3}$$

When the amplitude of the wave is small, the number of trapped particles is low because v_ϕ corresponds to a region of low density in the phase space. But if the

TABLE I
Parameters of Simulation

Simulation	B (nT)	T_i (eV)	T_e (eV)	Beams Velocities	m_i/m_e	$\Delta t\omega_{pe}$	Size, (λ_D)	Part./Cell
two-stream	20	400	100	$+2.5v_t$ and $-2v_t$	1836	0.1	512 × 16	400

instability does not saturate too early, the amplitude E reaches a high enough level to trap a few particles of the beam, or even particles of the core distribution. In that case, particles from the high density regions penetrate into the phase space domain of initially low density (the channel). The channel is destroyed (partially filled), it is replaced by a succession of vortices of dimensions Δz and Δv given by the equations 1 and 3. The middle of a vortice, centered at v_ϕ and propagating at this velocity, is in a low density area because it corresponds to the low density of particles trapped by the initially low amplitude wave. These vortices are electron phase space holes.

The case of the nonlinear evolution of an electron two-stream instability that leads to the growth of various kinds of electron holes has been studied in Mottez *et al.* (1997) and will be used as an example all along this paper. The parameters of the simulations are summarized in Table I. A magnetized plasma composed of an ion population and two electron populations with the same density and opposite velocities is unstable. The most unstable wave has a wave vector along the ambient magnetic field, its real frequency is equal to zero. Mottez *et al.* (1997) have shown that the growth rate is high, the spectrum is rather narrow, and strong coherent structures develop: electron phase space holes. Figure 3 shows a succession of large amplitude electron phase space holes created through the nonlinear evolution of the electron two-stream instability.

The growth of electron coherent structures in the magnetotail, interpreted as electron holes, has been modelized as the nonlinear evolution of an electron-beam plasma instability (Omura *et al.*, 1996) or in terms of an electron two-stream instability (Mottez *et al.*, 1997).

The growth of ion phase-space holes can also be explained as a product of the nonlinear evolution of a beam-plasma instability. For instance, ion phase space holes that propagate at the sound velocity can be produced in a plasma destabilized by an ion beam through an ion acoustic instability (Sato and Okuda, 1981; Chanteur *et al.*, 1983).

As already mentioned, the scenario described in this section, based on particle trapping by a wave, requires an almost monochromatic spectrum of unstable waves. If the spectrum is more spread in phase velocity $v_\phi(k)$, other effects (phase mixing, quasilinear diffusion) invalidate particle trapping and saturate the wave amplitude before its amplitude is high enough to catch particles of the beam and the core distribution. This is why the growth of coherent electrostatic structures is in general

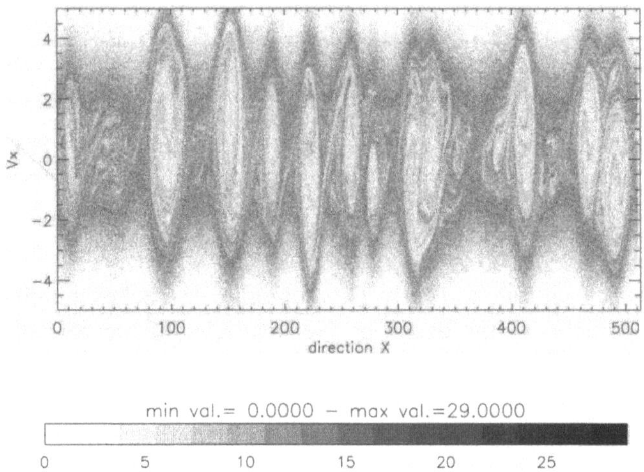

Figure 3. Two-stream instability, 1D simulation. Electron phase-space (z, v) density at time $t = 100\omega_{pe}$ showing a sucession of phase-space holes surrounded by trapped particles.

obtained from plasma instabilities with a narrow spectrum of unstable waves. In other words coherent structures require rather monochromatic waves. When monochromatic waves reach a high amplitude, they can trap particles and create vortices (phase space holes), but they can also couple nonlinearly with other waves. In that case, things can become more intricate, but it does not necessarilly rule out the possibility of a growth of coherent structures, as shown in Mottez *et al.* (1992).

4. How do Phase Space Holes Interact?

The above scenario, often observed in numerical simulations shows that phase space holes are generally not isolated at all, but that there is about one structure at every wavelength. They have the same velocity $\sim v_\phi$ and they can be close enough to each other to interact. Berk *et al.* (1970) have shown that phase space holes attract each other. The basic idea of their demonstration is very simple. Instead of considering a phase space of particles with charge q and mass m, that is roughly a band in the z, v space (stable), with holes, they consider a stable band of particles q, m and describes the holes as assemblies of particles of charge $-q$ and negative mass $-m$. These particles, that represent the holes, interact electrostatically. The electrostatic interaction between particles of the same charge and negative mass is attractive, therfore, each phase space holes remain coherent, and phase space holes attract each other.

This attractive interaction between phase space holes leads to collisions, and, unlike solitons, there is no reason to suppose that these collisions are elastic. On

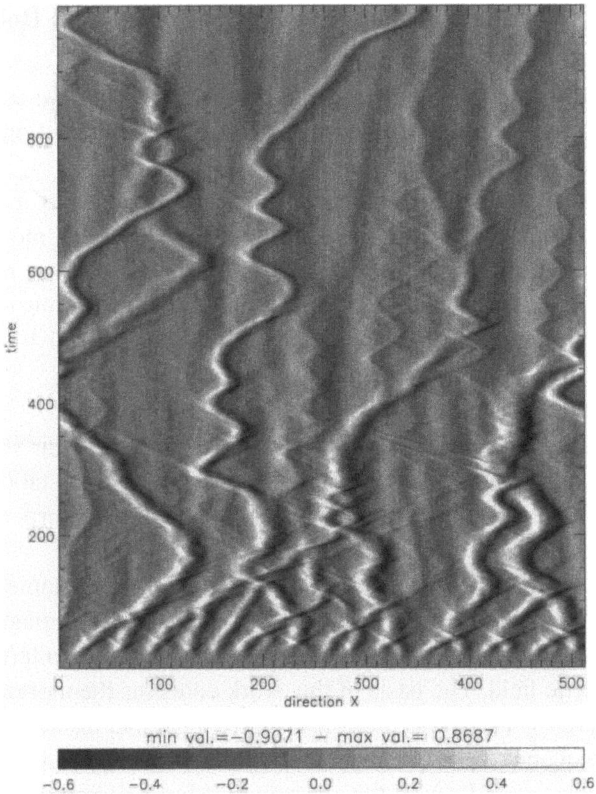

Figure 4. Two-stream instability, 1D simulation. Evolution of the parallel electric field E_z as a function of the position z and time t.

the contrary, simulations show that new kinds of holes are created during collisions, and some other are destroyed.

Figure 4 shows the space time evolution of the parallel electric field $E_z(z, t)$, from the onedimensional simulation of a two-stream instability (the same as for Figure 3). The regions of high electric field correspond to electron holes. We can see that the initial wave, with a null phase velocity, creates electron holes with a null velocity. But, after a few milliseconds, the holes start to attract each other. There are many kinds of interaction. Sometimes the holes cross each other, once or several times. Quite often, they merge and give a stronger electron hole. In most cases, the interactions give rise to one or two smaller and faster structures. The small and fast structures are stable, unless they are destroyed by a later collision with a large amplitude hole.

5. The Bidimensional Stability of Phase Space Holes

The theories presented up to now are onedimensional. We have seen that isolated space holes are very stable in many cases, and that their interactions tend to create new phase space holes, of various kinds.

Nevertheless, not all the 1D coherent structures are stable. For instance, the effect of the ions on electron holes must already be taken into account into a onedimensional theory. It has been shown that the ions that are reflected against an electron hole damp it; therefore, to be stable, a structure must be centered on a velocity that is quite different from that of the ion core distribution, so as to minimize the number of reflected ions.

Many structures that are stable in 1D are unstable in 2D.

From observations, we do not know what is the perpendicular size of the coherent electrostatic structures. Theoretical criteria about the 2D stability of coherent structures might help us to determine what are the maximum perpendicular lengths of these structures.

Very recent theoretical results on the question have been obtained by Muschietti (1999). These results concern electron phase space hole in a magnetized plasma. These structures are aligned with the magnetic field, and have a large extention across the magnetic field. The basis of this work concerns the motion of the trapped electrons (Figure 6). The trapped electrons have a parallel oscillatory motion, with the trapping frequency ω_t (Equation 2). Their perpendicular motion is cyclotronic, with a frequency $\omega_c = qB/m$. If the electron hole has a perpendicular perturbation, the shape of the hole is curved. Let us suppose first that there is no magnetic field, Muschietti has shown that the density of trapped oscillating electrons is increased in some places and lowered in others, in a way that reduces the electron density where the electron hole has already moved: the perturbation is amplified. Therefore, an electron hole in an plasma with a negligible magnetic field is unstable. Ultimately, the structure will be destroyed. If there is a magnetic field, the perpendicular gyromotion mix the electrons, and counteract this effect. Muschietti has established that the criterion of instabilty of the structure is $\omega_c < \omega_t$. This criterion is completed by a requirement about the perpendicular thermal velocity of the electrons.

The electron phase space, taken from the same simulation as Figures 3 and 4 is shown at time $t = 500\omega_{pe}$ on Figure 5. We can see that, because of the hole-hole interactions, the morphologies of the phase space electron holes are more varied than in Figure 3. We still can see the large amplitude and slow electron holes that appear in Figure 3; but there are also many structures centered on higher velocities (positive or negative) and of smaller amplitude. Mottez *et al.* (1997) have performed 2D simulations with the same parameters. It is shown that in the 2D simulations, the large amplitude electron holes are unstable, they are destroyed in 150 milliseconds. But before being destroyed completely, they have time to interact and to create the faster structures of smaller amplitude that we can see on

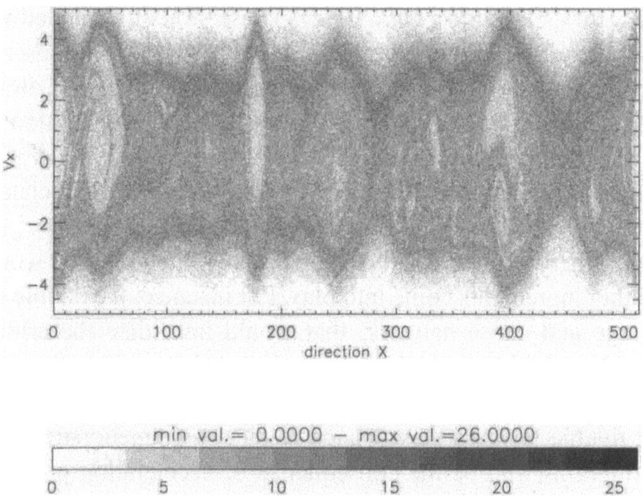

Figure 5. Two-stream instability, 1D simulation. Electron phase-space (z, v) density at time $t = 500\omega_{pe}$.

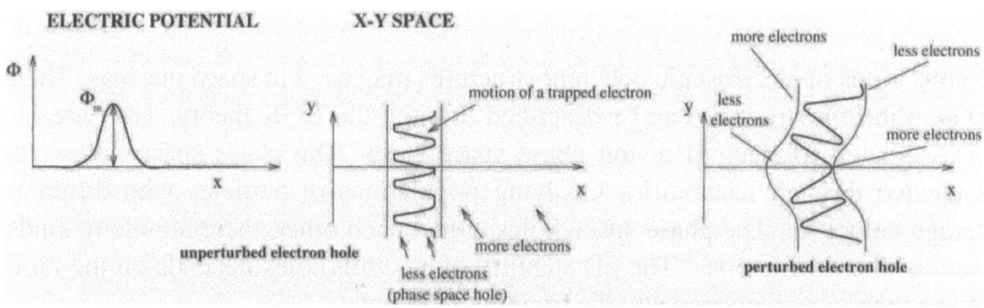

Figure 6. Transverse destabilization of an electron hole.

Figure 5. The fast and small amplitude structures are stable. The theory developped by L. Muschietti has been applied recently to the 2D simulation data shown in these simulations. It successfully explains the instability (in 2D simulations) of the large amplitude electron holes, and the stability of the small and fast electron holes (Muschietti, 2000). Moreover, contrary to the case of the 1D simulations where the small structures are destroyed by further interactions with large amplitude structures (Figure 4), in 2D simulations, the large structures are destroyed after 150 ms, and the small structures remain till the end of the simulations (that represent almost one second).

6. Can Double Layers Accelerate Particles?

Coherent structures focused attention when they were discovered in the acceleration regions of the auroral zone because most of them carry a net potential drop, of the order of 1 mV (Temerin *et al.*, 1982). These structures, called weak double layers, seem to be ion holes (Boström *et al.*, 1988). As they are observed in great numbers, they were suspected of carrying a large scale potential drop of several kV, that would be enough to explain the auroral particle acceleration. But, it was mentioned that such structures could not explain alone the acceleration (Bryant, 1990): they are electrostatic structures, and the charge density associated to them is very similar to an electrostatic dipole. Therefore, at large distances r, the potential decreases as r^{-2}, therefore, the resulting large scale potential drop would be negligible, unless other ingredients come into play. For instance, the double layers might constitute a large and dense network, that would invalidate the critics based on the electric dipole model because it refers to isolated structures. But experimental data does not seem to confirm this hypothesis (Malkki *et al.*, 1993). Another possiblilty is that double layers are associated to electromagnetic structures of larger scales. They might as well be the consequence of acceleration process, a kind of structuring of electric fields that are generated by other processes.

7. Conclusion

Various kinds of electrostatic coherent structures may exist in space plasmas. Their 1D equilibrium structure can be described through the BGK theory. They are often associated to electron or ion phase space holes. The phase space holes can be created through instabilities involving populations of particles with different average velocities. The phase space holes attract each other, there are many kinds of collisional interactions. The 2D stability of electron holes depends on the ratio between the gyrofrequency and the trapping frequency.

Double layers are observed in plasma acceleration regions. However, they cannot be the only cause of acceleration. They might be as well a consequence of other acceleration processes.

The developpment of coherent structures through homogeneous plasma instabilities cannot always be invoked. This is the case of the coherent ion acoustic structures observed in the solar wind, were no source of free energy that would destabilize the plasma appears clearly in the particle distribution functions (Salem *et al.*, 1999). Some other processes must be invoked. Verga *et al.* (1988) have shown that low current, under the threshold of linear instability, can trigger the growth of weak double layers. The effect of time varying boundary conditions (Mangeney, 2000) that could be induced by larger scale fluctuations could also provide an explanation to the weak double layers observed in the solar wind.

Acknowledgements

I wish to thank the organizing commitee of 'Les rencontres de l'Observatoire' for inviting me to give a talk entitled 'Instabilities and Formation of Coherent Structures' on which the present paper is based.

The numerical computations have been done on a C98 at the IDRIS/CNRS computational facility (France).

References

Al'thsul, L. and Karpman, V.I.: 1966, Theory of nonlinear oscillations in a collisionless plasma, *Soviet Phys. J.E.T.P.* **22**, 361.

Bale, S.D., Kellog, P.J., Larson, D.E., Lin, R.P. and Lepping, R.P.: 1998, Bipolar electrostatic structures in the shock transition: evidence of electron phase space holes, *Geophys. Res. Lett.* **25**, 2929–2932.

Berk, H.L., Nielsen, C.E. and Roberts, K.V.: 1970, Phase space hydrodynamics of equivalent nonlinear systems: Experimental and computational observations, *Phys. Fluids* **13**, 980.

Bernstein, I.B., Green, J.M. and Kruskal, M.D.: 1957, Exact nonlinear plasma oscillations, *Phys. Rev.* **108**, 546.

Boström, R., Gustafsson, G., Holback, B., Holmgreen, G., Koskinen, H. and Kintner, P.: 1988, Characteristics of solitary waves and weak double layers in the magnetospheric plasma, *Phys. Rev. Lett.* **61** (1), 82.

Bryant, D.A.: 1990, *Phys. Scripta* **T30**, 215.

Chanteur, G., Adam, J.C., Pellat, R. and Volokhitin, A.S.: 1983, Formation of ion acoustic double layers, *Phys. Fluids* **26**(6), 1584.

Dovner, P.O., Eriksson, A.I., Boström, R. and Holback, B.: 1994, Freja multiprobe observations of electrostatic solitary structures, *Geophys. Res. Lett.* **21**, 1827.

Dupree, T.H.: 1982, Theory of phase-space density holes, *Phys. Fluids* **25**, 277.

Ergun, R.E., Carlson, C.W., McFadden, J.P., Mozer, F.S., Delory, G.T., Peria, W., Chaston, C., Temerin, M., Roth, I., Muschietti, L., Elphic, R.C., Strangeway, R., Pfaff, R., Cattel, C.A., Klumpar, D., Shelley, E., Peterson, W., Moebius, E. and Kistler, L.: 1998, FAST satellite observations of large-amplitude solitary structures, *Geophys. Res. Lett.* **25**, 2041.

Mälkki, A., Eriksson, A.I., Dovner, P.O., Boström, R., Holback, B., Holmgren, G. and Koskinen, H.E.J.: 1993, A statistical survey of auroral solitary waves and weak double layers, occurence and net voltage, *J. Geophys. Res.* **98**, 15,521.

Mangeney, A., Salem, C., Lacombe, C., Bougeret, J.-L., Perche, C., Manning, R., Kellogg, P.J., Goetz, K., Monson, S.J. and Bosqued, J.-M.: 1999, WIND observations of coherent electrostatic waves in the solar wind, *Ann. Geophys.* **17**, 307–320.

Matsumoto, H., Kojima, H., Miyatake, T., Omura, Y., Okada, M., Nagano, I. and Tsutsui, M.: 1994, Electrostatic solitary waves (ESW) in the magnetotail: BEN wave forms observed by Geotail, *Geophys. Res. Lett.* **21**, 2915.

Mottez, F., Chanteur, G. and Roux, A.: 1992, Filamentation of plasma in the auroral region by an ion-ion instability: A process for the formation of bidimensional potential structures, *J. Geophys. Res.* **97**, 10,801–10,810.

Mottez, F., Roux, A., Perraut, S. and Louarn, P.: 1997, Coherent structures in the magnetotail triggered by a double electron beam, *J. Geophys. Res.* **102**, 11399.

Muschietti, L.: 1999, BGK electron holes and stability *AGU fall meeting*.

Omura, Y., Matsumoto, H., Miyake, T. and Kojima, H.: 1996, Electron beam instabilities as generation mechanism of electrostatic solitary waves in the magnetotail, *J. Geophys. Res.* **101**, 2685.

O'Neil, T.: 1965, Collisionless damping of nonlinear plasma oscillations, *Phys. Fluids* **8**, 2255.

Salem, C., Mangeney, A. and Bougeret, J.L.: 1999, Coherent Electrostatic Nonlinear Waves in Collisionless Space Plasmas, in: T. Passot and P.L. Sulem (eds.), *Nonlinear MHD Waves and Turbulence, Lecture Notes in Physics*, N, Springer-Verlag.

Sato, T. and Okuda, H.: 1981, Numerical simulations of ion acoustic double layers, *J. Geophys. Res.* **86**, 3357–3368.

Schamel, H.: 1982, Kinetic theory of phase space vortices and double layers, *Phys. Scripta* **T2/1**, 228–237.

Temerin, M., Cerny, K., Lotko, W. and Mozer, F.S.: 1982, Observations of double layer and solitary waves in the auroral plasma, *Phys. Rev. Lett.* **48**, 1175.

Tetreault, D.: 1991, Theory of electric fields in the auroral acceleration region, *J. Geophys. Res.* **96**, 3,549.

Verga, A.D., Chanteur, G. and Pellat, R.: 1988, Current driven double layers under linearly stable conditions, *Phys. Fluids* **31**, 2625.

MAGNETIC TURBULENCE AND ION DYNAMICS IN THE MAGNETOTAIL

A.L. TAKTAKISHVILI
Abastumani Astrophysical Observatory, Tbilisi, Georgia

A. GRECO, P. VELTRI and G. ZIMBARDO
Physics Department, University of Calabria and Istituto Nazionale di Fisica della Materia, Unita' di Cosenza, I-87036 Arcavacata di Rende, Italy

L.M. ZELENYI and A.V. MILOVANOV
Space Research Institute, Moscow, Russia

Abstract. The ion dynamics in the Earth's magnetotail is studied in the case when a cross tail electric field E_0 and reconnection-driven magnetic turbulence are present in the current sheet. The magnetic turbulence observed by the Interball spacecraft is modeled numerically by a power law magnetic fluctuation spectrum. A test particle simulation is performed for the ions, and the distribution function moments are obtained as a function of the magnetic fluctuation level, $\delta B/B_0$, and of the value of the normal component B_n. It appears that even in the presence of magnetic turbulence, the normal component has a marked influence on particle dynamics: the ion bulk velocity along E_y and ion temperature are almost inversely proportional to B_n. The magnetic turbulence causes the current to split in two layers, and the level of magnetic fluctuations needed to have splitting is roughly proportional to B_n. It appears that in the relevant range of parameters, B_n and $\delta B/B_0$ have opposite effects on the current structure and on ion heating.

1. Introduction

Recent observations of Geotail and Interball spacecrafts revealed the existence of a magnetic field turbulence in the distant and near Earth's magnetotail. The observed turbulence has a power law spectrum which may be interpreted as indication of the presence of tearing mode turbulence in the different parts of the tail. On the other hand, the triggering of magnetospheric substorms, the most energetic events in the Earth's magnetosphere, remains a much debated issue, and it would be interesting to understand the influence of magnetic turbulence on the magnetotail equilibrium structure during the growth and expansion phases of a substorm.

In a previous paper (Veltri *et al.*, 1998), we investigated the effect of such a turbulent magnetic field on the ion dynamics for the distant tail geometry, $x < -100R_E$ (we use the GSM coordinate system, where x is along the Earth-Sun direction, z is oriented in the direction of the Earth's magnetic moment and y accomplishes the right-handed system), where the normal component of the unperturbed magnetic field, $B_{0z} \equiv B_n$, is zero. This configuration is often called

Astrophysics and Space Science **277:** 71–79, 2001.

neutral current sheet. A cross tail electric field E_y was also taken into account. We obtained, by a test particle simulation, that the magnetic fluctuations play the role of effective scattering mechanism. This leads to the inflating of the current sheet: the larger the fluctuation level the larger the sheet thickness. We also obtained a current sheet splitting in two layers for relatively high perturbation levels, which is in agreement with the double humped profile of the current sheet deduced from the Geotail (Hoshino *et al.*, 1996) and ISEE 1 and 2 (Sergeev *et al.*, 1993) observations (see also Delcourt *et al.*, 1994). In our numerical model this profile of the current is a consequence of the braking due to the magnetic fluctuations, which is larger in the central region, $z = 0$, because the turbulence is concetrated in the neutral sheet.

In this paper we investigate ion dynamics in the presence of magnetic turbulence for closer regions ($x > -100R_E$) of the magnetotail, where the average B_n is not zero anymore and is northward directed. We study the influence of B_n on the distribution function moments, and we find that it heavily influences the dynamics: unless $\delta B/B_0$ is rather high, the current splitting is suppressed by B_n; the level of perturbations necessary to obtain splitting grows with increasing B_n. Also, the normal component strongly reduces the growth of bulk velocity and temperature with y. Indeed, it appears that ion heating is inversely proportional to B_n, while it grows with $\delta B/B_0$.

In the next section we set up the model geometry and write down the main equations. Then we present the simulation results and we discuss them in connection with the onset of the expansion phase of substorms.

2. Magnetic Field Model and Equations of Motion

The considered magnetic field configuration consists of an unperturbed, sign reversing component, directed along Earth-Sun (x) axis; an unperturbed, constant, normal to the sheet component $\mathbf{B}_n = B_n\mathbf{e}_z$, and of three dimensional stationary magnetic fluctuations $\delta\mathbf{B}$ having power law spectrum: $\mathbf{B} = \mathbf{B}_{0x}(z) + \mathbf{B}_n + \delta\mathbf{B}$. The reversing field component $\mathbf{B}_{0x}(z)$ is positive in the upper (northern) lobe of the tail and negative in the lower lobe. Following Veltri *et al.*, 1998, we use for it the expression of a modified Harris magnetic field reversal, setting it in such a way that the asymptotic value B_0 is reached at the edges of the simulation box at $z = \pm 0.5L$, where also the derivative of the field with respect to z becomes zero:

$$\mathbf{B}_{0x}(z) = B_{0x}(z)\mathbf{e}_x = B_0 \frac{\tanh(z/L) - (z/L)\cosh^{-2}(L/2\lambda)}{\tanh(L/2\lambda) - (L/2\lambda)\cosh^{-2}(L/2\lambda)}.$$

Here, L is the total thickness of the considered magnetic field configuration and λ is the current sheet half thickness. The normal component and the reversing component describe the parabolic field line geometry of the Earth tail from approximately 15 to $100R_E$ downtail. Since $B_n \ll B_0$, this configuration is referred to as quasineutral current sheet. The distinct power law spectrum of the fluctuations observed by the Geotail (Hoshino *et al.*, 1994) and Interball satellites (Peroomian *et al.*, 1999) in this region of the magnetotail is typical of a well developed turbulence, which means that a saturation level has been reached and suggests that magnetic turbulence is in a steady state, in statistical sense. To model these fluctuations we represent them as the sum of static magnetic perturbations, $\delta\mathbf{B}(\mathbf{r}) = \sum_{k,\sigma} \delta B_\sigma(\mathbf{k})\mathbf{e}_\sigma(\mathbf{k}) \exp(i(\mathbf{kr} + \phi_\mathbf{k}^\sigma))$, where

$$\delta B_\sigma(\mathbf{k}) = \frac{(2\pi/L)^{3/2}(1/R)\,C}{(k_x{}^2 l_x{}^2 + k_y{}^2 l_y{}^2 + k_z{}^2 l_z{}^2 + 1)^{\alpha/4+1/2}} \tag{1}$$

and where $\mathbf{e}_\sigma(\mathbf{k})$ are the polarization unit vectors. The fluctuating field components have specified parity with respect to the z direction, corresponding to the formation of the magnetic islands during the reconnection in the current sheet (details of the numerical model are given in Veltri *et al.*, 1998). Also the correlation lengths l_x, l_y, l_z of the turbulence reproduce the features of the actual magnetotail, with l_x,$l_y \gg l_z$, and the random phases $\phi_\mathbf{k}^\sigma$ are fixed in a way to concentrate the magnetic fluctuations in the quasi-neutral sheet, in agreement with the observations. In addition to this magnetic field configuration, we assume that there is a constant cross tail electric field, $\mathbf{E} = E_y\mathbf{e}_y$, oriented in the dawn to dusk direction and whose origin lays in the solar wind-magnetosphere coupling. For simplicity we assume that the fluctuating electric field of the magnetic perturbations has negligible impact on the particle dynamics, which is justified for ions with velocities $v > v_A$, where v_A is the Alfven velocity (see Veltri *et al.*, 1998). For the other ions we still consider that electric field fluctuations can be effective only for those particles which have velocities close to wave phase velocity, which again requires $v \approx v_A$. Since no resonance condition is required to obtain a significant acceleration by the steady field E_y, this will have a much stronger effect on acceleration of the bulk of particles than the fluctuating field, even when E_y is smaller than δE.

The equations of motion for particles are $md\mathbf{v}/dt = e(E + \mathbf{v} \times \mathbf{B}/c)$; these can be written in a dimensionless form by normalizing all length scales to the box size in z direction L, velocities to electric drift velocity $v_E = cE_y/B_0$, magnetic and electric fields to B_0 and E_y, respectively, and time to ion gyrofrequency $\omega_0 = eB_0/mc$. We also introduce $b_n \equiv B_n/B_0$. We fix for our unit length the value $L = 10^4$ km. Ion motion is studied in a box extending up to $20L$ in cross tail (y) direction from $y = -10L$ to $y = 10L$, which corresponds to the realistic physical dimension of the tail, $L_y = 2 \cdot 10^5$ km $\simeq 30R_E$, up to L in the vertical (z) direction, from $z = -0.5L$ to $z = 0.5L$ and unlimited in x direction.

2.1. Particle Injection Scheme

In this paper we consider that the source of the particles that are entering the current sheet from the lobes, $z = \pm 0.5L$, is located somewhere in the magnetospheric mantle. The relatively cold ion distribution in a strong mantle magnetic field can be described as a shifted Maxwellian (e.g Ashour-Abdalla *et al.*, 1994)

$$f(v_{\parallel}, v_{\perp}) = (\sqrt{2\pi} v_{th}^3)^{-1} \exp\left(-\frac{(v_{\parallel} - u)^2 + v_{\perp}^2}{2v_{th}^2}\right). \quad (2)$$

Here u is the streaming velocity and v_{th} is thermal velocity. Following Ashour-Abdalla *et al.*, 1994, we assume that $u = 200$ km s^{-1} and $v_{th} = 170$ km s^{-1}, which corresponds to a temperature $T = 0.3$ keV of the mantle ions. Other values of u and v_{th} have been used, too, the main difference with the results reported below being that when increases the number density becomes more peaked around $z = 0$. In agreement to the Liouville theorem, we will consider that the form of the particle distribution function does not change during propagation from the mantle source to the current sheet lobes, where they enter into our simulation box through the surfaces at $z = \pm 0.5L$. Thus, to describe this process, we require that particles are injected with random velocity components in such a way to reproduce a distribution function given by (2). Consequently, the particle flux probability density has to be proportional to $F(v_x, v_y, v_z) \sim v_z \cdot f(v_{\parallel}, v_{\perp})$. The details will be given elsewhere.

In Figure 1 we present the projection on the plane yz of sample trajectories of particles with starting points at $z = \pm 0.5L$, while x and y are randomly chosen. In this case the values of parameters are $b_n = 0.05$ and $E_y = 0.05$ mV/m, while the magnetic fluctuation level is null in the upper panel and is $\delta B/B_0 = 0.3$ in the lower panel. As it is clear from this plot, in the lobe regions (closer to $z = \pm 0.5L$), the magnetized particles move towards the central region $z = 0$, due to $\mathbf{E} \times \mathbf{B}_{0x}$ electric drift. Gradient drift, $\mathbf{B}_{0x} \times \nabla \mathbf{B}_{0x}$, in the negative y direction is also clear in this part of the particle trajectories. In the vicinity of the center of quasineutral sheet $z = 0$, where B_{0x} becomes small, particles start to bounce between the walls of stronger unperturbed magnetic field, being accelerated by the electric field E_y. Note that for $b_n = 0$, most of the particles remained in the magnetic 'trap' in z (that is, in the neutral sheet) all the time while traversing the whole tail in y direction. Now, as it is clear from Figure 1, a nonzero b_n plays the role of the guiding channel for particle and causes them to exit the quasineutral sheet in z direction. The effect of the magnetic fluctuations on the particle trajectories is clearly seen, too, and consists in widening in z of the region of particle bouncing and in multiple deflections which slow down the acceleration along y.

The detailed description of the way we calculate distribution function moments is given in Veltri *et al.*, 1998. The normalization for the number density is based on the consideration that particle current I_y has to be enough to reproduce the unperturbed magnetic field $B_{0x}(z)$ introduced in our model. Then, Ampere's law leads the following relation for considered magnetic field configuration $2B_0L_x = \frac{4\pi}{c} I_y$. Here I_y is the total current across the current sheet and L_x is the box length in the x direction.

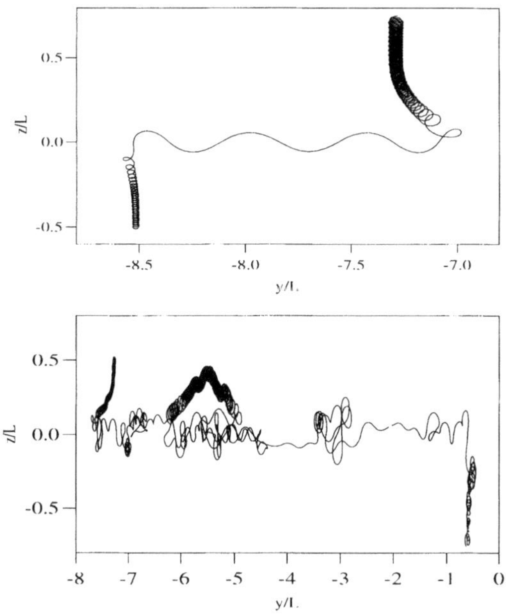

Figure 1. Sample trajectories of particle in the quasi neutral sheet with $b_n = 0.05$ and $E_y = 0.05$ mV m^{-1}. Top panel, $\delta B/B_0 = 0$. Bottom panel, $\delta B/B_0 = 0.3$.

3. Numerical Results

In this section we describe the main numerical results and present the vertical cross-section of some distribution function moments for different values of b_n and $\delta B/B_0$.

The simulation results for $b_n = 0.05$, $E_y = 0.1$ mV m^{-1} and $\delta B/B_0$ varying from 0.1 to 0.9 are shown in the top panels of Figure 2. When making the comparison with the results found for $b_n = 0$ by Veltri *et al.*, 1998, we can see that b_n modifies the magnetic field topology, and makes the motion along the y direction more difficult. However, for $b_n = 0.05$ ion motion is not adiabatic, and some current in the y direction is found. This current grows with $\delta B/B_0$, since the fluctuations bend the average field lines out of the $y =$ const planes, and the motion along y, also under the effect of the electric field E_y, becomes easier. This effect can be seen in the velocity profiles, as the velocity increase for $\delta B/B_0$ going to 0.1 to 0.5. Since number density is normalized in order to have a constant total current, a decrease in ion density corresponds to the increase of velocity (left upper panel). Also, the width of the current region increases with $\delta B/B_0$, as the magnetic

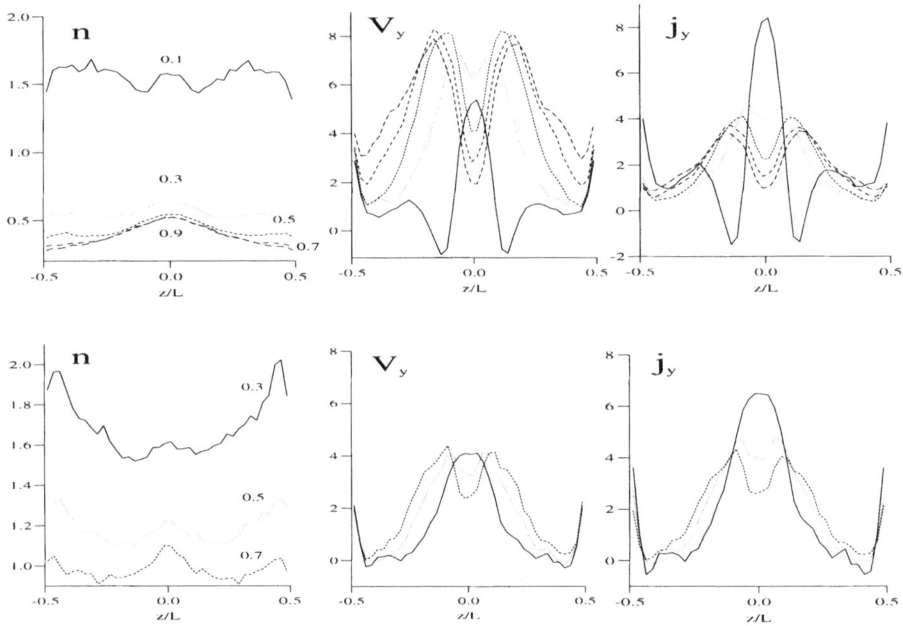

Figure 2. Top: density, bulk velocity and current density profiles for $b_n = 0.05$ and for 5 values of $\delta B/B_0$, indicated in the left panel close to the 5 different line styles. Bottom: the same profiles for $b_n = 0.1$ and for 3 values of $\delta B/B_0$, indicated in the left panel close to the 3 different line styles.

turbulence swings particles out of the central region. An evident feature shown in Figure 2 is the splitting of the velocity and current profiles in two peaks, that is, in a double current layers, when the fluctuation level is $\delta B/B_0 \geq 0.3$. This structure, which is also inferred from the observation (e.g., Hoshino *et al.*, 1996) is due in our numerical model to the concentration of turbulence in the quasineutral sheet, which, as explained earlier, corresponds to the observations. A similar splitting in a double humped current profile was also obtained by Delcourt *et al.*, 1994 by using a realistic magnetic configuration but without magnetic turbulence: in that case, the current splitting is due to the ion serpentine orbits, see the upper panel of Figure 1, which imply maximum v_y away from $z = 0$. However, in our model, thermal spread masks the effect of serpentine orbits, and no current splitting is obtained without magnetic fluctuations. It appears that the stronger the fluctuation level, the wider the separation between the two peaks in the current profile. The same features are observed for $b_n = 0.1$ (bottom panels of Figure 2), where the simulation results are reported for $\delta B/B_0 = 0.3$, 0.5 and 0.7. It is important to notice that for $b_n = 0.1$ many of the ions are magnetized even at $z = 0$, and a negligible V_y is obtained for $\delta B/B_0 = 0$. As shown by Figure 2, the velocity is generally smaller in this case, and the density higher. However, velocity and current splitting are obtained for for $\delta B/B_0 \geq 0.5$ (while for $b_n = 0.05$, splitting was obtained for $\delta B/B_0 \geq 0.3$). In general, we find that increasing b_n, a larger $\delta B/B_0$ is required in order to obtain the current splitting. This finding of current splitting is interesting for the onset of the substorm expansion phase, when the current is diverted from the current sheet into the ionosphere in the North and South auroral regions. The so called thinning of the current sheet corresponds to a decrease of b_n, so that, for a given $\delta B/B_0$, the condition where current splitting is found are met, and we can tentatively interpret the current splitting as the beginning of the current diversion process. This possibility is illustrated in Figure 3, where the density, velocity V_y, current density, and ion temperature are reported for $\delta B/B_0 = 0.3$, and $b_n = 0.1, 0.05$ and 0.02. It is clear that decreasing b_n, the current and velocity split in two layers, and that the temperature increases. The latter increase with the decrease of b_n is due to the weaking of the barrier against y motion due to b_n, so that the electric potential drop is converted into heating. In this connection, it is interesting to note that a pre-onset heating is often observed (Petrukovich, 2000). In many respects, we note that b_n and $\delta B/B_0$ have opposite effects on the current structure and ion heating (at least in the considered range of parameters). Also note that the number density profile is very flat, consistent with the current sheet being embedded in the plasma sheet. With the growth of the normal component, the current splitting becomes weaker and for $b_n = 0.1$ disappears completely. We note that the maximum temperature is almost inversely proportional to the value of the normal component.

We checked whether the current splitting effect is consistent with the structure of the average field and represents a real feature of the considered system or not: we derived a new profile of the average magnetic field $B_{0x}^{new}(z)$ by integrating the simulation particle current $j_y^{new}(z)$. We found that the double peak in the current density leads to a flattening of the magnetic field profile in the vicinity of $z = 0$ region. Then we introduced it into our model instead of the original modified Harris field $B_{0x}(z)$ and calculated the current density again. The numerical simulation showed a rapid convergence of the current and magnetic field profiles, which means that current splitting is a consistent and stable feature of the system under investigation.

4. Conclusions

We analyzed the ion dynamics in the presence of a stationary magnetic turbulence in the magnetotail and we studied the influence of the normal component of the magnetic field B_n on the particle density, bulk velocity, current density, and temperature. This study is relevant to the pre-substorm expansion phase of

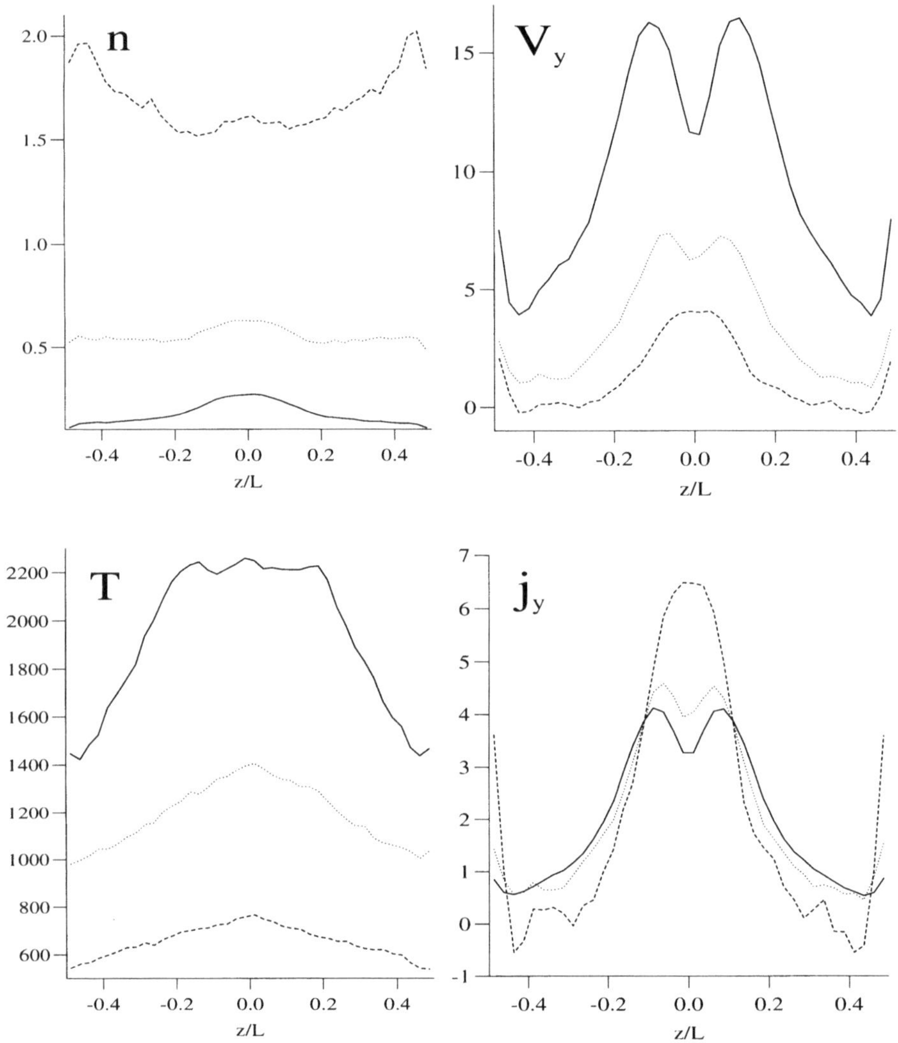

Figure 3. Top: density, bulk velocity profiles. Bottom: temperature and current density profiles for $\delta B/B_0 = 0.3$ and for 3 values of b_n (solid line $b_n = 0.02$, dotted line $b_n = 0.05$ and dashed line $b_n = 0.1$).

the magnetotail dynamics when, in the tail thinning phase, B_n becomes smaller while the tail stretches in the down tail direction.

The magnetic turbulence observed by the Interball satellite was modeled numerically, and a test particle simulation was performed in which ions are assumed to be injected into the quasineutral current sheet from the mantle source. We obtained the following results:

i) the normal component suppresses the current splitting for low perturbation levels $\delta B/B_0$;

ii) the level of $\delta B/B_0$ needed to have splitting is roughly proportional to that of B_n;

iii) for a given perturbation level $\delta B/B_0$, the ion heating is almost inversely proportional to B_n, which is in agreement with the observations prior to substorm;

iv) the temperature in the considered configuration grows toward the central plane which is also in agreement with the Interball observations.

Thus, our simple model proves the importance of magnetic fluctuations on the bulk features of the ions in the magnetotail and gives hints to explain some observed experimental data. This seems to be a promising field of investigation for magnetotail studies.

Acknowledgements

This work was supported by INTAS 97-1612 grant. During his stay in Italy A.L.T. was supported by grants of Italian INFM and MURST.

References

Ashour-Abdalla, M., Zelenyi, L.M., Peroomian, V. and Richard, R.: 1994, Consequences of magnetotail ion dynamics, *J. Geophys. Res.* **99**, 14,891.

Delcourt, D.C. and Belmont, G.: 1999, Ion dynamics at the Earthward termination of the magnetotail current sheet, *J. Geophys. Res.* **103**, 4605.

Hoshino, M., Nishida, A., Yamamoto, T. and Kokubun, S.: 1994, Turbulent magnetic field in the distant magnetotail: Bottom-up process of plasmoid formation?, *Geophys. Res. Lett.* **21**, 2935.

Hoshino, M., Yamamoto, T., Mukai, T., Nishida, A. and Kokubun, S.: 1996, Structure of plasma sheet current in distant magnetotail: Doubly humped electric current sheet, *Adv. Space Res.* **18**(8), 5.

Peroomian, V., Ashour-Abdalla, M., Zelenyi, L.M. and Petroukovich, A.: 1999, Influence of convection on magnetotail variability, in: D. Sibeck and K. Kudela (eds.), *Interball in ISTP program*.

Petrukovich, A.A.: 2000, The growth phase: comparison of small and large substorms, *Proceedings of the International Conference on Substorms*, St. Petersburg, 16–20 May, 2000, ESA SP-443, 9.

Sergeev, V.A., Mitchell, D.G., Russel, C.T. and Williams, D.J.: 1993, Structure of the tail plasma/current sheet at $\sim 11 R_E$ and its changes in the course of a substorm, *J. Geophys. Res.* **98**, 17,345.

Veltri, P., Zimbardo, G., Taktakishvili, A.L. and Zelenyi, L.M.: 1998, Effect of Magnetic Turbulence on the Ion Dynamics in the Distant Magnetotail, *J. Geophys. Res.* **103**, 14,897.

STATISTICAL MECHANICS OF STABLE STATES FAR FROM EQUILIBRIUM: THERMODYNAMICS OF TURBULENT PLASMAS

RUDOLF A. TREUMANN
Centre for Interdisciplinary Plasma Science, Max-Planck-Institut for Extraterrestrial Physics, D-85741 Garching, Germany
International Space Science Institute, CH-3012 Bern, Switzerland

Abstract. This paper reviews some recent developments in the theory of stationary states in collisionless media that are very far from thermal equilibrium. Such states may evolve under conditions when the binary collision time is much longer than any characteristic time of the processes under consideration. A typical example of such a system is collisionless turbulence in a plasma when the plasma evolves into a highly nonlinear state entirely dominated by wave generation, wave-wave and wave-particle interaction and generating a nearly stationary level of turbulence. Sometimes it is very difficult to describe the evolution of such a state. The present theory shows that it is nevertheless possible to develop a macroscopic picture in the framework of statistical mechanics and thermodynamics which allows for the macroscopic description of such states. This can be achieved introducing a control parameter κ. The equilibrium distribution which replaces the Maxwell-Boltzmann distribution is a generalized Lorentzian or κ-distribution. We sketch the underlying statistical mechanics and provide some arguments for the validity of this approach. On this level it is not possible to obtain a microscopic theory of κ, however, which must be constructed on the way of referring to the particular kind of turbulence. We note a number of unresolved problems.

1. Introduction

During the past four decades the theory of turbulent plasmas has been developed up to a certain stage of maturity which allows us to describe a large number of turbulent phenomena in great detail. Nevertheless it is still good for surprises as a susceptible amount of phenomena are still barely understood. Among those phenomena the behaviour of a collisionless strongly turbulent plasma far away from equilibrium is one of the most interesting. Collisionless plasmas can indeed assume long-lived quasi-stationary equilibria when the plasma is in a state of strong turbulence but otherwise behaves nearly stable. Such phenomena are frequently observed in nature and simulations. In some simple cases their description is possible in the framework of quasilinear theory or weak turbulence. In such cases the turbulence remains weak in the sense that a small expansion parameter exists which usually is taken as the ratio of turbulent wave energy density and kinetic energy contained in the plasma, $\epsilon = W_{tur}/nk_BT$. The assumption of weak turbulence implies that the expansion with respect to higher orders in ϵ rapidly converges, in other words, the turbulence must be renormalizable. This is the case for wave-

Astrophysics and Space Science **277:** 81–95, 2001.

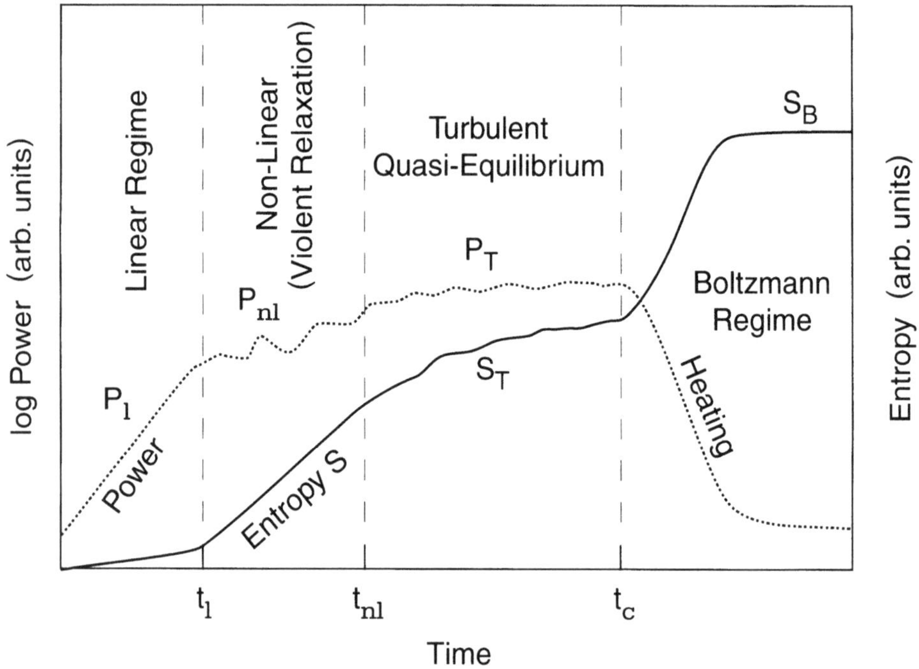

Figure 1. The schematic evolution of turbulent power and entropy in a collisionless plasma system after an initial disturbance has introduced free energy. This figure assumes that the nonlinear evolution time is considerably shorter than the collision time. Before collisions can set on the turbulence may evolve into a strong though quasi-stationary state with slowly growing entropy but high fluctuation level. This is the quasi-stationary state far from equilibrium which is referred to in the present paper. This state may or may not be reached in a smooth way without phase transition. It is not important for our discussion here in which way it is reached. It is only important *that* such a state can evolve and may exist.

wave and weak wave-particle interactions. In some cases, however, the turbulence violates this kind of expansion. When this happens, the wave spectrum evolves into localized wave packets, solitary waves or phase space holes with their own dynamics and the plasma becomes strongly turbulent. A description can then be obtained only in rather limited cases subject to some evolution equations like the Korteweg-de Vries equation, Burger's equation, Hasegawa-Mima equation and others more. A more adequate description can be given merely by numerical simulations. These do frequently show that after an initial state the plasma evolves into some quasi-stationary equilibrium where it may remain for a substantially long time as shown in Figure 1. This equilibrium is usually highly fluctuating but evolves slowly. This poses the question whether a thermodynamic description of such observed quasi-stationary states far from equilibrium can be made possible.

In the following we show that one can indeed derive a kind of thermodynamics that describes collisionless states far from equilibrium. We find an equilibrium distribution function that replaces the commonly used and celebrated Maxwell-

Boltzmann equilibrium distribution, and we derive a new kind of statistical mechanics that describes the states close to such stationary states. The distribution is of the family of generalized Lorentzian distributions or κ-distributions which have been detected experimentally in different regimes of collisionless plasmas and have so far been barely understood. However, a number of unsolved questions is raised by the present approach which we will pose at the end of this article without proposing a solution to most of them. Hence, this article ends with a kind of working program for future research.

Previous analysis on this subject has concentrated solely on the fitting of distributions to measured data (see e.g. Christon *et al.*, 1988, 1991), on the construction of κ-like distributions from special interaction models of waves and particles (see e.g. Hasegawa *et al.*, 1985; Reynolds *et al.*, 1998; Collier, 1993, 1999; Ma and Summers, 1999) and on the application of κ-like distributions to special problems in space plasma physics (see e.g. Scudder and Olbert, 1970a,b; Meyer-Vernet and Perche, 1989; Chateau and Meyer-Vernet, 1991; Summers and Thorne, 1991; Montcuquet *et al.*, 1995; Mace and Hellberg, 1995; Summers *et al.*, 1996; Mace *et al.*, 1998; Ma and Summers, 1998; Issautier *et al.*, 1999; and others). The applications to measurements of the distribution functions allowed to determine the power law index of the distributions. They, however, missed to infer about another quantity of interest which plays an important role in the theory of states far from equilibrium, the chemical potential. This will be elucidated in the section on statistical mechanics below. The applications of the κ-distribution to plasma theory suffer from the neglect of the chemical potential but otherwise are of interest as they show that the inclusion of Lorentzian distributions changes the character of the response of the plasma. The reasons for this become clear, however, only in a rigorous theory as that reviewed in the present communication.

2. The Turbulent Slab: An Example of Lévy Flight Statistics

Before starting let us consider a simple example which gives an indication of the possibility of a different kind of statistics than ordinary Boltzmann-Gibbs statistics. Let us assume that the turbulence is confined into a slab of length L. In addition assume that the medium is collisionless. A particle immersed into the slab at position x_0 at time t_0 may move some distance, will interact with the turbulence, is re-scattered or trapped for a while, moves, when released, a further distance where it is trapped again, and so forth until it escapes from the slab. During all this time it will have had a residence time $\tau = t_L - t_0$ inside the slab.

We attempt to describe the motion of a hypothetical particle in a very simple model map choosing $f(x_i) = ax_i^{1+1/\kappa} + g(x_i)$ such that

$$x_{i+1} - x_i = ax_i^{1+1/\kappa} + g(x_i) \tag{1}$$

where $\kappa > 1$ is a real number, and $g(x_i)$ is an arbitrary disturbance. Maps of this kind have been extensively investigated in the past (e.g. Hirsch *et al.*, 1982; Procaccia and Schuster, 1983, who investigated it for $g = 0$) and have been shown to exhibit intermittency and chaos in different regimes of the parameters a, κ and $g(x_i)$ (see for instance Geisel *et al.*, 1984, 1986; Barkai and Klafter, 1997). The continuous time version of the above model equations can be written

$$\dot{x} = ax^{1+1/\kappa} + g(x) \tag{2}$$

Solving for the time t yields

$$t - t_0 = \int_{x_0}^{x} \mathrm{d}x \left[ax^{1+1/\kappa} + g(x)\right]^{-1} \tag{3}$$

Of course, such maps are only rough approximations to the real interaction of the particle with the turbulent wave field. Their purpose is mainly to elucidate the behaviour of the particle under nonlinearity. The left-hand side can be interpreted as velocity. The right-hand side is then a nonlinear function of x. The velocity depends on time t only trough $x(t)$ and the inhomogeneity $g[x(t)]$. This map possesses an unstable origin which implies that particles close to the origin will experience long trapping times and will stay for long time in the slab. We do not go into the details of a discussion of the validity of the above map as it serves only as an illustration for our purposes.

Putting $t = t_L$ and taking the upper limit $x = L$ gives an expression for the 'residence time' of the hypothetical particle. The general solution is difficult to obtain, but it suffices for our purposes to set $g = 0$. The probability $P(\tau)$ of residence time a particle will experience in the slab is

$$P(\tau) \propto \left[1 + aL^{1/\kappa}\tau/\kappa\right]^{-(\kappa+1)} \tag{4}$$

which is precisely the form of the generalized Lorentzian or κ phase space distributions observed in the space measurements (see e.g. Christon *et al.*, 1989, 1991). This suggests that the particles which obey such distributions somehow undergo a collisionless chaotic motion of the kind of Lévy flights (Shlesinger *et al.*, 1987, 1993; Klafter *et al.*, 1996). They are thus in a state of collisionless quasi-stationarity far away from thermal equilibrium. We note that inclusion of the function g changes this distribution in as far as it causes an exponential cut-off of the residence time probability at large τ. This cut-off is due to the finite convection velocity introduced by g which when superimposed on the turbulence convects the particle out of the slab. When g is very large, the first term in Equation (1) can be neglected and the turbulence is unimportant. This case is of no interest. We have shown elsewhere what the use of the above distribution would be. One can, in particular, determine the mean square displacement $\langle(\Delta x)^2\rangle \propto t^\alpha$ of the particles which turns out to be proportional to a power of time different from one which indicates that the dynamics of the particles is highly nonlinear causing deviations from classical diffusion (Treumann, 1997).

3. Kullback's Contrast

What kind of statistical mechanics can be constructed to macroscopically describe either many particle Lévy flight dynamics or quasi-stationary behaviour of states far from thermal equilibrium? Treumann and Kull (2000) have recently shown that there is a direct formal relation between Boltzmann-Shannon information theory and such a statistical mechanics from Kullback's contrast (Kullback, 1951). Define the thermodynamic probability as p_i, with index i the running number of r states. Generalizing Kullback's contrast to include the order parameter κ yields in a straightforward way the following expression for the κ-entropy,

$$S_\kappa(P) = -\frac{\kappa k_B}{r^{1/(1+\kappa)}} \sum_{i=1}^{r} p_i \left[1 - p_i^{-\frac{1}{1+\kappa}}\right] \tag{5}$$

This entropy under same conditions and for $\kappa < \infty$ is always *larger* than the Boltzmann entropy S, and it is not difficult to demonstrate that it approaches S in the limit $\kappa \to \infty$. [Note that the last expression implies a redefinition of Boltzmann's constant $k_B \mapsto \kappa k_B / r^{1+1/\kappa}$.] An expression for entropy similar to the entropy derived above and different from Boltzmann's expression has first been given by Daróczy (1970) and revived by Tsallis (1988).

4. The Boltzmann κ Problem

We now turn to the problem of what an entropy different from that of Boltzmann's definition means. Boltzmann derived his entropy from the requirement that a system in thermal equilibrium will obey a general and unique phase space distribution function and at the same time will satisfy the condition that a slight deviation from thermal equilibrium will be extinguished by the system such that it turns back to a new equilibrium with a higher degree of disorder (or higher temperature) (see e.g., Huang, 1995). Hence, according to the second law of thermodynamics the entropy of the system will always increase. This raises the question if some argument similar to Boltzmann's may exist under conditions of collisionless quasi-steady states far from thermal equilibrium. Here we will show that the answer to this question is yes.

The physics of many body systems is properly described by the two lowest order kinetic equations of the BBGKY hierarchy which solve the exact Liouville equation. These equations are the kinetic equation for the one particle distribution function $f_1(1)$, and the kinetic equation for the two-particle distribution $f_2(1, 2)$

where the argument indicates the dependence of the distribution function on the momenta and positions of the particles. It can be shown that Boltzmann's ansatz for the two-particle distribution

$$f_2(1,2) = f_1(1)f_1(2) + correlations \tag{6}$$

solves the two-particle hierarchy equation with vanishing three-particle collision term (see Huang, 1995). Neglecting this collision term is a reasonable assumption because three-particle interactions should appear at far lower probability than two-particle interactions. On the other hand, the additional assumption of neglecting the correlations depends on the kind of interaction. In Boltzmann's case the interaction between two particles is restricted to a miniscule fraction of space such that the interaction volume V_{int} remains small. In collisional plasmas this implies that the Debye sphere is small. Even taking into account its finiteness does not change the assumption of neglecting the correlations. It merely produces a slight correction on the Boltzmann collision term known as the Landau-Lenard-Balescu equation.

In strong collisionless turbulence the argument of a small interaction volume may fail. In particular in scale invariant interaction the correlation length diverges and the interaction volume becomes the total volume occupied by the system. In this case the correlations in Equation (6) become dominant, and Boltzmann's ansatz must be replaced. A reasonable assumption in this case is to set

$$f_2(1,2) = g_\kappa[f_1(1)]g_\kappa[f_1(2)] + \dots \tag{7}$$

with $g_\kappa[f_1]$ a so far undetermined functional of the one-particle distribution function. For convenience we have introduced an arbitrary order parameter κ at this place. It is a lengthy procedure to solve the two-particle equation for $g_\kappa[f_1]$. It turns out that $g_\kappa[f_1]$ must be an *exponential* functional of f_1 in oder to become an exact solution of the BBGKY hierarchy.

$$g_\kappa = \exp\{\phi_\kappa[f_1]\} \tag{8}$$

The Boltzmann-like solution is the particular case when ϕ_κ is a logarithm. Otherwise, imposing the condition that $g_\kappa \mapsto f_1$ for $\kappa \to \infty$ is the identity functional and thus reproduces Boltzmann's solution, one finds that

$$\phi_\kappa[f_1] = \kappa\left[1 - f_1^{-\frac{1}{\kappa+1}}(\kappa, \mathbf{p})\right] \tag{9}$$

with $\kappa > \kappa_{min}$ where κ_{min} is determined from the condition of detailed balance in interaction. In our case this implies that total particle number and energy must be conserved in the interaction. Hence, at least the second moment of the distribution function should exist. Thus $\kappa_{min} = 3/2$.

As in Boltzmann kinetic theory one assumes that in equilibrium the new two-particle collision integral

$$\mathcal{C}_\kappa = \int \mathrm{d}\tau M_{12}\{g_\kappa[f_1(1')]g_\kappa[f_1(2')] - g_\kappa[f_1(1)]g_\kappa[f_1(2)]\} = 0 \tag{10}$$

vanishes identically. Here M_{12} is the interaction matrix which is always positive definite. In equilibrium it must not be specified as the necessary condition for the collision integral $\mathcal{C}_\kappa$ to vanish identically; it is simply that its integrand vanishes. Detailed equilibrium requires that

$$\ln g_\kappa[f_1] = -\beta\left(\frac{p^2}{2m} - \mu\right) \tag{11}$$

with μ the chemical potential. The solution for the average most probable one-particle distribution is thus given by

$$f_\kappa(\mathbf{p}) = A_\kappa\left[1 + \frac{\beta}{\kappa}\left(\frac{p^2}{2m} - \mu\right)\right]^{-(\kappa+1)} \tag{12}$$

The constant A_κ is a proper normalization constant which has to be determined assuming conservation of particle number. However, this constant is not entirely free as the chemical potential also is defined only through particle number.

The equilibrium distribution found in this rigorous way thus replaces Boltzmann's famous distribution function for collisionless states far away from equilibrium. Figure 2 shows a schematic of the plot of the κ distribution for different κ. The assumption that these states are far from equilibrium has been implicit to the above ansatz and has been explicated in the justification of our initial assumption. It is now obvious that when $\kappa \to \infty$ and collisions take over the system evolves into thermal equilibrium. Therefore our new distribution can be taken as the quasi-thermal equilibrium distribution far from equilibrium. It is now natural to ask for the sufficient condition that this distribution describes such a quasi-equilibrium. It has been shown (Treumann, 1999a,b) that such a proof follows Boltzmann's argument and works when the entropy is defined as

$$\mathcal{S} \propto -k_B \int \mathrm{d}\tau f_\kappa \ln\{g_\kappa[f_\kappa]\} \tag{13}$$

where $\mathrm{d}\tau$ is the volume of phase space. This is a definition close to that of Boltzmann and converges to Boltzmann's expression for $\kappa \to \infty$. It is thus a generalization of Boltzmann's concept of the H-theorem. This entropy does always increase as well which is in agreement with statistical mechanics, thermodynamics, and physical experience.

At this place we mention briefly that it is not difficult to extend the entire formalism to the domain of quantum statistics (Treumann, 1999b). To this end one redefines the integrand in the collision term Equation (10) according to

$$G_\kappa[f] = \left(\sum{}' - \sum\right)_{i=1,2} \frac{g_\kappa[f_1(i)]}{1 \mp g_\kappa[f_1(i)]} \tag{14}$$

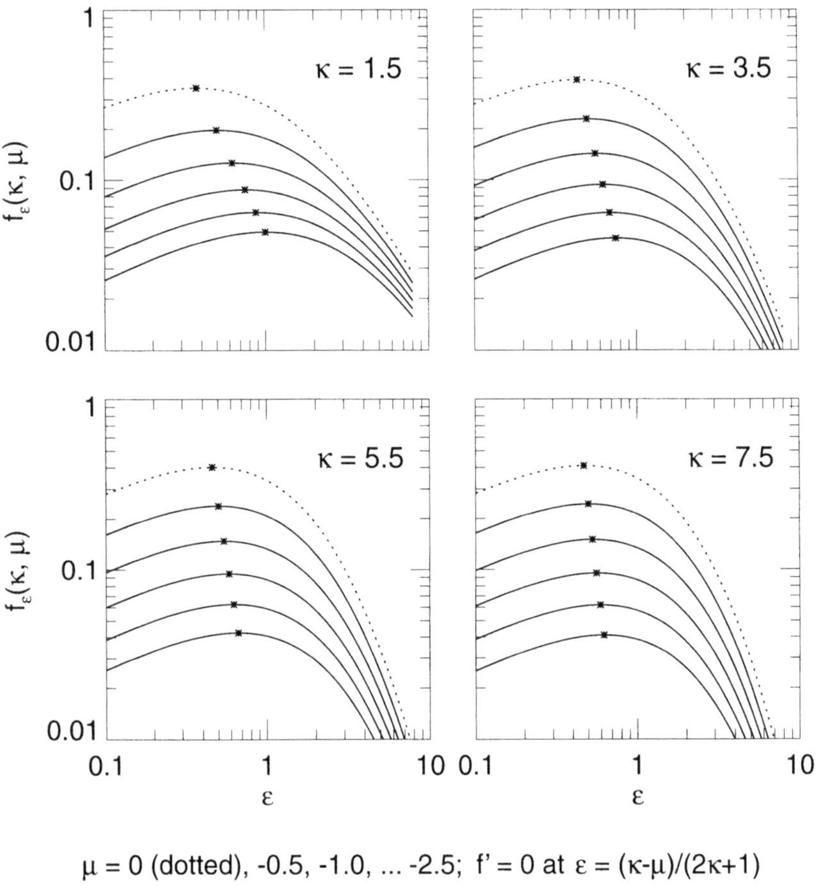

Figure 2. Schematic plot of the κ-distribution as function of particle energy $\epsilon_p = p^2/2m$. In energy space the distribution is multiplied by $\sqrt{\epsilon_p}$. The asymptotic power law decay of the distribution allows for a determination of κ. The maximum of the distribution is related to the temperature β^{-1} and the chemical potential μ.

Following the usual formalism explicated above this ansatz immediately leads to the two new average distribution functions

$$f_\kappa^{F,B} = A_\kappa \left\{ \left[1 + \frac{\beta}{\kappa}\left(\frac{p^2}{2m} - \mu\right)\right]^{\kappa+1} \pm 1 \right\}^{-1} \tag{15}$$

The positive sign corresponds to the κ-Fermi-Dirac, the negative sign to the κ-Bose-Einstein distribution, respectively. Investigation of the relevance of these distributions lies outside this review. We note in passing that both distributions explicitly exclude the possibility of κ-states at zero temperature. This at a first glance

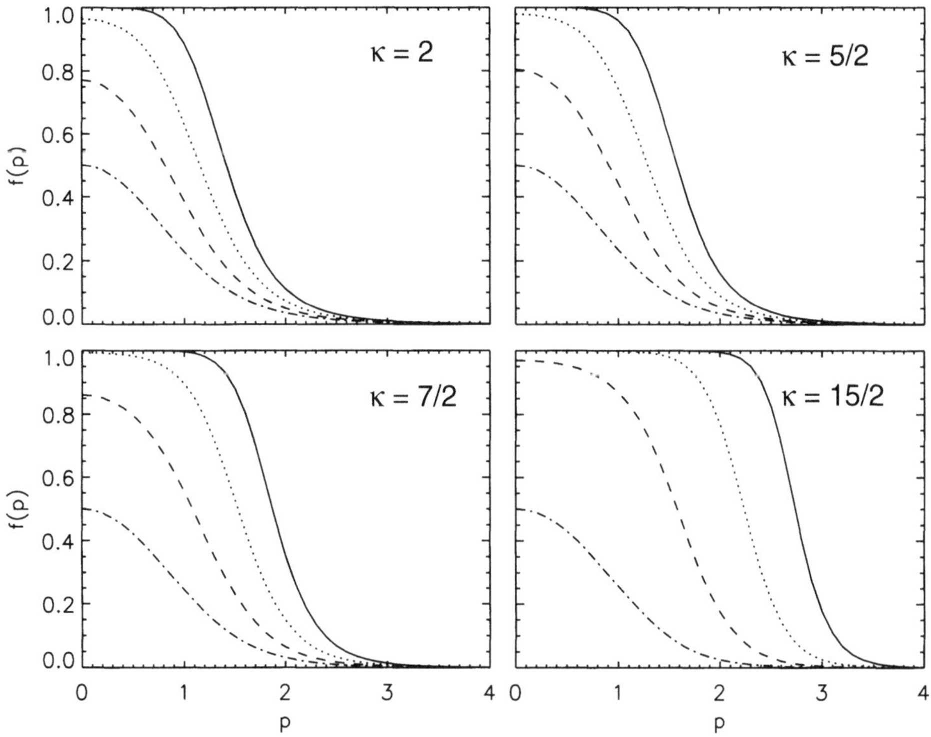

Figure 3. The κ-Fermi distribution as function of momentum p for four different κs. This distribution is very similar to the low temperature Fermi distribution but holds only for finite values of the temperature. The uppermost curve in each frame corresponds to the maximum possible value $\mu_m = \kappa k_B T$ of the chemical potential. The lowest curve is the curve with vanishing chemical potential $\mu = 0$.

surprising result has the very suggestive physical interpretation that not correlations can evolve at $T = 0$. Hence, at zero temperature the systems become ordinary Fermi or Bose gases. The two implications resulting from this insight are that, firstly, κ quantum gases exhibit macroscopic quantum states only and, secondly, $\kappa(T)$ must be a function of temperature which for $T \to 0$ turns to $\kappa \to \infty$ in order to reproduce classical quantum statistics. This conclusion is nothing but a different expression of our claim that κ is a function of the turbulent wave spectrum. A simple formula for the temperature dependence of κ for Fermi gases has been derived elsewhere (Treumann, 1999b). The κ-Fermi distribution is shown schematically in Figure 3.

5. Thermodynamics

The above formula for the entropy suggests that it should be possible to construct an equivalent thermodynamics for the quasi-stationary state far from thermal equilibrium. Such a thermodynamics can be considered as a formal theory which allows us to describe the macroscopic quasi-equilibrium properties of the system after it has settled for some possibly long time in the quasi-equilibrium state. Such a description is of high practical value because it frees one from knowing all the microscopic processes which proceed in the system. In particular, when the system, in our case the plasma, is highly turbulent, it is not necessary to investigate the microscopic interactions, wave transformation and particle heating processes in order to infer about the macroscopic behaviour of the plasma for time short with respect to the evolution time of the system.

The task of constructing such a kind of thermodynamics is non-trivial. A first attempt can be found in Treumann (1999a). The basic idea is that one defines a grand thermodynamic potential from which all other thermodynamic functions follow. Knowing the functional form of the entropy one can define this potential by

$$\mathcal{Q}(\beta, \mu, V) = -\frac{V}{\beta}\int \frac{\mathrm{d}^3 p}{h^3}\left[1 + \frac{\beta}{\kappa}\left(\frac{p^2}{2m} - \mu\right)\right]^{-\kappa} \tag{16}$$

where V is the volume. Comparing with ordinary thermodynamics it can be concluded that $\beta = 1/k_B T$ is the inverse temperature of the system. Interestingly, the power of the term in brackets inside the integral is less by one unit than that of the distribution function. This choice guarantees that thermodynamics can be constructed in the traditional way. Indeed, it can be shown that the equivalent to the Helmholtz free energy is

$$F = \mathcal{Q} + \mu N \tag{17}$$

which is the usual thermodynamic formula when N is particle number. From here we find that as usual $\mu = \partial F/\partial N$ is the chemical potential. Moreover,

$$E = \frac{\partial}{\partial \beta}(\beta F) \tag{18}$$

is just the definition of the average total energy in the system at quasi-stationarity. Carrying out these calculations (see Treumann, 1999a) with the above definition of $\mathcal{Q}$ one can show that the energy is actually the second moment of the κ distribution. Moreover, one also can recover the expression for the entropy. This completes the thermodynamic approach and formalism.

From this point of view the present theory is complete and to some extent satisfactory. a number of questions do, however, remain unanswered and unsolved and have to be treated in near future. The first of these questions is the question for the development of a relevant statistical mechanics. This can be answered based on the above grand thermodynamic potential. it suggests that the relevant canonical partition function can be constructed in the same way as $\mathcal{Q}$ has been constructed, when resolving the integral into a sum over the microstates. The general relation between $\mathcal{Q}$ and the partition function $\mathcal{Z}$ is

$$\mathcal{Q} = -\beta^{-1} \ln \mathcal{Z} \tag{19}$$

Hence the κ-partition function can be formally written as

$$\mathcal{Z}_\kappa = \prod_i \exp\left\{\left[1 + \frac{\beta}{\kappa}(\epsilon_i - \mu)\right]^{-\kappa}\right\} \tag{20}$$

Here we have defined the energy states ϵ_i of the system. The summation is over all such energy states ϵ_i to which the system has access. With the help of $\mathcal{Z}_\kappa$ one will be in the position to construct a statistical mechanics valid in the quasi-stationary state of the system far from thermal equilibrium. Clearly this definition implies a redefinition of the Gibbsian approach to statistical mechanics. The mathematical form of this partition function is however much more complicate than in ordinary statistical mechanics. It requires a new kind of analysis. In particular there is some similarity to quantum statistics in as far as the chemical potential is retained in the partition function. Remember that in classical statistical mechanics μ drops out of $\mathcal{Z}$ because the partition function is a product of simple exponentials. This is different in our case. Obviously the correlations prevalent on all scales in the turbulence act in a similar way as the chemical potential: they introduce an interactive work the particles must perform when changing their states.

6. The Problem of Hydrodynamics and Divergencies

The second problem concerns the formation of moments of the distribution function. It is clear that only a finite number of moments can be taken on the κ distribution function with all higher moments higher than a certain critical κ_c will diverge. Does this imply that no higher moments exist? Does it possibly imply the the hydrodynamic equations which ordinarily can be derived from the one-particle kinetic equation do not exist, in other words that the system can be described only by kinetic theory and no hydrodynamic theory is possible? Or does it mean that the hydrodynamic equations close by themselves and are automatically truncated? This is a difficult problem to answer.

Moments of the distribution function are taken in the usual way as averages over the distribution function. In fractal theory there have been proposals to take moments in the form of multifractal moments where the distribution function is taken to some power. This suggestion called *Rényi convention* has been used extensively in the past and has turned out to be formally very useful in the determination of the fractal dimensions of a system. In thermodynamics, it has also been used by Tsallis and his school but cannot be physically justified in an acceptable way. Our derivation of thermodynamics shows that in a consistent statistical mechanics of states fra from equilibrium moment formation remains the same as in conventional physics, and the Rényi-Tsallis convention is not needed. But the problem of the existence of hydrodynamic equations and the convergence of higher moments remains.

The divergence of higher physical moments is unsatisfactory. It may imply that a hydrodynamic description of strongly turbulent plasma is not possible in principle, and it is unavoidable to return to kinetic theory. On the other hand, the divergences may call for renormalization of the moments. Observations indicate that κ always assumes relatively large (fractal) values of the order of $\kappa > 5 - 6$ which suggests that the divergencies appear only at high moments. Since hydrodynamics is never stressed up to equations of order higher than the heat conduction equation, such moments are usually ignored anyway and do not hurt. However, their divergence remains principally disturbing. A possible recipe for their treatment can be suggested in the following way. Each moment satisfies its own conservation equation which depends on the next higher moment. For a given κ the lowest order conservation equations contain only finite moments but are not closed. When a diverging moment appears first in a high order conservation equation one may attempt to keep it including its diverging term and step to the next higher order. Here two diverging moments will appear, the former one and the higher order diverging moment. One may now imagine that part of the divergence of the next higher moment may be arranged to compensate for the former diverging moment thus closing the lower order equation. The remaining divergence of the higher order moment may be expected to be taken care of by the second next higher moment and so on. As result one expects that the 'naked' κ which has been assumed initially will change its value to become a 'dressed' renormalized κ. In such a theory it is not possible to neglect the interaction integral on the right hand side of the one-particle kinetic equation when deriving the moment conservation equations which affects the transport coefficients.

7. Conclusions

In the present communication we presented a theory of quasi-stationary many-body states far from equilibrium. Such a theory might be of importance in application to plasma turbulence and turbulence in astrophysical systems where long lasting stationary states are occasionally observed which nevertheless are far away from thermal equilibrium. Collisionless plasma turbulence is a good example for such systems. Other examples may be found in the behaviour of non-interacting dark matter like the matter in clusters of galaxies or the non-interacting dark matter in cosmological models of the universe which are all void of collisions.

Since no collisions act to thermalize the matter such systems are usually far from equilibrium. Whence they evolve into strong turbulence via excitation of a multitude of wave modes one may expect that they assume a state which is quasi-stationary for times much shorter than the interparticle collision time. Hence in this intermediate transitory state their phase space distribution functions will also differ from the Maxwell-Boltzmann distribution. We have demonstrated that they will belong to the family of generalized-Lorentzian or κ distributions. Application of the statistical mechanics reviewed in this paper is then of vital interest to such systems.

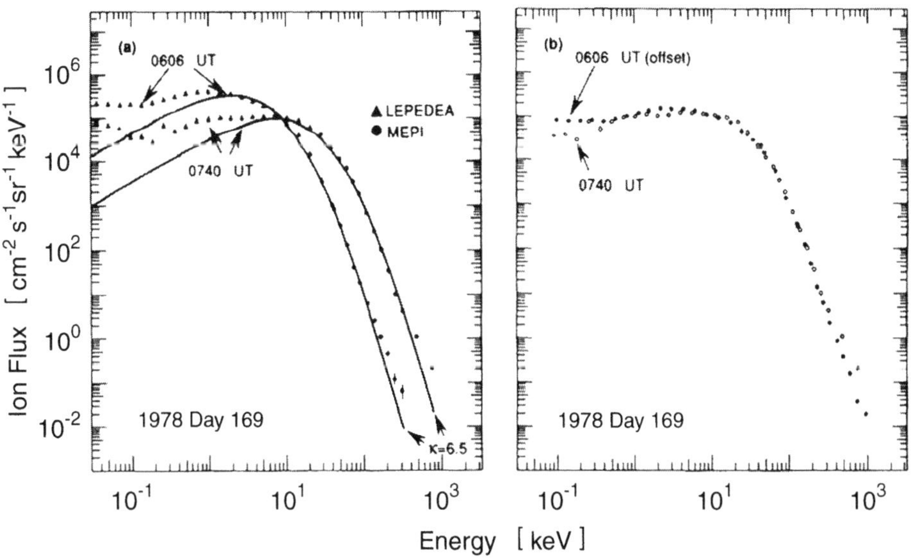

Figure 4. Actual measured energy distributions of magnetospheric ions. The distributions are typical for quiet times and clearly belong to the family of κ distributions. The power law tails allow for the determination of κ which is found to have values around 6. For such values of κ the most relevant physical moments do converge, thus allowing for the definition of average physical quantities like average energy, pressure, heat flux etc. The maximum of the distribution seems to be shifted to higher energies possibly suggesting that an acceleration mechanism were at work (after Christon *et al.*, 1991). During disturbed times this shift may even be larger, and in addition a beam component may be superimposed on the background distribution.

In near-Earth space (see e.g. Vasyliunas, 1968; Christon *et al.*, 1989, 1991, and others) there is a wealth of observations that support the presence of κ distributions under otherwise collisionless conditions (see Figure 4). The κ value in these observations is relatively large indicating convergence of the lower moments up to fifth order. Formal use of such distributions in equilibrium situations is possible. In cosmological structure formation theory it has been speculated for long time that non-Gaussian statistics should be applied to primordial density fluctuations in order to generate the observed structure in the Universe. The present theory provides the theoretical and physical though still formal background for their existence.

Classical systems subject to Lorentzian distributions will exhibit transport coefficients different from ordinary thermal equilibrium. Application to quantum theory implies that the quantum system enters a *macroscopic* quantum state with macroscopic wave function. One may speculate that such a theory may have some relevance in high-temperature superconductivity, the theory of white dwarfs or neutron stars.

Application to radiative processes is under progress. First estimates show that in bremsstrahlung emission the use of the generalized Lorentzian changes the spectrum. Interpreted in terms of the Boltzmann statistics causes overestimate of the temperature and underestimates the density. The main problem consists in the calculation of the proper interaction cross sections. These will be low as they depend on the distribution function. In future work some of these questions will have to be explored in greater depth.

Acknowledgements

The author thanks the organizers of OBS2000 for their invitation and hospitality. Discussions with M. Scholer, S. Christon, and R. Jokipii are gratefully acknowledged. The author thanks G. Haerendel, G. Morfill, J. Trümper (MPE), J. Geiss, G. Paschmann, and R. v. Steiger (ISSI) for interest and support as well as the ISSI staff for hospitality. This work has been initiated at STEL, Nagoya University, during a visiting professorship of the author. He thanks Y. Kamide, director of STEL, for his hospitality, and the Nagoya University for support. Finally, the author thanks the referee for his constructive comments.

References

Barkai, E. and Klafter, J.: 1997, *Phys. Rev. Lett.* **79**, 2245.
Chateau, Y.F. and Meyer-Vernet, N.: 1991, *J. Geophys. Res.* **96**, 5825.
Christon, S.P., Mitchell, D.G., Williams, D.J., Frank, L.A., Huang, C.Y. and Eastman, T.E.: 1988, *J. Geophys. Res.* **93**, 2562.
Christon, S.P., Williams, D.J., Mitchell, D.G., Huang, C.Y. and Frank, L.A.: 1991, *J. Geophys. Res.* **96**, 1.
Collier, M.R.: 1993, *Geophys. Res. Lett.* **20**, 1531.
Collier, M.R.: 1999, *J. Geophys. Res.* **104**, 28559.
Daróczy, Z.: 1970, *Inform. Control* **16**, 36.
Dubois, M., Rubio, M.A. and Berge, P.: 1983, *Phys. Rev. Lett.* **51**, 1446.
Dum, C.T.: 1995, in: *Cross-Scale Coupling in Space Plasmas, Geophys. Monogr.* **93**, p. 1, AGU, Washington, D.C.
Geisel, T. and Thomae, S.: 1984, *Phys. Rev. Lett.* **52**, 1936.
Geisel, T., Nierwetberg, J. and Zacherl, A.: 1985, *Phys. Rev. Lett.* **54**, 616.
Hasegawa, A., Mima, K. and Duong-van, M.: 1985, *Phys. Rev. Lett.* **54**, 2608.
Hirsch, J.E., Hubermann, B.A. and Scalapino, D.J.: 1982, *Phys. Rev. A: Gen. Phys.* **25**, 519.
Huang, K.: 1995, *Statistical Mechanics*, Wiley, New York.
Issautier, K., Meyer-Vernet, N., Montcuquet, M. and Hoang, S.: 1999, *J. Geophys. Res.* **104**, 6691–6704.
Johnson, J.R. and Cheng, C.Z.: 1998, *Geophys. Res. Lett.* **25**.
Klafter, J., Shlesinger, M.F. and Zumofen, G.: 1996, *Phys. Today* **49**, 33.
Kullback, S.: 1951, *Information Theory and Statistics*, Wiley, New York.
Ma, C. and Summers, D.: 1998, *Geophys. Res. Lett.* **25**, 4099.
Mace, R.L. and Hellberg, M.A.: 1995, *Phys. Plasmas* **2**, 2098.
Mace, R.L., Hellberg, M.A. and Treumann, R.A.: 1998, *J. Plasma Phys.* **59**, 393.
Meyer-Vernet, N. and Perche, C.: 1989, *J. Geophys. Res.* **94**, 2405.

Moncuquet, M., Meyer-Vernet, N. and Hoang, S.: 1995, *J. Geophys. Res.* **100**, 21697.
Reynolds, M.A., Fried, B.D. and Morales, G.J.: 1998, *Phys. Plasmas* **4**, 1286.
Scudder, J.D. and Olbert, S.: 1979a, *J. Geophys. Res.* **84**, 2755.
Scudder, J.D. and Olbert, S.: 1979b, *J. Geophys. Res.* **84**, 6603.
Shlesinger, M.F., West, B.J. and Klafter, J.: 1987, *Phys. Rev. Lett.* **58**, 1100.
Shlesinger, M.F., Zaslavsky, G.M. and Klafter, J.: 1993, *Nature* **363**, 31.
Summers, D. and Thorne, R.M.: 1991, *Phys. Fluids B* **3**, 1835.
Summers, D., Thorne, R.M. and Matsumoto, H.: 1996, *Phys. Plasmas* **3**, 2496.
Treumann, R.A.: 1997, *Geophys. Res. Lett.* **24**, 1727.
Treumann, R.A.: 1999a, *Phys. Scripta* **59**, 19 and 204.
Treumann, R.A.: 1999b, *Europhys. Lett.* **48**, 8.
Treumann, R.A. and Kull, A.: 2000, *Phys. Scripta* **621**, 403.
Treumann, R.A., LaBelle, J. and Bauer, T.M.: 1995, in: P. Song, B.U.Ö. Sonnerup, M.F. Thomsen (eds.), *Physics of the magnetopause*, AGU Washington D.C., p. 331.
Tsallis, C.: 1988, *J. Stat. Phys.* **52**, 479.
Vasyliunas, V.M.: 1968, *J. Geophys. Res.* **73**, 2839.
Winske, D., Thomas, V.A. and Omidi, N.: 1995, in: *Physics of the Magnetopause, Geophys. Monogr.* **90**, p. 321, AGU, Washington, D.C.

MAGNETIC FIELD LINE TRANSPORT IN THE HELIOSPHERE AND ENERGETIC PARTICLE PROPAGATION FROM COROTATING INTERACTION REGIONS TO HIGH HELIOGRAPHIC LATITUDES

GAETANO ZIMBARDO, PIERRE POMMOIS and PIERLUIGI VELTRI
Dipartimento di Fisica, Università della Calabria and Istituto Nazionale di Fisica della Materia, Unità di Cosenza, 87036 Arcavacata di Rende CS, Italia

Abstract. A numerical model of magnetic turbulence allows us to compute the diffusion coefficients for different degrees of anisotropy, and in particular for the cases corresponding to the solar wind. We consider here the problem of the propagation of the energetic particles accelerated by corotating interaction regions (CIRs) and observed by Ulysses at high heliographic latitudes. The excursion in latitude of a magnetic field line located at large heliospheric distance is evaluated with a Monte Carlo simulation. In the calculations, the random 'force' terms are proportional to the square root of the diffusion coefficient in each direction. By tracing a large number of field lines, we can devise how the particles would be seen by Ulysses. Good agreement with the actual observations is found.

1. Introduction

To lowest order in the drift approximation, charged particles move along the magnetic field lines, so that the magnetic field structure determines particle transport. Besides the global, average magnetic field configuration, a relevant level of magnetic turbulence is often present, which induces a random walk of magnetic field lines (Jokipii and Parker, 1968). This problem also attracted much interest in the plasma fusion field (Rechester and Rosenbluth, 1978).

In this paper we consider Ulysses recurrent observations of energetic particles at high southern heliographic (e.g., Simnett *et al.*, 1998). These particles are thought to be accelerated by corotating interaction regions (CIRs) which extend in latitude up to $\pm$ 30°–35°, while energetic particles enhancements were observed up to 70°–74°S for protons, and even 80°S for electrons. We show that latitudinal magnetic field line diffusion due to anisotropic magnetic fluctuations can indeed explain the observations of both ion and electron events (Kóta and Jokipii, 1995). The intensity of particles accelerated at CIRs is maximum at 3–4 AU, and then decreases with radial distance. In our model, we assume that particles are accelerated at heliographic distances larger than those explored by Ulysses, $R_{\mathrm{CIR}} \simeq$ 6–10 AU, and that they travel toward the inner heliosphere along the average magnetic field; at the same time, magnetic turbulence cause a random walk of particles. By means of a Monte Carlo simulation, we can trace a large number of field lines, and transport from a distant CIR can be evaluated.

Astrophysics and Space Science **277:** 97–100, 2001.

2. Monte Carlo Simulation of Field Line Transport

In the solar wind we adopt a local frame of reference where z is along the unperturbed magnetic field, x is normal to the solar wind speed-magnetic field plane, and y completes the right-handed system (Pommois *et al.*, 2001). l_x, l_y, and l_z represent the magnetic turbulence correlation length in each direction. Extensive numerical simulation of magnetic field line transport were performed by Pommois *et al.* (1999) in the case when $l_x > l_y = l_z$, and by Zimbardo *et al.* (2000) when $l_x = l_y \neq l_z$. We have computed the diffusion coefficient in the more general case when all the correlation lengths are different. For the fluctuation levels relevant to the solar wind, $\delta B/B_0 \simeq 0.5$–1, where $\delta B = \sqrt{\langle \delta \mathbf{B} \cdot \delta \mathbf{B} \rangle}$, and for the degree of anisotropy of interest, $l_x \gg l_y$, the diffusion coefficients can be modelled as

$$D_i = \mathcal{D} \left(\frac{\delta B}{B_0} \frac{l_z}{l_x} \right)^{\mu} \frac{{l_x}^2}{l_z} \left(\frac{l_i}{l_x} \right)^{\nu} \tag{1}$$

with $i = x, y$, and where a best fit yields $\mathcal{D} = 0.029$, $\mu = 1.57$ and $\nu = 0.69$ (these parameters are dimensionless, while D_i has the dimension of a length). The magnetic field is given by the sum of the average component $\langle \mathbf{B}(\mathbf{r}) \rangle$ and of the fluctuating component $\delta \mathbf{B}$, $\mathbf{B} = \langle \mathbf{B}(\mathbf{r}) \rangle + \delta \mathbf{B}$. The equations of the magnetic field lines which we have to integrate are

$$\frac{d\mathbf{r}}{d\xi} = \frac{\langle \mathbf{B}(\mathbf{r}) \rangle + \delta \mathbf{B}}{|\langle \mathbf{B}(\mathbf{r}) \rangle|} = \hat{e}_B + \frac{\delta \mathbf{B}(\mathbf{r})}{B_0(\mathbf{r})} \tag{2}$$

where $\mathbf{r}$ is the field line position, $\hat{e}_B$ is the unit vector along the unperturbed spiral structure, and $B_0(\mathbf{r}) \equiv |\langle \mathbf{B}(\mathbf{r}) \rangle|$. The fluctuating component of (2) is modelled as a random process, $\delta B_i(\mathbf{r})/B_0(\mathbf{r}) = \eta_i(\xi) A_i(\mathbf{r})$ where $i = x, y$ in a local frame of reference. Above, $\eta_x(\xi)$ and $\eta_y(\xi)$ are two uncorrelated random functions with no memory, and the random 'force' amplitudes $A_i(\mathbf{r})$ are related to the diffusion coefficients by $A_i(\mathbf{r}) = \sqrt{6 D_i(\mathbf{r})}$. Once D_x and D_y are specified by (1), we can integrate (2) to trace a magnetic field line in the heliosphere.

Several runs where done varying the parameters $\delta B/B_0$, l_y/l_x, l_z/l_x, and the heliospheric distance R_{CIR} of the magnetic connection to the CIR.

In Figure 1, we show the angular positions of field lines obtained at Ulysses for an anisotropy of the correlation lengths given by $l_y/l_x = l_z/l_x = 0.125$, a fluctuation level at 1 AU of $\delta B_{\mathrm{E}}/B_{\mathrm{E}} = 0.7$ and $R_{\mathrm{CIR}} = 8$ AU. It can be seen that some field lines travel until a latitude of 70° and more, therefore showing that magnetic field line diffusion can provide the required magnetic connection between CIRs and high heliographic latitudes. In order to make the comparison with the Ulysses observations we consider that the pattern in Figure 1 rotates over Ulysses every 26 days, while the spacecraft slowly changes latitude. This yields the histogram of field line transport as a function of time reported in Figure 2.

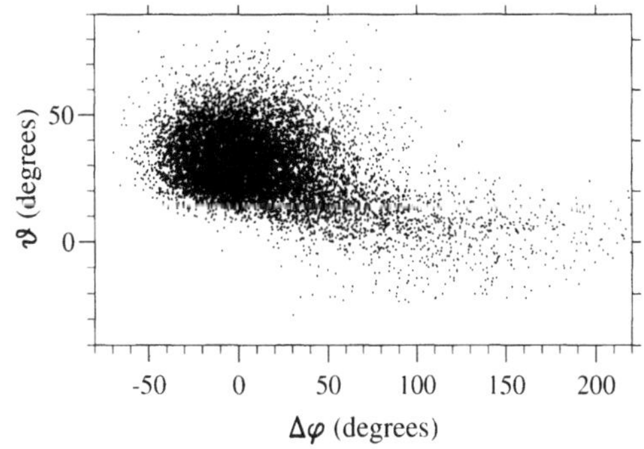

Figure 1. Values of the angle ϑ and $\Delta\varphi$ obtained at the end of the integration.

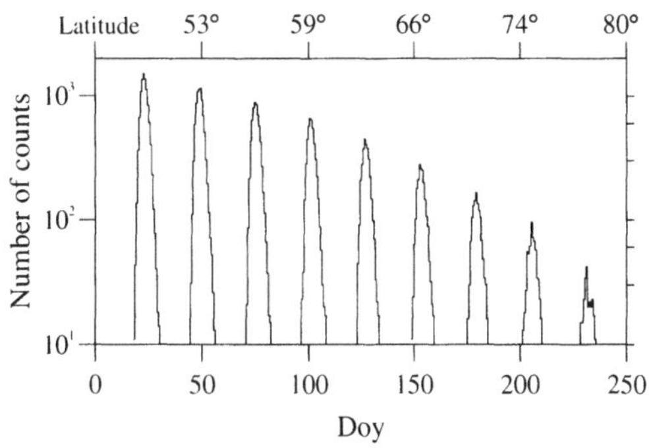

Figure 2. Histogram of the distributions of magnetic field lines with latitude obtained using the Monte Carlo simulation (arbitrary units). The lower (upper) horizontal axis reports the day of year 1994 (the latitude of Ulysses).

3. Discussion and Conclusions

We have shown in the previous section that magnetic field line transport due to anisotropic turbulence can provide the magnetic connection between high and low heliographic latitudes. We argue that this mechanism can explain both the electron and the ion observations reported, for instance by Simnett *et al.* (1998). Indeed, Figure 2 exhibits peaks decreasing in intensity with latitude, with a peak at nearly 80°, as observed for the electrons.

The electrons closely follow the magnetic field lines, since their Larmor radius is much smaller than the shortest wavelength of magnetic turbulence (see below). On the other hand, the ions diffuse *less* than the electrons because they have larger Larmor radii. In particular, we can estimate that the inertial range of turbulence at 3 AU goes from the radial correlation length $\lambda_r \approx 1.5 \times 10^7$ km, to the dissipation scale $\lambda_{\text{diss}} \equiv 2\pi V_A/\Omega_i \approx 2500$ km, assuming $B \simeq 1.3$ nT and number density $n \simeq 0.5$ cm^{-3}. In such a magnetic field, 1.2 MeV protons have gyroradius $\rho_i \simeq$ 120.000 km, and 50 keV electrons have gyroradius $\rho_e \simeq 600$ km. Hence, the electrons see all of the field line random walk as $\rho_e \ll \lambda_{\text{diss}}$, while the ions are not sensitive to the part of field line transport due to short wavelengths as $\rho_i \gg \lambda_{\text{diss}}$. Indeed, it is well known from the physics of plasma confinement in laboratory that magnetic field line transport is most effective on electrons, while for the ions the magnetic fluctuations with scales less than the ion gyroradius are averaged out in the gyromotion (e.g., Mynick and Krommes, 1979).

Acknowledgements

Work supported by the Italian MURST, CNR, and ASI.

References

Jokipii, J.R. and Parker, E.N.: 1968, Random walk of magnetic lines of force in astrophysics, *Phys. Rev. Lett.* **21**, 44–47.

Kóta, J. and Jokipii, J.R.: 1995, Corotating variations of cosmic rays near the south heliospheric pole, *Science* **268**, 1024–1025.

Mynick, H.E. and Krommes, J.A.: 1979, Particle diffusion by magnetic perturbations of axisymmetric geometries, *Phys. Rev. Lett.* **43**, 1506–1509.

Pommois, P., Veltri, P. and Zimbardo, G.: 1999, Anomalous and Gaussian transport regimes in anisotropic 3-D magnetic turbulence, *Phys. Rev. E* **59**, 2244–2252.

Pommois, P., Veltri, P. and Zimbardo, G.: 2001, Field line diffusion in the solar wind magnetic turbulence and energetic particle propagation across heliographic latitudes, *J. Geophys. Res.*, in press.

Rechester, A.B. and Rosenbluth, M.N.: 1978, Electron heat transport in a Tokamak with destroyed magnetic surfaces, *Phys. Rev. Lett.* **40**, 38–41.

Simnett, G.M., Kunow, H., Flückiger, E., Heber, B., Horbury, T., Kóta, J., Lazarus, A., Roelof, E.C., Simpson, J.A., Zhang, M. and Decker, R.B.: 1998, Corotating particle events, *Space Sci. Rev.* **83**, 215–257.

Zimbardo, G., Veltri, P. and Pommois, P.: 2000, Anomalous, quasilinear, and percolative regimes for magnetic-field-line transport in axially symmetric turbulence, *Phys. Rev. E* **61**, 1940–1948.

MAGNETIC FIELD LINE TRANSPORT IN ANISOTROPIC MAGNETIC TURBULENCE: ANOMALOUS, QUASILINEAR, AND PERCOLATIVE REGIMES VERSUS THE KUBO NUMBER

GAETANO ZIMBARDO, PIERLUIGI VELTRI and PIERRE POMMOIS

Dipartimento di Fisica, Università della Calabria and Istituto Nazionale di Fisica della Materia, Unità di Cosenza, 87036 Arcavacata di Rende CS, Italy

Abstract. The transport of plasma and of energetic particles because of magnetic turbulence is relevant to many space plasmas, ranging from the planetary magnetospheres to the solar corona and to the heliosphere. Various transport regimes for magnetic field lines can be obtained depending on the Kubo number. Here we show, by means of a numerical simulation, that the Kubo number also determines the level of chaos of the field lines. Weak chaos, closed magnetic surfaces, and anomalous transport regimes are obtained for $R \ll 1$; widespread chaos, destroyed magnetic surfaces, and quasilinear scaling of the diffusion coefficient for $R \gtrsim 0.3$; and global stochasticity as well as percolation scaling of the diffusion coefficient for $R \gg 1$.

The transport of magnetic field lines in magnetic turbulence strongly influences plasma transport. The Kubo number, which for magnetic turbulence can be defined as $R = (\delta B/B_0)(l_\parallel/l_\perp)$, where δB is the rms fluctuating magnetic field, B_0 is the average field, and $l_\parallel$ ($l_\perp$) is the turbulence correlation length parallel (perpendicular) to $\mathbf{B}_0$, is a parameter which characterizes the different regimes of transport. For $R \ll 1$ we can use a perturbative treatment and the quasilinear diffusion coefficient is obtained, with $D \sim (\delta B/B_0)^2 l_\parallel$ (Jokipii and Parker, 1968). Recently, Isichenko (1991) derived the scaling of the diffusion coefficient with the level of fluctuations in the percolation limit, $R \gg 1$, obtaining $D = l_\perp (l_\perp/l_\parallel)^{3/10} (\delta B/B_0)^{7/10}$.

On the other hand, Zimbardo *et al.* (1995) and Pommois *et al.* (1998) have shown by means of numerical simulations that in isotropic turbulence (that is when $l_\parallel = l_\perp$), at low fluctuation levels, $\delta B/B_0 \lesssim 0.2$, anomalous, i.e., superdiffusive and subdiffusive, regimes can be found in a periodic magnetic turbulence model. These regimes are characterized by a mean square deviation of the field line position $\langle \Delta x^2 \rangle$ which grows as $\langle \Delta x^2 \rangle = 2Ds^\alpha$, where s is the field line length (note that s takes the role of time for field line transport), D is the 'diffusion coefficient', and $\alpha \neq 1$ is the anomalous diffusion exponent. Increasing the level of fluctuations a regime of global stochasticity is reached, so that the K.A.M. tori are destroyed, and normal diffusion ($\alpha = 1$) is attained. In those works, a turbulent magnetic field is realized by adding a power law spectrum of Fourier modes upon a uniform magnetic field B_0.

Astrophysics and Space Science **277:** 101–102, 2001.

More recently, Pommois *et al.* (1999) and Zimbardo *et al.* (2000) performed numerical simulations of magnetic field line transport in a three dimensional anisotropic magnetic turbulence, in which the magnetic fluctuation level $\delta B/B_0$ and the correlation lengths l_x, l_y and l_z can be varied independently of each other. The case of anisotropy in the plane perpendicular to the average magnetic field, relevant to the solar wind magnetohydrodynamic turbulence, shows that diffusion is faster along the direction with the largest correlation length, with $D_x/D_y \approx l_x/l_y$ (see Pommois *et al.*, 1999). Three different magnetic field line transport regimes are obtained depending on the Kubo number R (Zimbardo *et al.*, 2000). For $R \ll 1$, weak chaos, closed magnetic surfaces, and anomalous transport regimes are obtained. For $R \gg 1$ we find global stochasticity and $D \simeq 0.06R^{0.7}l_x{}^2/l_z$, that is the percolation scaling for the diffusion coefficient (Isichenko, 1991). For $R \gtrsim 0.3$, a quasilinear regime is recovered where there is widespread chaos and the magnetic surfaces are destroyed; in this case the diffusion coefficient is of the form $D \simeq 0.015R^2l_x{}^2/l_z$. These findings are relevant for particle propagation in the solar wind, where R is of order 1 or more. Indeed, it appears that field line transport in the solar wind corresponds to a transition from the quasilinear to the percolative regime.

Acknowledgements

Work supported by the Italian MURST, CNR, and ASI.

References

Isichenko, M.B.: 1991, Effective plasma heat conductivity in a 'braided' magnetic field – II. Percolation limit, *Plasma Phys. Control. Fusion* **33**, 809–826.

Jokipii, J.R. and Parker, E.N.: 1968, Random walk of magnetic lines of force in astrophysics, *Phys. Rev. Lett.* **21**, 44–47.

Pommois, P., Zimbardo, G. and Veltri, P.: 1998, Magnetic field line transport in three dimensional turbulence: Lévy random walk and spectrum models, *Phys. Plasmas* **5**, 1288–1297.

Pommois, P., Veltri, P. and Zimbardo, G.: 1999, Anomalous and Gaussian transport regimes in anisotropic 3-D magnetic turbulence, *Phys. Rev. E* **59**, 2244–2252.

Zimbardo, G., Veltri, P., Basile, G. and Principato, S.: 1995, Anomalous diffusion and Lévy random walk of magnetic field lines in three dimensional turbulence, *Phys. Plasmas* **2**, 2653–2663.

Zimbardo, G., Veltri, P. and Pommois, P.: 2000, Anomalous, quasilinear, and percolative regimes for magnetic-field-line transport in axially symmetric turbulence, *Phys. Rev. E* **61**, 1940–1948.

DAMPING EFFECTS ON PARAMETRIC DECAYS OF ALFVÉN WAVES

L. GOMBEROFF[1] and J. A. ARANEDA[2]
[1] *Departamento de Física, Facultad de Ciencias, Universidad de Chile, Casilla 653, Santiago, Chile*
[2] *Departamento de Física, Facultad de Ciencias Físicas y Matemáticas, Universidad de Concepción, Concepción, Chile*

Abstract. It is shown that damping effects can destabilize regions that are stable under non-dissipative fluid theory. Damping effects on the ion-acoustic modes are simulated by a collisional-like term in the longitudinal component of the fluid equations. It is shown that when there is only a beat wave instability, damping effects can destabilize modulational and decay-type instabilities. These results are in agreement with hybrid kinetic-fluid approaches and hybrid simulation codes.

1. Introduction

In space plasmas large amplitude Alfvénic fluctuations have been observed in a number of space environments such as the Earth's bow shock (see e.g., Spangler, 1992), in comets (see e.g., Tsurutani, 1991), and in the solar wind (see, e.g. Marsch and Tu, 1990, 1993). The nature of the parametric decays of circularly polarized Alfvén waves depends on the ion-acoustic velocity. Thus, for $\beta < 1$ ($\beta = (v_s/v_A)^2$ is the ratio between the sound speed, v_s, and the Alfvén speed, v_A) there are three decay modes of the pump wave (Hollweg, 1994): An ordinary decay instability, a beat-wave decay, and a modulational instability. As shown by Hollweg (1994), when $\beta \geq 1$ the only parametric decay is the beat-wave instability. The decay instability involves backward propagating Alfvén waves. These waves are of particular interest in the solar wind because they can provide the backward propagating Alfvén waves which are thought to be necessary for the development of turbulence (see e.g., Dobrowolny *et al.*, 1980; Marsch and Tu, 1993). This instability might account for the observed decrease of normalized cross helicity in the solar wind at greater distances from the Sun (Marsch and Tu, 1990).

All these studies have used fluid theory and have ignored possible damping effects. Here we simulate Landau damping effects on the ion-acoustic modes by introducing a collision-like term in the longitudinal component of the fluid equations. This procedure introduces a phenomenological damping on the ion-acoustic modes while leaving the parent modes unaltered. Although it is not guaranteed that this is equivalent to kinetic Landau damping, we shall see that it leads to the same effects. We shall show that for $\beta \gtrsim 1$, where fluid theory predicts only the beat-wave decay, damping effects lead to new instabilities.

Astrophysics and Space Science **277:** 103–106, 2001.

2. The Dispersion Relation

The system consists of a plasma composed by protons and massless electrons in the presence of an external magnetic field B_0. Assuming that the system is current-free, the dispersion relation is given by,

$$y_0^2 = \frac{x_0^2}{(1 - x_0)}, \tag{1}$$

where $y_0 = k_0 v_A/\omega_{cp}$, $x_0 = \omega_0/\omega_{cp}$, $v_A = B_0/(4\pi n_p m_p)^{1/2}$, and $\omega_{cp} = qB_0/cm_p$ is the proton gyrofrequency. Eq. (1) is the dispersion relation of Alfvén/ion-cyclotron waves moving in the direction of the external magnetic field, which it has been chosen to be the x direction. Assuming now that the system consists of protons, electrons, and a large amplitude pump wave satisfying Equation (1) and propagating in the direction of the interplanetary magnetic field (the external magnetic field), the non-linear dispersion relation can be written in the following form (Hollweg, 1994),

$$M = L_- L_+ D + L_- R_- B_+ + L_+ R_- B_- = 0, \tag{2}$$

where

$$L_\pm = y_\pm^2 - x_\pm^2/\xi_\pm, \tag{3}$$

$$R_\pm = \frac{y_\pm}{2\xi_0}(x_0 - \frac{y x_0^2}{y_0 x} + \frac{x_\pm}{\xi_{pm}}), \tag{4}$$

$$D = r\beta_e y^2 - x^2 A - \Delta, \tag{5}$$

$$B_\pm = \pm\frac{A x \xi_\mp (y_\pm x_0^2 - y_0 \xi_0 x_\pm^2)}{y_0 y_\pm}, \tag{6}$$

$$\Delta = r(x^2 - \beta_p y^2), \tag{7}$$

with the definitions, $A = (B/B_0)^2$, where B is the magnetic field of the pump wave, $r = \xi_0\xi_+\xi_-$, $\xi_0 = 1 - x_0$, $\xi_\pm = 1 - x_\pm$, $x_\pm = x_0 \pm x$, $y_\pm = y_0 \pm y$, and $\beta_{e,p} = 4\pi n_p \gamma_{e,p} K T_{e,p}/B_0^2$. In the last relations K is the Boltzmann's constant, and $\delta p/p_0 = \gamma \delta n/n_0$ with γ the adiabaticity coefficient.

In order to simulate Landau damping in the ion-acoustic modes, we introduce a collisional-like term in the x-component of the fluid equations,

$$(\frac{\partial}{\partial t} + V \cdot \nabla)\vec{V} = \frac{q}{m}(\vec{E} + \frac{1}{c}\vec{V} \times \vec{B}) - \frac{\vec{\nabla} p}{mn} - m\nu V_x \hat{e}_x, \tag{8}$$

where ν is a phenomenological damping rate. The idea is to simulate ion-acoustic Landau damping through a collision-like term (Muñoz and Gomberoff, 1988, and references threin). After linearization of Equation (8), the new term adds a contribution $i\nu$ to ω in the left-hand side of the equation. It is simple to show that the

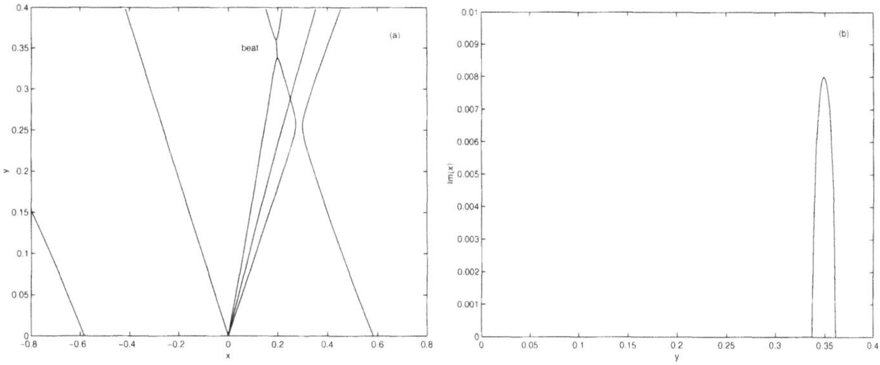

Figure 1. (a) Nonlinear dispersion relation, x vs. y, for $x_0 = 0.25$, $\beta = \beta_e + \beta_p = 1.0$, $A = 0.2$, and $\bar{\nu} = 0$. (b) $Im(x)$ vs. y for the beat wave instability.

effect of this term on the nonlinear dispersion relation, Equation (2), is equivalent to replace x^2 by $x(x + i\bar{\nu})$ in (7), where $\bar{\nu} = \nu/\omega_{cp}$ (Muñoz and Gomberoff, 1998).

3. Numerical Analysis

In Figure 1a we have solved Equation (2), for $x_0 = 0.25$, $\beta = 1.0$, $\bar{\nu} = 0$ and $A = 0.2$. From Figure 1 it follows that the only instability is a beat-wave instability (Hollweg, 1994). For the same parameters, the growth rate of the instability, $Im(x) = \gamma/\omega_p$, is shown in Figure 1b. In Figure 2, we have increased $\bar{\nu} = 0.08$. We see that the maximum growth rate of the beat-wave instability has decreased with respect to Figure 1b ($\bar{\nu} = 0$), but there are two new instabilities with one of them being clearly a modulational instability. It is possible to show that the growth rate of the new instabilities increases until a critical $\bar{\nu}$-value, such that for values larger than this critical value, the growth rate of the instabilities decreases. By plotting $Re(M) = 0$ and $Im(M) = 0$ as a function of $Im(x)$ and $Re(x)$, it is possible to show that the modulational-like instability involves a forward propagating sideband wave, and the decay-like instability (i.e., the one around $y \simeq 0.32$, see Figure 2) gives rise to a backward propagating sideband wave.

4. Summary

By simulating Landau damping effects on the ion-acoustic modes through a collisional-like term in the longitudinal component of the fluid equations, we have shown that for $\beta \gtrsim 1$, there are two new instabilities (see Figure 2). One of them is a modulational-like instability, and the other occurs in the region where a decay instability could have been expected in the absence of damping (see Figure 1). These results are in good agreement with the simulation of Vasquez (1995) based

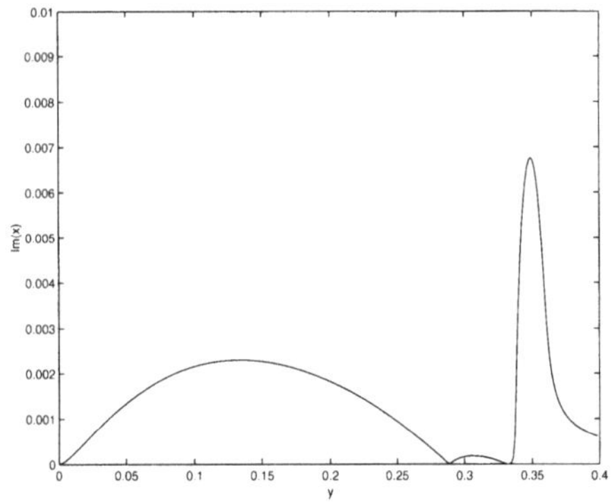

Figure 2. Same as Figure 1b, but for $\bar{\nu} = 0.08$.

on a 1-D hybrid simulation code, and with Araneda (1998) who used a hybrid approach to the problem.

Acknowledgements

This work has been partially supported by FONDECYT grant N°1990047, and Dirección de Investigación, Universidad de Concepción.

References

Araneda, J.A.: 1998 *Phys. Scripta* **T57**, 164.
Dobrowolny, M., Mangeney, A. and Veltri, P.: 1980, *Astron. Astrophys.* **83**, 26.
Hollweg, J.V.: 1994, *J. Geophys. Res.* **99**, 23,431.
Marsch, E. and Tu, C.Y.: 1990, *J. Geophys. Res.* **95**, 8211.
Marsch, E. and Tu, C.Y.: 1993, *J. Geophys. Res.* **98**, 21045.
Muñoz, V. and Gomberoff, L.: 1998, *Phys. Plasmas* **5**, 3171.
Spangler, S.R.: 1992, The evolution of large-amplitude MHD waves near quasi-parallel shocks in the solar wind, in: E. Marsch and R. Shwenn (eds.), *Solar Wind Seven*, p. 539, Pergamon, New York.
Tsurutani, B.T.: 1991, Nonlinear low frequency waves : Comets and foreshockphenomena, in: *Physics and space Plasmas*, p. 91, Scientific, Cambridge, Mass.
Vasquez, B.J.: 1995, *J. Geophys. Res.* **100**, 1779.

RADIO WAVE PROPAGATION IN THE CORONA AND THE INTERPLANETARY MEDIUM

T.S. BASTIAN*
National Radio Astronomy Observatory, Charlottesville, VA 22903, U.S.A.
E-mail: tbastian@nrao.edu

Abstract. Radio wave propagation through an inhomogeneous, random plasma produces a variety of observable phenomena group delay, Faraday rotation, refraction, angular broadening, spectral broadening, and scintillations in phase, amplitude, and frequency. These may be exploited to constrain the mean and fluctuating properties of the medium through a variety of remote sensing techniques. In the case of the solar corona and the solar wind, the mean density, magnetic field, solar wind speed, and the spatial spectrum of the density fluctuations can all be constrained in regions that are inaccessible to *in situ* measurements.

1. Introduction

The interaction of radio waves with the tenuous plasma in which they propagate in the solar corona, the interplanetary medium (IPM), and the interstellar medium (ISM) has been studied for several decades. When certain properties of the emitting source are known, coupled with a theoretical understanding of propagation phenomena, the source may be used as a probe of the foreground medium. Unlike studies of the ISM, for which only pulsars and certain compact extragalactic sources may serve as probes, transillumination of the corona and IPM is possible using spacecraft radio beacons, radar signals, in addition to cosmic sources. Hence, a large variety of powerful experiments are possible, experiments which provide unique constraints on physical parameters of the corona and IPM.

A phenomenological overview of propagation phenomena and a sketch of the theoretical elements used to describe them has been presented by Bastian (2000). A more detailed discussion, with particular reference to spacecraft observations, may be found in Mullan and Yakovlev (1995) and Yakovlev and Mullan (1996). Thorough theoretical treatments may be found in Ishimaru (1978) or in Rytov *et al.* (1989). The present paper is confined to outlining some preliminary concepts in §2. A discussion of propagation phenomena involving interaction with a spectrum of plasma inhomogeneities is presented in §3. The diagnostic uses of these phenomena is outlined in §4. Large scale refraction and large-angle scattering in the context of coronal and interplanetary radio bursts is briefly discussed in §5.

* Currently at DESPA, Observatoire de Paris, Section de Meudon, 92195 Meudon Cedex, France

Astrophysics and Space Science **277:** 107–116, 2001.

2. Preliminaries

Propagation phenomena result from the interaction between radio waves and the medium in which they propagate – the tenuous, magnetized plasma of the solar corona and the interplanetary medium (IPM) in the present case. The dispersive properties of the medium are described by the dielectric constant $\epsilon = \mu^2$, where μ is the refractive index. For our purposes it is sufficient to consider an unmagnetized plasma for which $\mu = (1 - \nu_{pe}^2/\nu^2)^{1/2}$, where ν is the (cyclic) frequency of the wave, $\nu_{pe} = (e^2 n_e/\pi m_e)^{1/2}$ is the electron plasma frequency, and n_e is the electron number density. Expressing n_e and μ in terms of average and fluctuating parts, such that $n_e = n_{e\circ} + \delta n_e$ and $\mu = \mu_\circ + \delta\mu$, one sees that fluctuations in the electron number density δn_e produce fluctuations in refractive index

$$\delta\mu = \frac{1}{2\mu_\circ}\frac{\nu_{pe}^2}{\nu^2}\frac{\delta n_e}{n_{e\circ}} = \frac{1}{2\pi}\frac{\lambda^2 r_e}{\mu_\circ}\delta n_e$$

where $r_e = e^2/m_e c^2$ is the classical electron radius. A number of basic phenomena then result from variations in the refractive index $\delta\mu$ or, equivalently, in the electron number density δn_e.

Consider a uniform slab of fully ionized hydrogen of thickness Δz and refractive index μ. The group velocity is $v_{gr} = \partial\omega/\partial k = \mu c < c$. The difference in the time it takes a radio signal to propagate across Δz in the plasma slab and to propagate the same distance *in vacuo* is the *group delay* $\tau_{gr} = \Delta z(v_{gr}^{-1} - c^{-1})$ or, more generally,

$$\tau_{gr} \approx \frac{\lambda^2 r_e}{2\pi}\int n_e ds.$$

Propagation through a plasma inhomogeneity also leads to a change in the phase of the incident wave. Consider again the plasma slab. The number of wavelengths across the plasma slab is $\Delta z\mu/\lambda$ whereas the number of wavelengths across Δz *in vacuo* is $\Delta z/\lambda$. With $\delta\mu = 1 - \mu$, a plasma inhomogeneity then leads to a *phase fluctuation* given by

$$\delta\phi = 2\pi(\Delta z/\lambda - \Delta z\mu/\lambda) = k\Delta z\delta\mu = \lambda r_e \Delta z \delta n_e$$

The direction in which a wave propagates in a medium is given by the normal to the phase front. It is clear that a gradient in n_e – e.g., a plasma wedge – yields a phase gradient and changes the direction of the incident wave. In other words, a density gradient *refracts* the wave.

When a wave is incident on a plasma inhomogeneity with a transverse dimension $l_\circ$ the wave is forward scattered into a cone of angular size $\theta \sim \lambda/l_\circ$ (when $\lambda \ll l_\circ$); i.e., it is *diffracted.* In the presence of many such inhomogeneities, the diffractive cones overlap after propagating some distance L, and interference phenomena occur. Such diffractive effects can often be ignored, at least to first order.

However, an example where diffraction plays a dominant role is in strong intensity scintillations, described briefly in §3.5.

Finally, if the plasma is magnetized a linearly polarized wave experiences Faraday rotation. For the above plasma slab, now threaded by a magnetic field, the change in angle of the polarization plane relative to that *in vacuo* is

$$\Delta\chi = \frac{r_e^2\lambda^2}{2\pi e} n_e B_{\parallel} \Delta z$$

where $B_{\parallel}$ is the component of the magnetic field vector along the line of sight.

3. Propagation Phenomena

The basic effects described in the previous section lead to a number of observable propagation phenomena that may be exploited as probes of the physical properties of the medium. In many cases of interest, the effects of large scale density gradients are negligible compared to those due to small scale density inhomogeneities. We therefore defer discussion of large scale refraction to §5 and briefly outline the most important propagation phenomena caused by *scattering* on small scale inhomogeneities.

A plasma such as the corona or the IPM can be described in terms of a mean background with large scale gradients and a broadband spectrum of random inhomogeneities. It is convenient to describe the medium in terms of the spatial correlation functions $B_\epsilon = \langle \delta\epsilon(\mathbf{r}_1)\delta\epsilon(\mathbf{r}_2)\rangle$, $B_\mu = \langle \delta\mu(\mathbf{r}_1)\delta\mu(\mathbf{r}_2)\rangle$, or $B_{n_e} = \langle \delta n_e(\mathbf{r}_1)$ $\delta n_e(\mathbf{r}_2)$, where

$$B_\epsilon(\mathbf{r}_1, \mathbf{r}_2) = 4B_\mu(\mathbf{r}_1, \mathbf{r}_2) = (r_e^2\lambda^4/\pi) B_{n_e}$$

For a statistically homogeneous, isotropic medium the spatial correlation functions depend only on $r = |\mathbf{r}_1 - \mathbf{r}_2|$, while temporal correlation functions likewise depend only on time differences $\tau = t_1 - t_2$. The 3D Fourier transform of the spatial correlation functions $B_\epsilon(r)$, $B_\mu(r)$, and $B_{n_e}(r)$ yield the spatial power spectra $\Phi_\epsilon(q)$, $\Phi_\mu(q)$, and $\Phi_{n_e}(q)$, respectively, where $q = 2\pi/r$ is the wave number. Similarly, the Fourier transform of a temporal correlation function yields the temporal power spectrum (see below).

To achieve an intuitive understanding of scattering phenomena, it is useful to characterize the action of the 3D plasma medium in terms of a thin, non-absorbing, phase-changing screen in the $x - y$ plane located at $z = 0$. We take the observer to be at a distance $z = L$ from the screen. The action of the thin screen is to randomly perturb the phase as a function of position (x, y). In the case of the solar wind, there is reasonable justification to employ the thin-screen model of the medium in practice. Scattering phenomena depend on density as $(\delta n_e/n_{eo})^2$. Beyond a few solar radii, the solar wind density declines as $n_e \propto (R/R_\odot)^{-2}$. For an observation with an impact parameter b from the Sun, the bulk of the scattering occurs along a length $\sim b \ll L$.

The statistics of the phase fluctuations can be described by a correlation function $B_\phi(\rho, z) = (4\pi^2/\lambda^2)B_\mu(\rho, z)$ or, more conveniently, by the phase structure function $D_\phi(\rho, z) = \langle|\phi(\rho+\rho', z)-\phi(\rho', z)|^2\rangle$, where ρ is the transverse coordinate in the $x-y$ plane. The correlation scale, or coherence scale, $\rho_\circ$ is the transverse scale on which $D_\phi(z, \rho_\circ) = 1$. The transverse scale over which the solar wind density changes significantly is $l_{sw} \sim n_e/\nabla n_e = R_\odot/2$. The transverse scales of interest here are $\rho \ll l_{sw}$, thus ensuring the approximate validity of the homogeneous, thin-screen model.

Both the amplitude and phase of a wave propagating in a random medium are perturbed by the spectrum of plasma inhomogeneities. The wave is often represented as $u(\mathbf{r}) = A(\mathbf{r})\exp[i\phi(\mathbf{r})] = A_\circ \exp[\chi(\mathbf{r}) + i\phi(\mathbf{r})]$, where $\chi = \ln(A/A_\circ)$ is the log-amplitude and ϕ is the phase. The quantities of interest are the fluctuating parts of χ and ϕ.

3.1. Angular and Temporal Broadening

Consider a plane wave incident on the thin screen (corresponding to waves emitted by a point source at $z = -\infty$). After passing through the thin screen, the phase of a plane wave has been randomly perturbed as a function of transverse position (x, y). The phase front is now corrugated and can be characterized by a spectrum of plane waves whose wave normals encompass a range of directions. An observer therefore receives radiation from a range of angles. In other words, the source has suffered *angular broadening* and is no longer point-like. Angular broadening is easily measured using modern interferometric radio arrays, which measure the spatial coherence function $\Gamma(\rho) = \exp[-D_\phi(\rho)/2]$ of the incident radiation. The width of the angular spectrum is conveniently characterized as $\theta_S = 1/k\rho_\circ$. Angular scattering of this order clearly leads to a geometrical delay $\tau_D \sim L\theta_S^2/2c$. An impulse source of radio waves therefore experiences *temporal broadening* of this order.

3.2. Spectral Broadening

If the thin screen moves at a constant speed v_S transverse to the observer, a source observed at a frequency ν will experience *spectral broadening* of order $\Delta\nu/\nu \sim 2v_\parallel/c \approx 2v_S\theta_S/c$. The spectral broadening is a result of Doppler shifts caused by the radial projection of the screen velocity over the characteristic angular width θ_S.

3.3. Phase Scintillations

Consider once again a plane wave propagating through the phase-changing screen. On the observer's plane the phase varies randomly in space. The variations can be described by the spatial correlation function $B_\phi(q)$. If we now suppose the random medium moves transverse to the observer, as before, and that the random plasma inhomogeneities are frozen into the medium, the phase measured at a fixed point in the observer's plane varies randomly in time. These *phase scintillations* can be described by a temporal correlation function $B_\phi(\tau)$, where τ is the difference of two times: $\tau = t_1 - t_2$. The temporal power spectrum $\Phi_\phi(\omega)$, with $\omega = 1/\tau$, is the Fourier transform of $B_\phi(\tau)$ and is given by

$$\Phi_\phi(\omega) = 8\pi^2 k^2 D \int_{\omega/v_S}^{\infty} q\, dq\, \Phi_\mu(q)\, (q^2 v_S^2 - \omega^2)^{-1/2} f_\phi(q)$$

where $f_\phi(q) = cos^2(q^2 r_F^2/2)$. Assuming $\Phi_\mu(q) \propto q^{-\alpha}$, with $2 < \alpha < 4$, one finds that $\Phi_\phi(\omega) \propto k^2 v_S^{\alpha-2} \omega^{-(\alpha-1)}$. In this case, the power spectrum of phase scintillations is dominated by the largest spatial scales in the spectrum of random inhomogeneities. Note that if one assumes more complex forms for the spatial power spectrum $\Phi_\mu(q)$ – the presence of an inner scale, for example – additional features appear in the spectrum (e.g., Woo, 1975).

3.4. Frequency or Doppler scintillations

Spectral broadening occurs as a result of scattering on a moving screen. A related phenomenon is that of *frequency scintillations*, also referred to as *Doppler scintillations*, where the center frequency of the scatter-broadened line fluctuates in time. If a monochromatic signal of frequency ν is transmitted through the moving, phase-changing screen, we have already seen that an observer sees phase scintillations. The frequency fluctuates about ν as $\delta\nu \sim \delta\phi/\delta t$. Since the largest spatial scales, and smallest wave numbers, of $\Phi_\mu(q)$ dominate phase scintillations, so too do they dominate frequency scintillations. Indeed, the temporal power spectrum of the phase scintillations is related to that of frequency fluctuations as $\Phi_\nu(\omega) = \omega^2 \Phi_\phi(\omega) \propto \omega^{-(\alpha-3)}$.

3.5. Intensity scintillations

An important spatial scale for intensity scintillations is the Fresnel scale: $r_F = \sqrt{\lambda L/2\pi}$. If $r_F \ll \rho_\circ$, intensity variations are caused by weak focusing and defocusing by inhomogeneities with spatial scales $l \sim r_F$. The time scale for these *weak* scintillations (weak scattering regime) is $t_W \sim r_F/v_S$. If, on the other hand, $r_F \gg \rho_\circ$, the incident radiation is scattered into a diffractive cone of angular size $\theta_S \sim 1/k\rho_\circ$. At a distance D from the screen, an observer is illuminated by all cones lying within a patch of sky of size $\sim D\theta_S = r_F^2/\rho_\circ$, or $N \sim (r_F/\rho_\circ)^4$ cones. These constructively and destructively interfere with one another, producing *strong* scintillations for which the intensity modulation is 100%. The time scale of strong scintillations (strong scattering regime) is $t_S \sim \rho_\circ/v_S \ll t_W$. Observations in the weak scattering regime are of the most practical interest because they can be related most easily to physical quantities of interest: e.g., the screen velocity and the spatial spectrum of density inhomogeneities.

As is the case for phase scintillations, a temporal correlation function $B_\chi(\tau)$ describes the log-amplitude scintillations, the Fourier transform of which yields the temporal power spectrum of the log-amplitude fluctuations:

$$\Phi_\chi(\omega) = 8\pi^2 k^2 D \int_{\omega/v_S}^{\infty} q\, dq\, \Phi_\mu(q)\, (q^2 v_S^2 - \omega^2)^{-1/2} f_\chi(q)$$

where $f_\chi(q) = \sin^2(q^2 r_F)/2$. When $\Phi_\mu(q)$ is a power law the solution for $\Phi_\chi(\omega)$ is a complex function of ω. For $\omega \gg v_S/r_F$, corresponding to spatial scales $l \ll r_F$, $f_\chi \approx 1$ and $\Phi_\chi \propto \omega^{-(\alpha-1)}$, the same dependence as $\Phi_\phi(\omega)$. At frequencies $\omega < v_S/r_F$, the spectrum saturates and becomes independent of ω. The break in the spectrum depends on v_S and can therefore be used to determine v_S. It is illuminating to consider the product $\omega\Phi_\chi(\omega)/v_S$, the fluctuation power per log-interval of ω. The maximum occurs at $\omega_{max} = v_S/r_F$, corresponding to a spatial scale $v_S\tau_{max} = r_F$. Weak intensity scintillations are therefore dominated by spatial scales r_F as noted above.

4. Diagnostic Uses of Propagation Phenomena

In this section, the manner in which propagation phenomena are exploited in practice is outlined. Experimental configurations are first described and then a small sample of the diagnostic uses of propagation phenomena discussed. A tabulation of the phenomena discussed in previous sections is given in Table I, along with a summary of the plasma properties that can be inferred through careful observations of the specific phenomenon.

4.1. Experimental Configurations

A large number of experimental configurations can be employed to observe propagation phenomena in the corona and IPM. I mention four broad classes here and reference early examples: 1) distant cosmic sources of radio waves – e.g., pulsars and certain quasars and radio galaxies – appear point-like (spatially coherent) to an observer on Earth and can therefore be used as a probe of the foreground medium (e.g., Erickson, 1964); 2) a spacecraft can carry a transmitter on board which can broadcast a monochromatic (temporally coherent) signal through the coronal or IP medium to receiving apparatus on Earth (e.g., Woo *et al.*, 1976); 3) a spacecraft can transmit a temporally coherent signal through the medium to a receiver on board another spacecraft (e.g., Celnikier *et al.*, 1983); 4) a radar transmitter can transmit a temporally coherent signal from Earth to a planet and the reflection can be observed back on Earth after a double-passage through the medium (e.g., Harmon and Coles, 1983). Each of these experimental configurations, or variants thereof, have been used for many years.

4.2. Mean Properties of the Outer Corona

While direct radio, soft X-ray, UV, EUV, and white light observations have been highly successful in probing the nature of the inner corona, these are largely ineffective in studying the outer corona and the IPM (the outstanding exception being the relatively recent SOHO/LASCO experiment, which has been useful in studying aspects of the corona/IPM out to 30 $R_\odot$). Observations of group delay and Faraday rotation (§2) have been successfully exploited to constrain the mean plasma density and longitudinal magnetic field strength in the outer corona during spacecraft occultations. A recent example is found in Pätzold *et al.* (1996) who used dual-band transmitters on board the *Ulysses* spacecraft to measure the coronal electron column density between $20 - 30\ R_\odot$ over a wide range of position angle. Pätzold *et al.* (1987) used a linearly polarized signal from a transmitter on board the *Helios* spacecraft to constrain the mean coronal magnetic field in the range $3 - 10\ R_\odot$.

4.3. The Nature of $\Phi_{n_e}(q)$

Interest in the nature of inhomogeneities in the corona and IPM has been intense for several decades. The full range of propagation experiments possible has been brought to bear on this problem. A complete discussion of the experiments and their results is far beyond the scope of this short paper.

On the basis of observations of phase scintillations, Doppler scintillations, spectral broadening, and angular broadening, the spatial spectrum $\Phi_{n_e}(q)$ is well- constrained. Briefly, the spatial spectrum of electron density inhomogeneities is complex (e.g., Coles and Harmon, 1989). On spatial scales $l \gtrsim 1000$ km, it is 'Kolmogorov-like'; i.e., the spatial spectrum is $\Phi_{n_e}(q) \propto q^{-\alpha}$, with $\alpha \approx 11/3$. On spatial scales of a few $\times 10$ to $\lesssim 1000$ km, the spectrum flattens significantly, with $\alpha \sim 1$. On the smallest spatial scales it becomes quite steep, with $\alpha \approx 4$. The latter is interpreted as the 'inner scale' or 'dissipation scale' l_{in} of a turbulent spectrum. Interestingly, $\Phi_{n_e}(q)$ is highly anisotropic on intermediate spatial scales close to the Sun (Armstrong *et al.*, 1990). The inhomogeneities are elongated in the direction of the local magnetic field. All of these features must be accommodated and explained by a theory of energy transport in the corona and IPM. Recently, attempts have been made to exploit fluctuations in the Faraday rotation to constrain the nature of Alfvén waves in the solar wind, as well as large scale structures (Mancuso and Spangler, 1994).

4.4. The Solar Wind Speed

While spacecraft in the IPM can perform *in situ* measurements of various plasma parameters, including the solar wind speed, obviously they can only do so at the particular location of the spacecraft. Spacecraft have rarely ventured significantly inside of 1 AU and, until recently, remained in the ecliptic. Remote sensing observations have therefore played an important role in measuring or otherwise con-

TABLE I

Diagnostic uses of radio propagation phenomena

Observation/Technique	Plasma Property
Group delay	Mean electron number density n_e
Refraction	Gradients in electron number density ∇n_e
Faraday rotation	Mean magnetic field B
Faraday fluctuations	Magnetic field fluctuations δB
Spectral broadening	Index α of spatial spectrum Φ_{n_e}
	(on spatial scales $<$ few $\times 100$ km)
	The inner scale of the turbulence l_{in}
	Dispersion in electron number density $\delta n_e / n_e$
Angular broadening	Index α of spatial spectrum Φ_{n_e}
	(on spatial scales $<$ few $\times 100$ km)
	The inner scale of the turbulence l_{in}
	Degree of anisotropy a/b
	Dispersion in electron number density $\delta n_e / n_e$
Phase scintillations	Index α of spatial spectrum Φ_{n_e}
	(on spatial scales $>$ few $\times 100$ km)
	Outer scale of the turbulence l_{out}
Frequency scintillation	Solar wind velocity v_S
(Doppler scintillations)	Index α of spatial spectrum Φ_{n_e}
	on spatial scales $>$ few $\times 100$ km
Intensity scintillations	Solar wind velocity v_S
(weak)	Index α of spatial spectrum Φ_{n_e}
	on spatial scales $< r_F \sim$ few $\times 100$ km
	Dispersion in electron number density $\delta n_e / n_e$

straining plasma parameters in regions where no *in situ* measurements are possible. Particularly important have been measurements of solar wind speed and its acceleration profile with radius.

Solar wind speed measurements are most commonly performed by fitting to the temporal power spectrum of weak amplitude scintillations (§3.5; e.g., Manoharan *et al.*, 1994). More recently, measurements of the moving scintillation pattern on the ground by two widely-spaced antennas have been used to infer the solar wind speed and acceleration profile (Grall, 1995). This technique has been used to suggest that the solar wind is accelerated to high speeds within 2 $R_\odot$ (Grall *et al.*, 1996). Spacecraft measurements of spectral broadening, Doppler scintillations, and phase scintillations (Woo, 1978) have also been used to constrain the solar wind speed and acceleration profile in radial ranges inaccessible to direct measurement.

5. Solar and Interplanetary Radio Bursts

The bulk of this short review has been given over to the uses of propagation phenomena as probes of the corona and IPM. Such uses rely on the fact that the properties of the radio source are well-known and that the all distortions to the signal can be attributed to propagation in the medium. In the case of solar and IP radio bursts, the intrinsic properties of the burst are only crudely known. The corona is an extremely complex and structured plasma environment. Furthermore, coronal and IP radio bursts typically occur near ν_{pe} or its harmonic. Hence refraction and reflection, as well as the scattering effects described in §3 all act to distort and/or destroy information about the source.

General refraction was included in radiative transfer calculations from the very earliest days of radio astronomy in which ray tracing calculations were employed to follow ray trajectories. Beginning with the work of Steinberg *et al.* (1971), the effects of random inhomogeneities were also incorporated into a ray treatment of radio emission from coronal radio bursts. This, and many subsequent studies have been useful in outlining the qualitative effects of propagation. Implicit to existing treatments is the assumption of small-angle scattering relative to the unperturbed ray trajectory. Since this is not always the case, alternative methods must be explored. A promising approach is that proposed by Arzner and Magun (1999) where photon scattering is treated in the context of the geometrical optics approximation and a Fokker-Planck transport equation.

References

Armstrong, J.W., Coles, W.A., Kojima, M. and Rickett, B.J.: 1990, *Astrophys. J.* **358**, 685.

Arzner, K. and Magun, A.: 1999, *Astron. Astrophys.* **351**, 1165.

Bastian, T.S.: 2000, in: R.G. Stone, K. Weiler, M. Goldstein, J.-L. Bougeret (eds.), *Space Based Radio Observations at Long Wavelengths*, Proc. Chapman Conf., 19–23 October 1998, Paris.

Celnikier, L.M., Harvey, C.C., Jegou, R., Kemp, M. and Moricet, P.: 1983, *Astron. Astrophys.* **126**, 293.

Coles, W.A. and Harmon, J.K.: 1989, *Astrophys. J.* **337**, 1023.

Erickson, W.C.: 1964, *Astrophys. J.* **139**, 1290.

Grall, R.R.: 1995, Ph.D Thesis, 141 pp., Univ. of California, San Diego.

Grall, R.R., Coles, W.A., Klinglesmith, M.T., Breen, A.R., Williams, P.J., Markkanen, J. and Esser, R.: 1996, *Nature* **379**, 429.

Harmon, J.K. and Coles, W.A.: 1983, *Astrophys. J.* **270**, 748.

Ishimaru, A.: 1978, *Wave propagation and scattering in random media*, 2 vol., Academic Press, New York.

Mancuso, S. and Spangler, S.R.: 1999, *Astrophys. J.* **525**, 195.

Manoharan, P.K., Kojima, M. and Misawa, H.: 1994, *J. Gen. Relativ.* **99**, 23411.

Pätzold, M., Bird, M.K., Volland, H., Levy, G.S., Seidel, B.L. and Stelzried, C.T.: 1987, *Sol. Phys.* **109**, 91.

Pätzold, M., Karl, J. and Bird, M.K.: 1996, *Astron. Astrophys.* **316**, 316.

Rytov, S.M., Kravtsov, Yu.A. and Tartarskii, V.: 1989, *Principles of Statistical Radiophysics*, 4 vol., Springer-Verlag, New York.

Steinberg, J.-L., Aubier-Giraud, M., Leblanc, Y. and Boischot, A.: 1971, *Astron. Astrophys.* **10**, 362.
Woo, R.: 1978, *Astrophys. J.* **201**, 238.
Woo, R.: 1978, *Astrophys. J.* **219**, 727.
Woo, R., Yang, F.-C. and Ishimaru, A.: 1976, *Astrophys. J.* **210**, 568.
Mullan, D.J. and Yakovlev, O.I.: 1995, *Irish Astron. J.* **22**, 119.
Yakovlev, O.I. and Mullan, D.J.: 1996, *Irish Astron. J.* **23**, 7.

SUMMARY OF SESSION 1: FUNDAMENTAL PROCESSES IN SPACE PHYSICS FROM MACROSCOPIC TO MICROSCOPIC SCALES

ANDRÉ MANGENEY
DESPA, Observatoire de Paris, 92195 Meudon, France

Abstract. I present a rapid overview of the first session of the conference, during which have been discussed basic astrophysical phenomena covering a wide range of spatial scales, as well as several level of descriptions from the macroscopic MHD level, to the microphysics of wave particle interactions.

The first thing I would like to stress in this overview is how well a crude parametrization of stellar convection performs in the understanding of the internal structure of stars. As discussed by E.N. Parker in his review, the theory of the solar interior is firmly established: the sun is a normal star which is burning its central hydrogen and may be considered as spherically symmetric, in a stationary state governed by hydrostatic and thermal equilibrium, energy being transported by means of radiation except in a convective zone covering about 30% of the radius while other effects like those due to magnetic fields, rotation and other macroscopic motions can be ignored. Then, the internal structure is fully determined by the microscopic properties of the solar material, as the equation of state, the opacity, the nuclear reaction rates and the microscopic diffusion parameters, except in the convective zone where fluid motions are essential for the energy transport. A very limited understanding of convection proves to be sufficient to take into account this region in the description of stellar interiors: average hydrostatic equilibrium, the use of a simple Schwarzschild criterion, adiabatic stratification and a rough parametrization of energy transport using an empirical parameter: the mixing length. The first success of the theory was to explain the statistical properties of stars, like the accumulation of stars along the main sequence in the H.R. diagram, i.e. in the plane (luminosity,effective temperature). More recently, a beautiful confirmation came from the confrontation with the results from helioseismology showing that there is a remarkably good agreement between the sound speed profile predicted by the standard theory and that deduced from inversion of the helioseismological observations. The largest deviations occur at the solar surface and at the base of the convection zone, where a larger than predicted sound speed, indicating mixing below the convection zone: as a rule, the solar interior revealed by seismology is smoother than predicted by models pointing towards the necessity of including some, moderate, amount of mixing in the models. This high level of success of the

Astrophysics and Space Science **277:** 117–122, 2001.

mixing length theory is surprising when one recalls that, in the stellar context, convection occurs in a strongly stratified atmosphere, a highly compressible medium, with rotation affecting at least the largest scales, and in a regime characterized by very large Reynolds numbers; it is also surprising in view of the picture given by numerical simulations where fully compressible convection appears as highly intermittent with strong, longlived downwards flows and slower random upwards motion. Clearly, the highly turbulent convection expected in the stellar context must reach some sort of 'asymptotic' state, which cannot be attained by present day numerical simulations.

Beyond the fact that it is an essential ingredient in the working of what E.N. Parker has called the 'Thermal Sun', the solar convective zone is also the basic engine which maintains an extended heliosphere against radiative or mass losses. It is to be noted that the situation is similar for the outer atmospheres of other solar type stars, in the range of masses where spectroscopic signatures of convection are detected. The efficiency of the convective engine for the sun is relatively small. For example, if we take the ratio between the power emitted by the nonradiatively heated outer atmosphere and the total luminosity of a star as a measure of this efficiency, it is about 10^{-5} for the sun , while for fast rotating solar-type stars it can reach as much as 10^{-2}!

To understand how the convective zone can provide the 'mechanical' power needed in the outer atmosphere, one requires a deeper understanding of stellar convection. For example, it is thought that the convective motions produce compressible waves, part of which are reflected low in the solar atmosphere and observed as trapped 'p-modes' and part of which are able to propagate upwards and dissipate by shock formation in the solar chromosphere; they are also able to maintain at the basis of the convective zone a magnetic field which is itself partially unstable to a convective type of instability so that some of the flux tubes are intermittently expelled in the solar atmosphere and heat the solar corona.

There have been numerous attempts to estimate the power emitted upwards in the form of sound (or other) waves; most of them consider the convective motions as prescribed random motions characterized by a small number of statistical parameters, corresponding to a parametrisation of turbulent convection which is subtler but similar in nature to the mixing length theory. It is then found that the efficiency of conversion of convective energy into sound waves, either trapped or propagating is proportional to a high power of the velocity of the energy bearing convective eddies. However, recent observations seem to suggest that, at least, part of this wave power may be due to superficial 'seismic events' resulting from the complex interaction between the convective flow and photospheric radiative losses, which is not usually taken into account.

Similarly, a satisfactory nonlinear dynamo model for solar type stars is still missing. The consensus now is that a hydrodynamic dynamo operates driven by differential rotation and some effect of the cyclonic turbulence; here also, at the present stage of modelling some 'empirical' parametrization of small scale effects, the so called Λ, α,..., effects is necessary. The dynamo amplification process must saturate: either when magnetic buoyancy makes flux tubes of the toroidal to move outwards and emerge above the photosphere, or by the back reaction of the generated fields on the fluid motions. There is yet no satisfactory model predicting the amplitude of the field which can be achieved nor its highly intermittent spatial distribution at the photospheric level. While some global features like the observed cyclic solar dynamo can be understood on the basis of kinematic models of the '$\alpha\Omega$' type, there remains considerable difficulties, as discussed in Professor Parker's review.

The magnetic field, once emerging into the solar atmosphere, appears to determine its dynamical and thermal structure, which may vary on all observationally accessible time and spatial scales, as beautifully shown by the recent observations from the Yokoh, Soho and Trace spacecrafts. A considerable amount of work has been devoted to the aim of understanding the physical effects involved, in particular in producing and shaping the high observed coronal temperatures. Most of the usual scenarios make reference to some form of a turbulent cascade, from the energy containing scales – which may vary from a typical loop size to that of a X-ray bright point – to the extremely small scales where resistive effects may transform macroscopic motions into heat, and/or accelerated particles. Without going into the details let us point some characteristics which make this 'turbulent' atmosphere so different from the usual model of incompressible, statistically homogeneous MHD turbulence. First, it is strongly stratified, with a strongly varying plasma β; radiative losses may be very important, introducing a dissipation which is not limited to small scales and specific nonlinearities. Second, the forcing occurs at the bottom, taken at the photospheric level, through the emergence of new field and the stirring of magnetic structures by the photospheric and subphotospheric motions. It appears that fine characteristics of the forcing play an important role in the evolution of the atmospheric structures, corresponding to the natural invariants, like magnetic helicity, of the ideal MHD equations governing the large scale behavior.

Does there exist some universal behavior? That it is so, is suggested by the observation of atmospheric activity in solar like stars. A very short summary can be done in the following way: there appears to be, for each spectral type, a minimum level of atmospheric activity, the 'basal' level, depending on the corresponding effective temperature and may correspond to the dissipation of the sound waves produced directly by the convective zone. Above this level, an excess of atmospheric emissions apparently depending on the star's rotation velocity is observed. A striking point is that the excess fluxes coming from various altitudes, from the chromosphere to the corona, are well correlated across the relevant region of the H.R. diagram, irrespective of the details of the atmospheric structure. This confirms that a dynamo process of the '$\alpha\Omega$' type is operating in the convective envelope of these stars, and that the details of the heating processes are not so important for the global energy budget of the atmosphere (see the recent book by Schrijver and Zwaan, 2000, for a detailed discussion).

Some attempts in this direction are based on a concept proposed in 1988 by Bak, Tang and Wiesenfeld who showed that fractal and universal behavior, that is power law distributions, can be observed in simple dissipative systems with many degrees of freedom, without fine tuning of the parameters characterizing the system, what they called self organized criticality (SOC). Using very simple equations, which hopefully capture qualitatively the non linear and dissipative aspects of the MHD equations, these models aim to describe how the solar atmosphere may develop intermittently the amount of large electric currents or strong gradients necessary to provide the magnetic dissipation required to maintain its thermal state.

Another facet of the turbulence in the solar atmosphere is that it is highly compressible so that the formation of supersonic flows and shocks are expected, which may reach the solar wind and perturb it significantly, as observed in CME's etc. ...; besides in situ evolution in the solar wind may lead also to shock formation, like forward and reverse shocks at the streams interface, or the solar wind termination shock. These shocks interact with charged particles and accelerate them, through the process called 'diffusive shock acceleration'. As discussed by J.R. Jokippii in his review, this acceleration mechanism is the most probable one in many astrophysical contexts as a source of cosmic rays; the heliosphere has proven to be a very good laboratory in this respect. There is yet no simple and general theory for the 'injection' process: how thermal particles can be accelerated up to the energies where the diffusive shock acceleration process works. More generally, a still open question concerns the partition of dissipated energy into heating (thermal energy deposited in the core of the particle distribution functions) and acceleration (energy deposited in the suprathermal tails). Note that these problems arise only when collisions become unsufficient to ensure the Maxwellian character of the particle distribution functions.

This occurs somewhere in the corona where a transition must occur from a collisional to a collisionless state of the plasma. In the first one, macroscopic information is transmitted by MHD waves, which can be viewed as macroscopic degrees of freedom. In the collisionless region, there is a priori an enormous increase in the number of degrees of freedom; simultaneously the electric fields needed to ensure electric neutrality are increased. This transition from a collisional to a collisionless state of the solar plasma may have some relation to a long standing problem of the theory of the solar wind. Indeed, hydrodynamic theories explain nicely the transition from a high pressure corona to the low pressure interstellar medium through a transsonic flow and a termination shock. However, they have difficulties to explain the large velocities observed in the fast wind, unless some ad-hoc physics is included. On the other hand, if one takes into account the fact that the particle distributions are not maintained close to local thermodynamical equilibrium by binary collisions, one may play with the electrostatic potential needed to keep charge neutrality, which depends in a sensitive way on the properties of these distribution functions. For example, a moderate increase in the number of suprathermal electrons, as observed in the solar wind close to the earth leads to an increase of the potential sufficient to account for the largest wind velocities, as shown by a simple argument based on energy conservation. However existing models of the solar wind use generally extreme assumptions: either fully collisional or fully collisionless with rather peculiar boundary conditions far from the sun, so that one has still to wait for a convincing answer to be given.

Another peculiarity of the collisionless situation is that the structure of the large scale fields become dominant to determine the shape of the distribution functions. An illustrative example is provided by the paper by A.L. Taktakishvili *et al.* on the structure of the distant earth magnetotail, where the distribution of the plasma is entirely shaped by the superposition of large scale magnetic structure and a – stationary – level of magnetic fluctuations. In many cases, however, there is a sharp separation of scales with a very slowly varying background magnetic field and magnetic fluctuations with a short correlation length. In that case quasilinear theory can be used to describe the motion of charged particles which takes on a diffusive character. G. Laval' s review presents a nice review of the associated diffusion processes, covering the collisional and collisionless regimes.

Along the same lines, one may speculate that a sufficiently complex structure of the boundaries where particle are injected will result in a highly intermittent distribution of particles in their respective phase space leading to a significant wave activity. A related problem concerns the dissipation of large scale MHD like motions in collisionless plasmas, which presumably occurs through the deformation of particle distributions leading to the excitation of plasma waves etc. ... Such a constrained plasma, will exhibit short periods of fast evolution (called 'violent relaxation' in stellar dynamics), until a state close to mechanical equilibrium ('virialisation') is achieved followed by a quasistationary, or slowly evolving, evolution far from equilibrium, with relatively strong fluctuations in phase space coupled to electrostatic fields. Is it possible to give a description of such states in a manner which does not depend too much on the fine details of the particular situation at hand, something like a thermodynamical description where everything is determined by a small number of macroscopic parameters. R.A. Treumann argues in his discussion that a $\varkappa$-distribution (may be suitably cut-off) is, in some sense, the most probable one, in agreement with some observations of particle distribution functions in the magnetosphere.

The virialisation process mentioned above, often results in particle trapping. As revealed by the high time resolution observations which are now routinely made, this is a very general phenomenon, producing electrostatic coherent structures like solitary waves, double layers etc. ..., associated with electron or ion holes. As discussed by Omura *et al.* and F. Mottez in their respective reviews such structures have been observed in many regions of the earth environment as well as in the solar wind, suggesting that they may play a significant role in the chain of processes involved in the dissipation of macroscopic energy and its transformation into heat, even if this concept may be misleading in a collisionless plasma.

My final comment will concern the coronal heating by ion cyclotron resonance and let us come back to the turbulent low solar atmosphere. A very popular scenario assumes presently that its spectrum of fluctuations extends to the high frequency region ($f \gtrsim 100$ Hz) in the form of Alfvén waves which can then propagate upwards and interact resonantly with the coronal protons, and other charged particles, what would explain nicely a number of observational facts: acceleration of the solar wind relatively close to the solar surface, anisotropy of the particle distribution functions, etc. ... In the present state of the art, the quasilinear theory of the resonant interaction is used without question. It is however not clear not clear that other processes cannot compete. For example, in the low beta corona, these high frequency Alfvén waves appear to be nonlinearly unstable, at least for parallel propagation, as discussed in the paper by Gomberoff and Araneda. Then the question is: what is the higher frequency to expect, and will it remain in the resonance domain for protons? For oblique propagation, as evidenced by observations in the solar wind, the tip of the wave magnetic field vector appear to perform a a random walk on the sphere, suggesting that the wave polarization reverses abruptly with a sort correlation time. If that is true also in the corona, resonance times between waves and charged particles can be severely diminished.

References

Bak P., Tang, C. and Wiesenfeld, K.: 1988, *Phys. Rev.* **A38**, 364.

Schrijver, C.J. and Zwaan, C.: 2000, *Solar and Stellar Magnetic Activity*, Cambridge University Press, Cambridge, U.K.

II

DUSTY AND EXOTIC PLASMAS

PLASMAS POUSSIÉREUX ET EXOTIQUES

SOME PHYSICAL PROCESSES IN DUSTY PLASMAS

M. ROSENBERG
Department of Electrical and Computer Engineering, University of California, San Diego, La Jolla, CA 92093

Abstract. Ionized gases containing fine (μm to sub-μm sized) charged dust grains, referred to as dusty plasmas, occur in diverse cosmic and laboratory environments. Dust occurs in many space and astrophysical environments, including planetary rings, comets, the Earth's ionosphere, and interstellar molecular clouds. Dust also occurs in laboratory plasmas, including processing plasmas, and crystallized dusty plasmas. Charged dust can lead to various effects in a plasma. In this review, some physical processes in dusty plasmas are discussed, with an emphasis on applications to dusty plasmas in space. This includes theoretical work on several wave instabilities, the role of dust as an electron source, and Coulomb crystals of positively charged dust.

1. Introduction

Ionized gases containing fine (μm to sub-μm sized) charged dust grains, referred to as dusty plasmas, occur in diverse cosmic and laboratory environments. Dust is an ubiquitous component of space and astrophysical environments, occurring for example in planetary rings, comets, the Earth's ionosphere, interstellar molecular clouds, and circumstellar disks (e.g. Goertz, 1989; Mendis and Rosenberg, 1994; Horanyi, 1996). Dust also occurs in laboratory plasmas, including those used for materials processing (e.g. Selwyn, 1993; Bouchele, 1993), and the novel crystallized dusty plasmas (e.g. Thomas *et al.*, 1994; Chu and I, 1994; Hayashi and Tachibana, 1994). Although spatial dimensions and time scales in these plasmas differ by orders of magnitude, there is a commonality in the basic physical processes in space and laboratory dusty plasmas.

Dust in plasma and/or radiative environments can get charged by various processes, such as electron and ion current collection from the plasma (e.g. Whipple, 1981; Whipple *et al.*, 1985), photo-electric emission in the presence of a flux of sufficiently energetic ultraviolet (UV) photons (e.g. Goertz, 1989), thermionic emission when the grain is heated (e.g. Sodha and Guha, 1971), or secondary electron emission in the presence of higher energy ($\gtrsim$ 50 eV) electrons (e.g. Meyer-Vernet, 1982; Chow *et al.*, 1993).

When a grain is charged only by plasma collection currents, the grain charges negatively owing to the larger electron mobility, much like a floating probe in a plasma. For an 'isolated' grain, with $a \ll \lambda_D \ll d$ (where a is the grain radius, λ_D is the plasma Debye length, and the intergrain spacing is $d = (3/4\pi n_d)^{1/3}$, where

Astrophysics and Space Science **277:** 125–133, 2001.

n_d is the dust density), the magnitude of the grain's equilibrium surface potential ϕ (with respect to the background plasma potential) is of the order of a few times the electron temperature T_e (Whipple, 1981; Whipple *et al.*, 1985). The grain charge is given by $Q_d = C\phi$ (where $C \sim a$ is the grain capacitance); the charge state can be estimated as $Z_d \sim 700\,a(\mu m)\,\phi(V)$. For example, a 1 μm 'isolated' grain in a $T_e \sim 1$ eV argon plasma can have a charge state $Z_d \sim 2000$.

In the presence of a flux of UV radiation radiation with energy $h\nu$ greater than the work function W of the grain material, the grains can get charged positively, when the photoemission current is dominant. The maximum grain charge estimated from the energetics of the photoemission process (i.e., $h\nu > W + e\phi$) gives $Z_d^{max} \sim (h\nu - W)a/e^2$ (Mohr *et al.*, 1996). Grains can also become charged positively by thermionic emission when the grain is sufficiently heated, as for example, due to immersion in a thermal plasma (e.g. Fortov *et al.*, 1996).

The charging of dust leads to a plethora of consequences both for the plasma and the dust. In this review, some physical processes in dusty plasmas are discussed, with an emphasis on applications to dusty plasmas in space. This includes theoretical work on several instabilities that may be relevant to space plasmas, the role of dust in enhancing the electron density in several environments, and a scheme for forming Coulomb crystals with positively charged dust.

2. Instabilities in Dusty Plasmas

The presence of charged dust in a plasma can both modify the behavior of standard ion waves and instabilities and lead to the appearance of new low frequency dust waves and instabilities. For weakly correlated dusty plasmas, in the regime where $\lambda_D \gg d \gg a$, the charged dust may be considered to be massive charged point particles, similar to heavy, multiply charged, negative (or positive) ions. Assuming uniform dust size (and thus uniform $Z_d \propto a$ and uniform $m_d \propto a^3$), the condition of equilibrium charge neutrality in the plasma is

$$\epsilon Z_d n_d + n_e = n_i \quad (1)$$

where $\epsilon = 1\,(-1)$ for negative (positive) dust.

Dust can affect ion waves through the charge neutrality condition, Equation (1). This can modify the phase speed of the wave and thus the conditions for instability. An example is the effect of dust on the Farley-Buneman instability, which is a two-stream (ion wave) instability, applicable to the E region of the ionosphere at altitudes of about $h \sim 90 - 130$ km (see e.g. Kelley, 1989). The instability is driven by an electron $E \times B$ drift (at these altitudes the electrons are magnetized, but the ions are unmagnetized since the ion collision frequency is much larger than the ion gyrofrequency). The effect of dust on this instability may be of interest, because charged dust (ice) may be present in the summer polar mesopause. For example, it is thought that charged dust (aerosols) plays a role in explaining the strong radar echoes referred to as PMSE (polar mesosphere summer echoes) that occur at altitudes of $h \sim 80 - 93$ km (reviews by Cho and Kelley, 1993; Cho and Rottger, 1997). Recently, both negatively and positively charged dust in the mesosphere has been detected by rocket probe measurements (Havnes *et al.*, 1996a) and and it was reported that dust can be a major negative charge carrier during PMSE conditions (Havnes *et al.*, 1996b). Rosenberg and Chow (1998) investigated the effect of dust on the Farley-Buneman instability and showed that if the dust is negatively charged, the phase velocity of the wave increases, and the critical drift decreases, for parameters that may be representative of the $h \sim 90 - 95$ km altitude range. It may be interesting to investigate further whether this instability could generate small scale structures in dusty mesospheric regions which may be relevant to PMSE.

Ion waves could also be excited by charged dust beams. Bernhardt *et al.* (1995) conjectured that water vapor of the expanding space shuttle exhaust plume in the ionosphere might condense into ice grains and get charged in the background plasma, creating a dusty plasma in which instabilities leading to radar backscatter might occur. Since the dust would have a directed speed V_{d0} of the order of the exhaust velocity with respect to the background plasma, which is $\gg$ the ion thermal speed, the dust comprises a charged beam. If the beam streams across the magnetic field, it can excite lower hybrid waves in the background F region plasma (Rosenberg *et al.*, 1999). In the lower hybrid frequency range, $\Omega_d \ll \Omega_i \ll \omega \ll \Omega_e$ (Ω_j is the gyrofrequency of species j, where $j = e, i$ and d for electrons, ions, and dust), the electrons are magnetized, while the ions and dust are non-magnetized. The dispersion relation for this instability is

$$1 - \frac{\omega_{LH}^2}{\omega^2} - \frac{\omega_{pd}^2}{A(\omega - kV_{d0})^2} \approx 0 \tag{2}$$

where $A = 1 + (\omega_{pe}/\Omega_e)^2$, $\omega_{LH} = \omega_{pi}/\sqrt{A}$ is the lower hybrid frequency, ω_{pj} is the plasma frequency of species j, and k is in the direction of V_{d0}. Maximum growth occurs at $\omega \approx \omega_{LH} \approx kV_{d0}$, with a growth rate $\gamma \sim \omega_{LH}\,(\omega_{pd}/\omega_{pi})^{2/3}$. The condition for maximum growth of the instability, using parameters for the space shuttle exhaust as given in Bernhardt *et al.* (1995) and representative parameters for the background plasma at altitudes of $\sim 300 - 400$ km, gives an unstable wavenumber k_s that approximately satisfies the Bragg condition for radar backscatter, $k_s = 2k_0$, for the wavenumber k_0 corresponding to the Arecibo 430-MHz radar (Rosenberg *et al.*, 1999). Thus this lower hybrid instability may be relevant to the observations of enhanced backscatter from space shuttle exhaust.

The presence of charged dust in a plasma can also lead to the appearance of new low frequency dust waves and instabilities associated with the dust dynamics. Dust acoustic waves are the low frequency analog of ion acoustic waves, in which the lighter plasma electrons and ions provide the pressure to sustain the wave, while the heavier dust grains provide the inertia (Rao *et al.*, 1990). The real part of the frequency is

$$\omega_r^2 \approx k^2\lambda_D^2\omega_{pd}^2 = k^2 Z_d \left| \frac{\delta - 1}{1 + \delta\frac{T_e}{T_i}} \right| \frac{T_e}{m_d} \tag{3}$$

where $\delta = n_i/n_e$, and for $k\lambda_D \ll 1$ (here $\lambda_D = [\lambda_{De}^{-2} + \lambda_{Di}^{-2}]^{-1/2}$). Dust acoustic waves have been observed in dedicated laboratory experiments (e.g. Barkan, *et al.* 1995; Thompson *et al.*, 1997). Since the phase speed of the wave ω/k is $\ll$ the ion thermal speed v_i, the wave can be excited by weak ion and electron drifts $u_0 < v_i$. For a collisionless plasma, dust acoustic instability is driven primarily by ion drift, with growth rate $\gamma \sim (u_0/v_i)\omega_r$ (Rosenberg, 1993). The dust acoustic instability may have application to planetary rings (Rosenberg, 1993; Melandso *et al.*, 1993), where there may be a relative drift between the plasma ions/electrons and the charged dust, because the ions/electrons tend to co-rotate with the planet, while the massive dust grains orbit the planet with speed near the Kepler speed. In Saturn's outer rings, $u_0 \gtrsim v_i$; numerical simulations of the saturation of the dust acoustic instability in this ion drift regime show that the instability saturates by trapping some of the ions, leading to ion heating (Winske *et al.*, 1995). This heating mechanism may have potential application to ion heating in Saturn's E ring (Winske *et al.*, 1995), where $T_i > T_e$ is observed (Richardson and Sittler, 1990).

Dust acoustic waves may also be excited by ion drift in laboratory dusty plasmas. For example, dust grains that have been observed in processing plasmas are localized near plasma sheath interface regions where there is a directed ion flow toward the electrode. The ion drift speed u_{0i} in these regions can be of the order of the ion sound speed $\sim (T_e/m_i)^{1/2}$ according to the Bohm sheath criterion. Since typically $T_e \gg T_i$ in these rf or dc glow discharges (see Winske and Jones, 1994), this implies $u_{0i} \gg v_i$, a condition for strong wave growth. However, laboratory dusty plasmas are generally collisional, since gas pressures are relatively high ($> 10-100$ mtorr); collisional effects reduce the growth rate, leading to to a dissipative type ion-dust streaming instability (Rosenberg, 1996). It was speculated that, if the nonlinear development of this instability results in ion heating, particularly in directions oblique to the ion flow, this may have implications for plasma etching (Rosenberg, 1996).

It has recently been shown that dust acoustic waves may be excited in dusty meteor trails by electron $E \times B$ drift, under certain conditions (Rosenberg and Shukla, 2000). Rosinski and Snow (1961) and Hunten *et al.* (1980) have shown theoretically that when meteors ablate, they can form dust by accretion. Recently, Kelley *et al.* (1998) reported on the detection of a meteor contrail and meteoric dust ($a \sim 50$ nm) in the Earth's upper mesosphere. If dust is indeed present in significant density in some meteor trails, dust acoustic waves might be excited by electron $E \times B$ drifts, in analogy to the excitation of ion waves by the Farley-

Buneman instability. A condition for instability is that the dust acoustic wave is weakly damped. Dust Landau damping is small when the wave phase speed is much larger then the dust thermal speed, which generally requires $Z_d \gg 1$ (viz. Equation 3). But if the dust is negatively charged by plasma collection currents, Z_d would be small in the low temperature (~ 200 K) upper mesosphere. However, if the dust grain material has low work function $\lesssim 4-5$ eV, the dust might be charged positively by photoemission in the presence of solar UV radiation, that is, under daytime conditions. Rosenberg and Shukla (2000) worked out some conditions under which dust acoustic type waves might be excited in dusty meteor trails at altitudes of $\sim 95 - 100$ km. Generally, these conditions involve the presence of sufficient density of positively charged dust with $Z_d \gg 1$, and weak collisionless and (dust-neutral) collisional damping. Critical electron drifts were found to be smaller than the ion thermal speed, which implies electric field strengths as small as several mV/m, representing the lower range of ionospheric electric field strengths. It may also be interesting to investigate in further studies whether the possible increase of electron density due to photoemission from dust in a meteor trail could be relevant to mechanisms for long duration meteor echoes, since the dust would be a source of ionization.

3. Dust as Source of Electrons

If dust is negatively charged by plasma collection currents, this reduces the background electron density (viz. Equation 1). However, dust can also act as a source of electrons, owing to electron emission processes such as photoemission or thermionic emission, as discussed in an early review by Sodha and Guha (1971).

For example, photoelectric emission from fine dust in an interstellar FUV radiation field may be a source of electrons in the warm photo-dissociation regions between molecular clouds and HII regions (Rosenberg *et al.*, 1994). Cosmic rays have long been considered a major source of ionization in molecular clouds relevant to star forming regions in the galaxy (see Norman and Heyvaerts, 1985), and it was subsequently shown (McKee, 1989) that photoionization of trace metal atoms by an interstellar FUV field (6 eV $< h\nu <$ 13.6 eV) could lead to larger ionization fractions in molecular clouds. The work function of many grain materials, however, is typically lower than the ionization potential of the trace metals. Rosenberg *et al.* (1994) found that photoelectric emission from dust could provide even higher degrees of ionization under certain conditions. The degree of ionization has implications for plasma transport processes, such as diffusion of plasma across the magnetic field, which is an issue in star formation (e.g. Nishi *et al.*, 1991). As another example of dust as an ionization source, rocket probe measurements of charged dust in the Earth's mesosphere show that on occasions the dust, when charged positive (perhaps due to photoemission under solar UV radiation) can be be a substantial source of electrons (Figure 3, Havnes *et al.*, 1996a).

Thermionic emission from dust may also enhance the electron density in certain space environments. For example, preliminary results show that thermionic emission from fine carbonaceous dust (soot) in the circumstellar shells around carbon-rich red giants could increase the electron density in those regions (Mendis *et al.*, 1998). Recently we (Sorasio *et al.*, 2000) have begun investigating the role of thermionic emission in meteor physics. Existing studies of meteoroids mainly concerns their deceleration and heating, their consequent ablation, and the ionization of the atmospheric molecules by the fast moving ablated molecules, typically having kinetic energies $\sim$ 100 eV (e.g see Bronshten, 1983). However, before the grains are hot enough to ablate (for carbonaceous particles ablation requires temperatures $\sim$ 3000 K, while metallic particles require temperatures of $\sim$ 2000 K), they would be hot enough to emit electrons thermionically (see also Bronshten, 1991). Preliminary results (Sorasio *et al.*, 2000) show this could be a major source of meteoric electrons, particularly if the work function of the meteoroid is sufficiently small (i.e. $\leq$ 2 eV). Our further work in this area includes studies of the spatial distribution of the thermionically emitted electrons, to investigate relevance to meteor head echoes.

4. Coulomb Crystals of Positively Charged Dust

A consequence of grain charging is that dust can form crystalline structures for sufficiently large coupling parameter

$$\Gamma_d = \frac{Q_d^2}{dT_d} exp(-d/\lambda_D) \tag{4}$$

where Q_d is the dust charge, d is the intergrain spacing, T_d is the dust thermal energy, and λ_D is the plasma screening length (Ikezi, 1986). Dusty plasma systems in which negatively charged, micron sized grains form lattice structures, referred to as Coulomb crystals or 'plasma crystals', have been produced in the laboratory by a number of experimental groups (e.g., Thomas *et al.*, 1994; Chu and I, 1994,; Hayashi and Tachibana, 1994; Melzer *et al.*, 1994; Fortov *et al.*, 1997; Mohideen *et al.*, 1998). Structures, dynamics, and phase transitions of these strongly coupled systems have been studied (e.g. review by Thomas and Morfill, 1997). Recently, 'space plasma crystal' experiments under microgravity conditions on sounding rockets have been reported (Morfill, *et al.*, 1999).

While the dust Coulomb crystals discussed above consist of Coulomb lattices of negatively charged dust in a background plasma, we (Rosenberg and Mendis, 1995; Rosenberg *et al.*, 1996), recently suggested a theoretical scheme for forming a Coulomb lattice of positively charged dust. This involves dispersing dust grains of low work function material in an inert gas in the presence of a flux of UV radiation with sufficient energy (e.g. $h\nu \gtrsim 6$ eV) to charge the grains positive by photoemission, yet with low enough energy (e.g. $h\nu < 12$ eV) to avoid photoionization of the gas.

This driven system would thus be composed of two charged components, positively charged grains and photo-emitted electrons, in a background gas to cool the grains; the dust itself provides the source of ionization and thereby controls the (electron) screening length. The positive equilibrium charge can be estimated by balancing the photoemission current with the electron recollection current, assuming overall charge neutrality $Z_d n_d = n_e$. It was found that that the condition for Coulomb crystallization, roughly $\Gamma_d > 170$, could be satisfied over a wide range of dust sizes in the range of microns to tens of microns, and interdust spacings in the range of tens to hundreds of microns, for a UV intensity of 8-9 eV photons of ~ 5 W cm^{-2}, and reasonable estimates for the photoelectric yield (Rosenberg *et al.*, 1996). We are also investigating the possible use of such crystals as filters of electromagnetic waves in the far infrared or submillimeter wavelength regimes, in analogy with colloidal crystals which have application as filters of visible light owing to their Bragg diffraction properties (e.g. Rundquist *et al.*, 1989).

Lattices of dust charged positively by photoemission have not yet been produced in the laboratory. However, recently experimental studies have been reported on liquid-like ordered formations of dust particles charged positively by photoemission in the presence of solar UV radiation, under microgravity conditions on the Mir orbiting space station (Fortov *et al.*, 1998).

Acknowledgements

The author would like to thank D. A. Mendis for helpful discussions. Work supported in part by DOE Grant no. DE-FG03-97ER54444 and by LANL/IGPP no. 1022.

References

Barkan, A., Merlino, R.L. and D'Angelo, N.: 1995, Laboratory observation of the dust-acoustic wave mode, *Phys. Plasmas* **2**, 3563–3565.

Bernhardt, P.A., Ganguli, G., Kelley, M.C. and Swartz, W.E.: 1995, Enhanced radar backscatter from space shuttle exhaust in the ionosphere, *J. Geophys. Res.* **100**, 23811–23818.

Bouchoule, A.: 1993, Dusty plasmas, *Physics World* **6**, 47–51, Aug. 1993.

Bronshten, V.A.: 1983, *Physics of meteoric phenomena*, D. Reidel, Dordrecht.

Bronshten, V.A.: 1991, Electrical and electromagnetic phenomena associated with meteor flight, *Sol. System Res.* **25**, 93–104.

Cho, J.Y.N. and Kelley, M.C.: 1993, Polar mesosphere summer echoes: observations and current theories, *Rev. Geophys.* **31**, 243–265.

Cho, J.Y.N. and Rottger, J.: 1997, An updated review on polar mesosphere summer echoes: Observation, theory, and their relationship to noctilucent clouds and subvisible aerosols, *J. Geophys. Res.* **102**, 2001–2020.

Chow, V.W., Mendis, D.A. and Rosenberg, M.: 1993, Role of grain size and particle velocity distribution in secondary electron emission in space plasmas, *J. Geophys. Res.* **98**, 19065–19076.

Chu, J.H. and Lin, I.: 1994, Direct observation of Coulomb crystals and liquids in strongly coupled rf dusty plasmas, *Phys. Rev. Lett.* **72**, 4009–4012.

D'Angelo, N. and Merlino, R.L.: 1996, Current-driven dust-acoustic instability in a collisional plasma, *Planet. Space Sci.* **44**, 1593–1598.

Fortov, V.E., Nefedov, A.P., Petrov, O.F., Samarian, A.A. and Chernyschev, A.V.: 1996, Particle ordered structures in a strongly coupled classical thermal plasma, *Phys. Rev. E* **54**, R2236–R2239.

Fortov, V.E., Nefedov, A.P., Torchinsky, V.M., Molotkov, V.I., *et al.*: 1997, Crystalline structures of strongly coupled dusty plasmas in dc glow discharge strata, *Phys. Lett. A* **229**, 317–322.

Fortov, V.E., Nefedov, A.P., Vaulina, O.S., Lipaev, A.M., Molotkov, V.I., Samaryan, A.A., *et al.*: 1998, Dusty plasma induced by solar radiation under microgravitational conditions: an experiment on board the Mir orbiting space station, *J. Exp. Theor. Phys.* **87**, 1087–1097.

Goertz, C.K.: 1989, Dusty plasmas in the solar system, *Rev. Geophys.* **27**, 271–292.

Havnes, O., Naesheim, L.I., Hartquist, T.W., Morfill, G.E., Melandso, F., *et al.*: 1996b, Meter-scale variations of the charge carried by mesospheric dust, *Planet. Space Sci.* **44**, 1191–1194.

Havnes, O., Troim, J., Blix, T., Mortensen, W., *et al.*: 1996a, First detection of charged dust particles in the Earth's mesosphere, *J. Geophys. Res.* **101**, 10839–10847.

Hayashi, Y. and Tachibana, K.: 1994, Observation of Coulomb-crystal formation from carbon particles grown in a methane plasma, *Jpn. J. Appl. Phys.* **33**, L804–L806.

Hill, J.R. and Mendis, D.A.: 1982, On the dust current of Saturn's F ring, *Geophys. Res. Lett.* **9**, 1069–1071.

Horanyi, M.: 1996, Charged dust dynamics in the solar system, *Annu. Rev. Astron. Astrophys.* **34**, 383–418.

Hunten, D.M., Turco, R.P. and Toon, O.B.: 1980, Smoke and dust particles of meteoric origin in the mesosphere and stratosphere, *J. Atmos. Sci.* **37**, 1342–1357.

Ikezi, H.: 1986, Coulomb solid of small particles in plasmas, *Phys. Fluids* **29**, 1764–1766.

Kelley, H.C.: 1989, *The Earth's Ionosphere: Plasma Physics and Electrodynamics,* Academic Press, San Diego.

Kelley, H.C., Alcala, C. and Cho, J.Y.N.: 1998, Detection of meteor contrail and meteoric dust in the Earth's upper mesosphere, *J. Atmos. Solar-Terr. Phys* **60**, 359–369.

Mathews, J.D., Meisel, D.D., Hunter, K.P., Getman, V.S. and Zhou, Q.: 1997, Very high resolution studies of micrometeors using the Arecibo 430 MHz radar, *Icarus* **126**, 157–169.

McKee, C.F.: 1989, Photoionization regulated star formation and the structure of molecular clouds, *Astrophys. J.* **345**, 782–801.

Melandso, F., Aslaksen, T.K. and Havnes, O.: 1993, A kinetic model for dust acoustic waves applied to planetary rings, *J. Geophys. Res.* **98**, 13315–13323.

Melzer, A., Trottenberg, T. and Piel, A.: 1994, Experimental determination of the charge on dust particles forming Coulomb lattices, *Phys. Lett. A* **191**, 301–308.

Mendis, D.A. and Rosenberg, M.: 1994, Cosmic dusty plasma, *Annu. Rev. Astron. Astrophys.* **32**, 419–463.

Mendis, D.A., Rosenberg, M. and Chow, V.W.: 1998, Ionization equilibria in dusty plasma environments, in: M. Horanyi, S. Robertson and B. Walch (eds.), *Physics of Dusty Plasmas*: 7th Workshop, AIP, New York.

Meyer-Vernet, N.: 1982, 'Flip-flop' of electric potential of dust grains in space, *Astron. Astrophys.* **105**, 98.

Mohideen, U., Rahman, H.U., Smith, M.A., Rosenberg, M. and Mendis, D.A.: 1998, Intergrain coupling in dusty plasma Coulomb crystals, *Phys. Rev. Lett.* **81**, 349–352.

Mohr, M.R., Matter, D. and Burtscher, H.: 1996, Efficient multiple charging of diesel particles by photoemission, *Aerosol. Sci. Technol.* **24**, 14–20.

Morfill, G.E. and Thomas, H.: 1996, Plasma crystal, *J. Vac. Sci. Technol. A* **14**, 490–495.

Morfill, G.E., Thomas, H.M., Konopka, U., Rothermel, H., Zuzic, M., Ivlev, A. and Goree, J.: 1999, Condensed plasmas under microgravity, *Phys. Rev. Lett.* **83**, 1598–1601.

Nishi, R., Nakano, T. and Umebayashi, T.: 1991, Magnetic flux loss from interstellar clouds with various grain size distributions, *Astrophys. J.* **368**, 181–194.

Norman, C. and Heyvaerts, J.: 1985, Anomalous magnetic diffusion during star formation, *Astron. Astrophys.* **147**, 247–256.

Rao, N.N., Shukla, P.K. and Yu, M.Y.: 1990, Dust-acoustic waves in dusty plasmas, *Planet. Space Sci.* **38**, 543–546.

Richardson, J.D. and Sittler, E.C.: 1990, A plasma density model for Saturn based on Voyager observations, *J. Geophys. Res.* **95**, 12019–12031.

Rosenberg, M.: 1993, Ion- and dust-acoustic instabilities in dusty plasmas, *Planet. Space Sci.* **41**, 229–233.

Rosenberg, M.: 1996, Instabilities in dusty plasmas, in: P.K. Shukla, D.A. Mendis and V.W. Chow (eds.), *The Physics of Dusty Plasmas*, World Scientific, Singapore.

Rosenberg, M.: 1996, Ion-dust streaming instability in processing plasmas, *J. Vac. Sci. Technol. A* **14**, 631–633.

Rosenberg, M.: 2000, Dusty plasmas: a topical review, *J. Phys. IV France* **10**, Pr5-73.

Rosenberg, M. and Chow, V.W.: 1998, Farley-Buneman instability in a dusty plasma, *Planet. Space Sci.* **46**, 103–108.

Rosenberg, M. and Mendis, D.A.: 1995, UV-induced Coulomb crystallization in a dusty gas, *IEEE Trans. Plasma Sci.* **23**, 177–179.

Rosenberg, M., Mendis, D.A. and Chow, V.W.: 1994, Weakly ionized cosmic gas: ionization and characterization, *Astrophys. Space Sci.* **222**, 247–253.

Rosenberg, M., Mendis, D.A. and Sheehan, D.P.: 1996, UV-induced Coulomb crystallization of dust grains in high-pressure gas, *IEEE Trans. Plasma Sci.* **24**, 1422–1430.

Rosenberg, M. and Shukla, P.K.: 2000, Low frequency Hall current instability in a dusty plasma, *J. Geophys. Res.* **105**, 23135–23139.

Rosenberg, M., Salimullah, M. and Bharuthram, R.: 1999, Lower hybrid instability driven by charged dust beam, *Planet. Space Sci.* **47**, 1517–1519.

Rosinski, J. and Snow, R.H.: 1961, Secondary particulate matter from meteor vapors, *J. Meteor.* **18**, 736–745.

Rundquist, P.A., Photinos, P., Jagannathan, S. and Asher, S.A.: 1989, Dynamical Bragg diffraction from crystalline colloidal arrays, *J. Chem. Phys.* **91**, 4932–4941.

Selwyn, G.S.: 1993, A phenomenological study of particulates in plasma tools and processes, *Jpn. J. Appl. Phys.* **32**, 3068–3073.

Sodha, M.S. and Guha, S.: 1971, Physics of colloidal plasmas, in: A. Simon and W.B. Thompson (eds.), *Adv. Plasma Phys.* **4**, Interscience, New York.

Sorasio, G., Mendis, D.A. and Rosenberg, M.: 2000, *The role of thermionic emission in meteor physics*, poster presented at *8th Workshop on the Physics of Dusty Plasmas*, Sante Fe. NM, April 26–28.

Thomas, H.M. and Morfill, G.E.: 1997, The processes involved in the solid-to-liquid phase transition, *Endeavour* **21**, 148–153.

Thomas, H., Morfill, G.E., Demmel, V., Goree, J., Feuerbacher, B. and Mohlmann, D.: 1994, Plasma crystal: Coulomb crystallization in a dusty plasma, *Phys. Rev. Lett.* **73**, 652–655.

Thompson, C., Barkan, A., D'Angelo, N. and Merlino, R.L.: 1997, Dust acoustic waves in a direct current glow discharge, *Phys. Plasmas* **4**, 2331–2335.

Whipple, E.C.: 1981, Potentials of surfaces in space, *Rep. Prog. Phys.* **44**, 1197.

Whipple, E.C., Northrop, T.G. and Mendis, D.A.: 1985, The electrostatics of a dusty plasma, *J. Geophys. Res.* **90**, 7405–7413.

Winske, D., Gary, S.P., Jones, M.E., Rosenberg, M., Chow, V.W. and Mendis, D.A.: 1995, Ion heating in a dusty plasma due to the dust/ion acoustic instability, *Geophys. Res. Lett.* **22**, 2069–2072.

Winske, D. and Jones, M.E.: 1994, Particulate dynamics at the plasma-sheath boundary in DC glow discharges, *IEEE Trans. Plasma Sci.* **22**, 454–464.

SHEAR INDUCED PHENOMENA IN DUSTY PLASMA FLOWS

GEORGE R. KHUJADZE
Department of Theoretical Astrophysics, Abastumani Astrophysical Observatory, Tbilisi, Georgia

STEFAAN POEDTS* and ANDRIA D. ROGAVA**
CPA, K.U. Leuven, Celestijnenlaan 200B, 3001 Leuven, Belgium

Abstract. It is found that velocity shear enables the extraction of kinetic energy from the background flow by Dust-Acoustic waves. It is also shown that the velocity shear leads to the appearance of a *new mode* of the dust particles collective behaviour, called *shear dust vortices*.

The majority of dusty space plasmas are flows with spatially inhomogeneous velocities, i.e. *shear flows.* Shear flows are known to host a number of so-called 'non-modal' phenomena, arising due to the non-self-adjointness of shear flow systems. One of these effects associated with compressible wave modes is that acoustic waves acquire the ability to extract energy from the flow (Chagelishvili *et al.*, 1994). Another phenomenon is that the velocity shear gives birth to a unique non-periodic mode of plasma collective behaviour – 'shear vortices' (Rogava *et al.*, 1997, 1998) – which eventually acquire wave-like features ('convert' into waves (Chagelishvili *et al.*, 1997) in flows with moderate or high shearing rates. In this paper we investigate the manifestation of these effects in dusty plasma shear flows. Are Dust-Acoustic waves (DAW) able to extract energy from the background flow? Does shear generate any new mode of the dust component collective behaviour? It is found that the answers on both questions are affirmative.

Consider a uniform plasma consisting of electrons, singly charged positive ions and negatively charged dust particles. We assume a Boltzmann distribution for both the electron number density: $N_e = \mathcal{N}_e \exp[e\phi/kT_e] \approx \mathcal{N}_e(1 + e\phi/kT_e)$, and the ion number density: $N_i = \mathcal{N}_i \exp[-e\phi/kT_i] \approx \mathcal{N}_i(1 - e\phi/kT_i)$, with k the Boltzmann constant. This assumption is standard (Rao *et al.*, 1990) for the study of the dust-acoustic mode. The particles have masses, number densities, temperatures, thermal pressures and velocities denoted by m_j, N_j, T_j, P_j, and $\mathbf{V}_j$ $(j \equiv e, i, d)$, respectively. Number densities of electrons, ions and dust particles $\mathcal{N}_j$ $(j \equiv e, i, d)$ satisfy the quasi-neutrality condition: $\mathcal{N}_i = \mathcal{N}_e + Z\mathcal{N}_d$. The steady state dust flow is: $\vec{\mathcal{V}}_d = (Ay, 0)$, with A a constant determining the shear of the flow.

The system is studied within the framework of the multicomponent fluid plasma model. The governing equations for perturbations are $[\mathcal{D}_t \equiv \partial_t + Ay\partial_x]$:

* Research Associate of the Belgian National Fund for Scientific Research (FWO-Vlaanderen)
** On leave from Abdus Salam International Centre for Theoretical Physics, Trieste 34100, Italy

Astrophysics and Space Science **277:** 135–138, 2001.

$$\Delta\phi = 4\pi e\,[n_e + Zn_d - n_i], \tag{1}$$

$$\mathcal{D}_t n_d + \mathcal{N}_d(\partial_x v_{dx} + \partial_y v_{dy}) = 0, \tag{2}$$

$$\mathcal{D}_t v_{dx} + Av_{dy} = -C_d^2\partial_x(n_d/\mathcal{N}_d) + (Ze/m_d)\partial_x\phi, \tag{3a}$$

$$\mathcal{D}_t v_{dy} = -C_d^2\partial_y(n_d/\mathcal{N}_d) + (Ze/m_d)\partial_y\phi, \tag{3b}$$

while from the Boltzmann condition for electrons we have: $n_e/\mathcal{N}_e = e\phi/k\mathcal{T}_e$ and for the ions $n_i/\mathcal{N}_i = -e\phi/k\mathcal{T}_i$. For the separation of variables we use the ansatz: $F(x, y, z; t) = \hat{F}(t)\exp[i(k_x x + k_y(t)y)]$, with $k_y(t) \equiv k_y(0) - Atk_x$, which reduces mathematics to the solution of an initial value problem. Using the dimensionless parameters: $\mu \equiv \mathcal{T}_i/\mathcal{T}_e$, $\nu \equiv \mathcal{T}_d/\mathcal{T}_e$, $\alpha \equiv \mathcal{N}_i/\mathcal{N}_e$, $\varepsilon \equiv m_i/m_d$, $\mathcal{C} \equiv (kT_e/m_i)^{1/2}$, $C_i \equiv (kT_i/m_i)^{1/2}$, $C_d \equiv (kT_d/m_d)^{1/2}$, the following dimensionless notation is introduced: $\beta_0 \equiv k_y(0)/k_x$, $R \equiv A/\mathcal{C}k_x$, $\tau \equiv \mathcal{C}k_x t$, $\beta(\tau) \equiv \beta_0 - R\tau$, $\mathbf{U}_d \equiv \hat{\mathbf{v}}_d/\mathcal{C}$, $n \equiv i\hat{n}_e/\mathcal{N}_e$, $d \equiv i\hat{n}_i/\mathcal{N}_i$, $D \equiv i\hat{n}_d/\mathcal{N}_d$, $\xi \equiv k_x\lambda_D$, $\lambda_D \equiv (\varepsilon_0 k\mathcal{T}_e/e^2\mathcal{N}_e)^{1/2}$, $\mathcal{K}^2(\tau) \equiv 1 + \beta^2(\tau)$, $\Phi \equiv ie\hat{\phi}/k\mathcal{T}_e$. For electrons and ions we have: $n = \Phi$ and $d = -\Phi/\mu$, which allows to derive from (1): $D = \left(1 + \alpha/\mu + \xi^2\mathcal{K}^2(\tau)\right)\Phi/(1-\alpha)$.

A remarkable feature of the above system is that there exists a linear combination of the perturbation functions, which is constant in time, viz. $U_{dy} - \beta(\tau)U_{dx} - RD = C$, with C a constant. Using this relation we can reduce the initial system of equations to the following second-order ODE in terms of $Y \equiv D/\mathcal{K}(\tau)$:

$$Y^{(2)} + \Omega^2(\tau)Y = \mathcal{F}_d(\tau), \quad \text{with } \mathcal{F}_d(\tau) \equiv -\frac{2RC}{\mathcal{K}^3(\tau)} \tag{4}$$

$$\text{and } \Omega^2 \equiv \varepsilon\nu\mathcal{K}^2(\tau) + \frac{(\alpha-1)Z\varepsilon\mathcal{K}^2(\tau)}{1 + \alpha/\mu + \xi^2\mathcal{K}^2(\tau)} + \frac{3R^2}{\mathcal{K}^4(\tau)}. \tag{5}$$

Equations (4)–(5) bear a significant similarity with the analogous equations describing the evolution of electrostatic ion perturbations (Rogava *et al.*, 1997) and the behaviour of electron periodic and non-periodic Langmuir perturbations (Chagelishvili *et al.*, 1997) in standard plasma shear flows.

When the initial perturbations are 'homogeneous' (i.e., when $C = 0$ and thus $\mathcal{F}_d = 0$) the system sustains shear-modified dust-acoustic waves. The velocity shear makes the effective frequency of the wave time-dependent ('adiabatic' when $R \ll 1$). The adiabaticity implies that the total energy of the perturbations stays proportional to the effective frequency during the whole evolution: $E(\tau) \simeq \Omega(\tau)$. Figure 1a displays the time evolution of the density perturbation $D(\tau)$ which shows the dust-acoustic wave with the variable effective frequency $\Omega(\tau)$ and a variable amplitude $a(\tau) \sim \Omega^{-1/2}(\tau)$ (for $\mu = 10^{-1}$, $\nu = 10^{-2}$, $Z = 10^3$, $\varepsilon = 10^{-6}$, $\alpha = 1.694$, $\xi = 10^{-3}$, $\beta_0 = 10$, $R = 10^{-2}$). The plot for the total energy is shown in Figure 1b, where the solid line denotes the energy, $E(\tau)$, while the dashed line

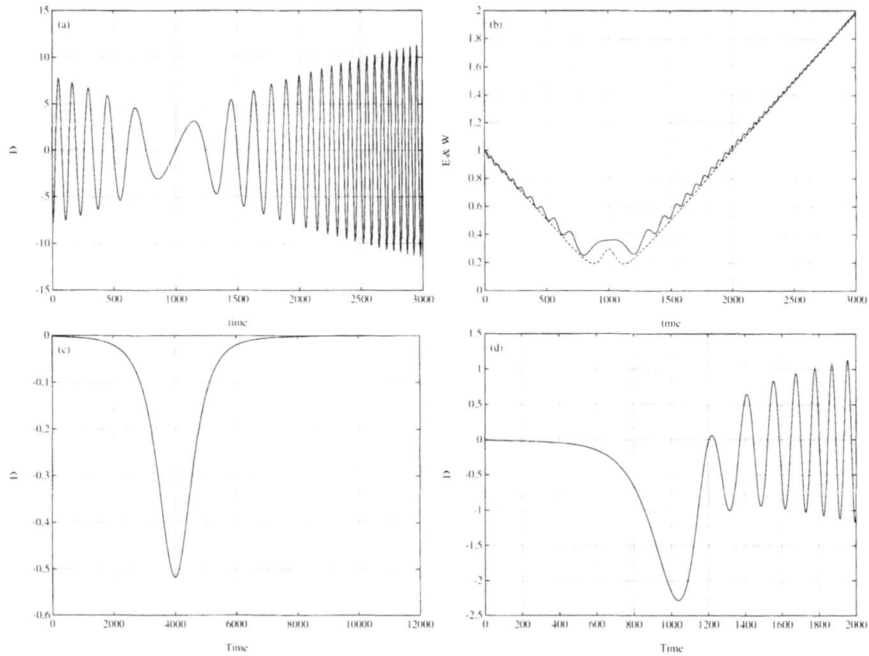

Figure 1. Some examples of dust dynamics obtained by nonmodal analysis.

displays the 'dispersion curve' of the dust-acoustic wave $\Omega(\tau)$. This shows that the dust-acoustic waves can 'extract' energy from the steady background flow. When the initial perturbations are 'inhomogeneous' ($C \neq 0$) the collective behaviour of the dust particles becomes diverse: together with the shear-modified dust-acoustic waves the system sustains also 'dust-acoustic vortices'. The zeroth-order approximate solution for the dust-acoustic vortex is given by: $Y_v(\tau) \simeq \mathcal{F}_d(\tau)/\Omega^2(\tau)$, which is a good approximation for low shearing rate flows ($R \ll 1$). Figure 1c shows the corresponding solution for $D(\tau)$ for the case $R = 10^{-3}$. The values of the other parameters are: $\mu = 10^{-1}$, $\nu = 10^{-2}$, $Z = 10^4$, $\varepsilon = 10^{-6}$, $\alpha = 1.7$, $\beta_0 = 4$, $\xi = 10^{-3}$, and $C = 10^{-1}$. The numerical solution is almost identical to the approximate solution.

For higher values of the shearing rate, however, these perturbations acquire wave-like features in the course of time. This tendency is well-pronounced even in rather weakly sheared dusty plasma flows, as can be clearly seen on Figure 1d, where $R = 4 \times 10^{-3}$, while all other parameters are the same as for Figure 1c. It is easy to see that if the function $Y(\beta)$ is the solution of Equation (4), then the function $Y(-\beta)$ is also a solution. Thus together with initially non-periodic and eventually oscillating solutions the system also sustains initially oscillating ('wave-like') solutions, which become non-periodic (vortical) and 'evanescent'. In general, an arbitrary initial perturbation exhibits some combination of the above types of temporal behaviour.

Thus our results indicate that dust-acoustic modes are seriously affected by the presence of the velocity shear. They acquire a shear-induced temporal variation of their basic frequency and become able to extract energy from the mean flow. Moreover, the shear 'drives' a new kind of collective behaviour of dust particles – *'Shear-Dust-Vortices'* – which are akin to analogous patterns in hydrodynamics (Chagelishvili *et al.*, 1994, 1997) and plasma physics (Rogava *et al.*, 1997, 1998).

Acknowledgement

This work was supported in part by INTAS grant 97-0504.

References

Chagelishvili, G.D., Rogava, A.D. and Segal, I.N.: 1994, *Phys. Rev. E* **50**, R4283.
Chagelishvili, G.D., Tevzadze, A.G., Bodo, G. and Moiseev, S.S.: 1997, *Phys. Rev. Lett.* **79**, 3178.
Rao, N.N., Shukla, P.K. and Yu, M.Y.: 1990, *Planet. Space Sci.* **38**, 543.
Rogava, A.D., Chagelishvili, G.D. and Berezhiani, V.I.: 1997, *Phys. Plasmas* **4**, 4201.
Rogava, A.D., Chagelishvili, G.D. and Mahajan, S.M.: 1998, *Phys. Rev. E* **57**, 7103.

GENERATION OF SUPERCOOLED STRONGLY-COUPLED PLASMA BY ARTIFICIAL INJECTION INTO SPACE

YU. V. DUMIN
IZMIRAN, Russian Academy of Sciences, Troitsk, Moscow region, 142190 Russia,
E-mail: dumin@cityline.ru; dumin@yahoo.com

Abstract. Attention is drawn to the possibility of transition of an expanding plasma cloud into the state with extremely high Coulomb's coupling. Conditions for such transition are studied, and properties of the resulting strongly-coupled plasma (particularly, the effective concentration of free-charge carriers) are briefly outlined.

1. Introduction

The experiments on artificial plasma injection into space are conducted since the early 1960s (see, for example, review by Haerendel, 1987). They were aimed both at diagnostics of the near-Earth plasma environment and studying dynamics of the ejected plasma itself. Aim of the present report is to show that such experiments may be a unique tool for studying a radically new physical system – a metastable supercooled plasma with extremely strong Coulomb's coupling.

2. Time Behavior of the Coulomb's Coupling Parameter

Time behavior of the concentration of charged particles in the expanding cloud of weakly-ionized plasma can be described by the continuity equation

$$\frac{\partial N}{\partial t} + \nabla(N\mathbf{v}_e) = -\frac{N_0}{\tau}\left(\frac{N}{N_0}\right)^{\kappa}\left(\frac{n}{n_0}\right)^{\tilde{\kappa}}\left(\frac{T_0}{T}\right)^{\eta}\left(\frac{T_{e0}}{T_e}\right)^{\tilde{\eta}}, \qquad (1)$$

where N is the concentration of charged particles (i.e., electrons and positive ions), n is the concentration of neutral gas, T is the temperature of heavy particles (i.e., ions and neutrals), T_e is the electron temperature, $\mathbf{v}_e$ is the velocity field of charged particles, κ, $\tilde{\kappa}$, η, and $\tilde{\eta}$ are the parameters characterizing the recombination channel under consideration, and τ is the time coefficient, which can be derived from the coefficient of recombination. For example, in the most typical case of two-electron recombination, $A^+ + e + e \rightarrow A + e$, we have

$$\kappa = 3\,,\ \tilde{\kappa} = 0\,,\ \eta = 0\,,\ \tilde{\eta} = 9/2\,,\ \tau = C\,\frac{m_e^{1/2}\,(k_B T_{e0})^{9/2}}{e^{10} N_0^2}\,, \qquad (2)$$

Astrophysics and Space Science **277:** 139–142, 2001.

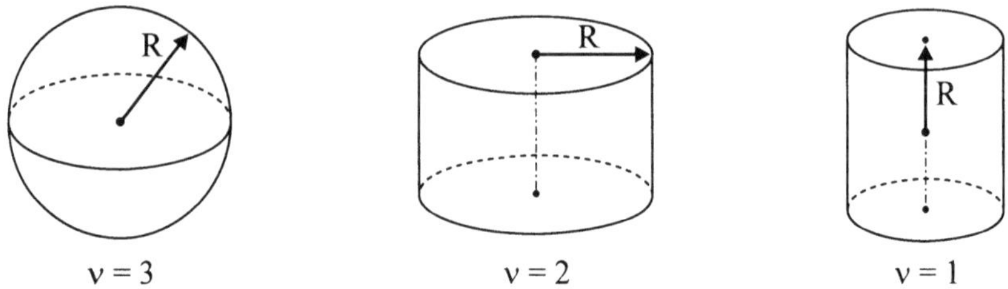

Figure 1. Geometric configuration of the cloud.

where e and m_e are the electron charge and mass, k_B is Boltzmann constant, and C is the dimensionless coefficient (e.g., Smirnov, 1985).

As was shown in our previous paper (Dumin, 2000a), Equation (1) can be solved in explicit form in the case of adiabatic expansion of a uniform cloud whose boundary moves by the law $R(t) = R_0 + u_0 t$. At $R \gg R_0$, such linear time dependence corresponds to the asymptotic stage of inertial expansion, when the most part of the initial thermal energy was transformed into kinetic energy of macroscopic motion (for more details, see Zel'dovich and Raizer, 1967).

By introducing the Coulomb's coupling parameter for electrons Γ_e^*, the final result can be written as*

$$\Gamma_e^*(t) = \frac{e^2 N^{1/3}}{k_B T_e} =$$

$$\Gamma_{e0}^* \left(1 + \frac{u_0 t}{R_0}\right)^{\frac{\nu}{3}} \left\{1 + \frac{R_0}{u_0 \tau} \frac{\kappa - 1}{\xi} \left[\left(1 + \frac{u_0 t}{R_0}\right)^{\xi} - 1\right]\right\}^{-\frac{1}{3(\kappa - 1)}}, \qquad (3)$$

$$\xi = \tilde{\nu}\Big[(\gamma - 1)\,\eta - \tilde{\kappa}\Big] + \nu\Big[(2/3)\,\tilde{\eta} - \kappa + 1\Big] + 1\,, \qquad (4)$$

where γ is the adiabatic index for heavy particles, ν and $\tilde{\nu}$ are the geometric factors of expansion for the charged and neutral particles respectively (see Figure 1): $\nu = 3$ corresponds to a spherical cloud; $\nu = 2$, to a cylindrical cloud expanding along its radius; and $\nu = 1$, to the cylindrical cloud expanding along its axis. From the physical point of view, the most interesting cases are $\nu = \tilde{\nu} = 3$ (3-dimensional free expansion of the unmagnetized cloud) and $\nu = 1$, $\tilde{\nu} = 3$ (1-dimensional expansion of the magnetized plasma cloud driven by 3-dimensional expansion of the neutral gas).

As follows from (3), there are four characteristic types of behavior of $\Gamma_e^*(t)$, drawn in Figure 2:

* Formula (4) represents generalization of the respective expression from our previous papers (Dumin, 2000a, 2000b), where $\tilde{\nu}$ was assumed to be equal to ν.

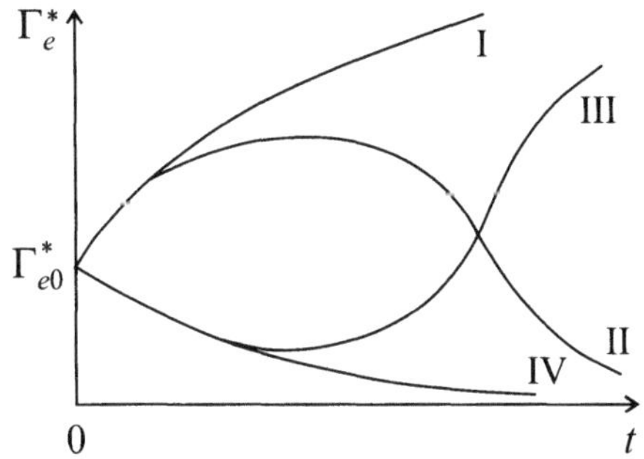

Figure 2. Four characteristic types of behavior of the function $\Gamma_e^*(t)$.

$$
\begin{array}{llll}
\text{I.} & \nu(u_0/R_0) > 1/\tau, & \xi < \nu(\kappa - 1); \\
\text{II.} & \nu(u_0/R_0) > 1/\tau, & \xi > \nu(\kappa - 1); \\
\text{III.} & \nu(u_0/R_0) < 1/\tau, & \xi < \nu(\kappa - 1); \\
\text{IV.} & \nu(u_0/R_0) < 1/\tau, & \xi > \nu(\kappa - 1).
\end{array}
$$

Of particular interest are types I and III**, which correspond to the asymptotic behavior $\Gamma_e^* \to \infty$ at $t \to \infty$. In such case, since potential energy of electrons decreases less rapidly than their kinetic energy, in certain time interval (when equilibrium with respect to elastic collisions is established but with respect to inelastic ones is not) Γ_e^* can reach considerable values (an order of magnitude greater than in laboratory devices for studying the nonideal plasmas).

3. Electric Properties of the Strongly-Coupled Plasma

In fact, when Γ_e^* reaches a value on the order of unity, the above formulas, referring to an ideal gas, are no longer applicable; and the charged particles will be in the strongly-coupled state. In general, an extensive numerical modeling is required to describe thermodynamic and electric properties of such plasma. Nevertheless, by using (a) the virial relations, (b) the assumption of ergodicity, and (c) the quasi-adiabatic approximation for the motion of electrons in the field of nearby ions, we succeeded to develop an efficient method for reduction of the many-particle distribution function for the system of strongly-coupled electrons to the effective one-particle function, whose most essential features turned out to be: (a) it does not depend on the temperature but is determined, to a first approximation, only by the electron concentration; (b) the most part of electrons has velocities considerably greater than the 'classical' thermal velocity; and (c) only a small part of electrons, corresponding to the tail of the distribution function, takes part in the transport

** For example, they take place for a free 3-dimensional expansion ($\nu = \tilde{\nu} = 3$) of the plasma cloud whose recombination is governed by the channel (2).

processes, for example, in electric conductivity (for details, see Dumin, 2000a, 2000b).

A relative concentration of the free charge carriers turns out to be

$$\frac{N_f}{N} \sim \frac{3\exp(3/C_u)}{2\pi C_u^2 \mu_0^{3/2}} \left(\frac{N_*}{N}\right)^{1/2} \exp\left\{-\frac{3}{2C_u^2\mu_0}\left(\frac{N_*}{N}\right)^{1/3}\right\} \tag{5}$$

at $(N/N_*) \ll 1$, where N_* is the electron concentration at the instant of plasma transition into the strongly-coupled state, C_u and μ_0 are the dimensionless coefficients on the order of unity, describing the many-particle correlations. The exponential suppression of N_f/N should lead to a new mechanism of the anomalous electric resistance (i.e., distinct from the usual ones, based on some kinds of the plasma turbulence).

Acknowledgements

I am grateful to V.I. Badin, M.G. Deminov, G.V. Kolmakov, Yu.B. Kudasov, B.V. Litvinov, I.I. Litvinov, S.A. Maiorov, Yu.P. Maltsev, S.A. Markovskii, V.B. Shikin, I.S. Veselovsky, V.S. Yarunin, B.N. Zakhariev, and L.M. Zelenyi for fruitful discussions and comments.

This work was supported by RFBR (RFFI) grant no. 99-05-65080.

References

Dumin, Yu.V.: 2000a, Transition of Plasma into a Strongly-Coupled State as a Possible Reason for Anomalous Resistance in Active Space Experiments, *Phys. Chem. Earth. (C)* **25**, 71–74.

Dumin, Yu.V.: 2000b, Studying the Properties of Supercooled Strongly-Coupled Plasma Created by Artificial Injection into Space, *J. Low Temp. Phys.* **119**, 377–383.

Haerendel, G.: 1987, Active Plasma Experiments, in: S.-I. Akasofu and Y. Kamide (eds.), *The Solar Wind and the Earth*, Terra Scientific, Tokyo, 215–242.

Smirnov, B.M.: 1985, *Physics of Weakly-Ionized Gas*, Nauka, Moscow (in Russian).

Zel'dovich, Ya.B. and Raizer, Yu.P.: 1967, *Physics of Shock Waves and High-Temperature Hydrodynamic Phenomena*, Academic Press, N.Y.

SUPERNOVAE RAYLEIGH-TAYLOR INSTABILITY EXPERIMENTS ON THE CEA-PHÉBUS LASER FACILITY

A. BENUZZI-MOUNAIX[1], S. BOUQUET[2], J.-P. CHIÈZE[1], F. MUCCHIELLI[2], R. TEYSSIER[1] and F. THAIS[1]

[1] *CEA/DSM, Orme-des-Merisiers, 91191 Gif-sur-Yvette Cedex, France*

[2] *CEA/DIF, BP 12, 91680 Bruyères-le-Châtel, France*

Abstract. Hydrodynamic instabilities play a key role in several astrophysical phenomena. We present a laser experiment related to the evolution of Rayleigh-Taylor instabilities (RTI) due to the passage of the shock wave through the He/H interface in a type II supernova. The experiment is performed on the CEA-Phébus laser facility and the RTI are studied using a bi-layer Cu/CH_2 target with sinusoidal ripple at the interface.

1. Introduction

Many astrophysical events involve gas flows which can be simulated by laboratory plasma experiments. We show that high energy lasers can be used to simulate the Rayleigh-Taylor instabilities (RTI) occuring during the explosion of type II Sne (Arnett, 1996; Kumagai *et al.*, 1989 and Baclet *et al.*, 1999) and we focus on the evolution of the RTI produced when the bouncing shock wave crosses the He/H interface.

2. Experimental Set-Up and Results

The experiment has been performed at the Commissariat à l'Energie Atomique on the Phébus laser facility. In order to simulate the transit of the shock wave through the He/H interface and the growth of the RTI, we used Cu/CH_2 targets (Baclet *et al.*, 1999; Remington *et al.*, 1997; Remington *et al.*, 1999). Two different initial conditions with a sinusoïdal (1D) interface ripple with $\lambda = 30\ \mu$m (resp. $\lambda = 125\ \mu$m) and an amplitude $a_0 = \pm 3\ \mu$m (resp. $a_0 = \pm 12.5\ \mu$m) have been taken. The shock is generated in an indirect drive way. We used the two beams of the Phébus laser at 3ω ($\lambda = 0.35\ \mu$m), with a squared pulse of 1.3 ns. Each beam delivered 2–2.8 kJ. The first beam has been focalized into a gold cavity to create a thermalized radiation at 170 eV and the second one has been focalized on a titanium target to obtain an X-rays backlighter source at 4.7 keV (providing a 10 μm-resolution). The 2D transverse radiography (second beam delayed of 30 ns)

Astrophysics and Space Science **277:** 143–146, 2001.

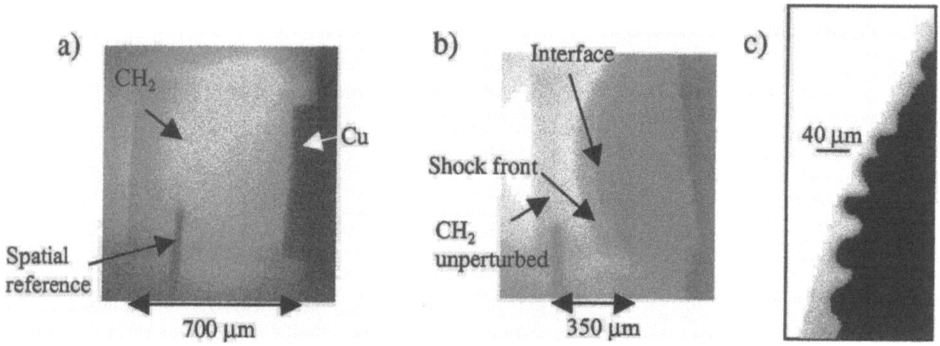

Figure 1. a) Radiography obtained with the Wolter microscope of an unshocked target at $t = 0$; b) Radiography of a shocked target at $t = 30$ ns with initial perturbation $\lambda = 125\ \mu$m and $a_0 = \pm 12.5\ \mu$m; c) Blow-up of the Radiography of the interface Cu/CH$_2$ at $t = 30$ ns (the initial modulation was $\lambda = 30\ \mu$m and amplitude $a_0 = \pm 3\ \mu$m).

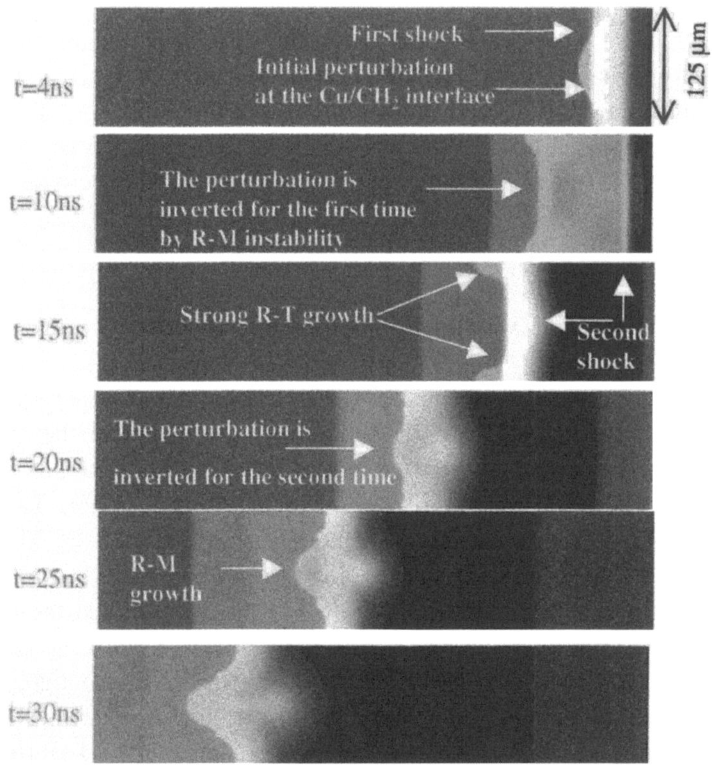

Figure 2. Time sequence of the 2D density profile.

has been performed by using a Wolter microscope coupled with a CCD camera. Taking into account the thickness of the plastic (cold CH_2), the transmission has been estimated, approximately, to 25%; on the contrary, the copper is completely opaque. On Figure 1a we show a radiography of a target, before being shocked by the laser, with no modulated interface, to be used as a test.

On Figure 1b, we present a typical image of a shocked target (the initial modulation was $\lambda = 125\ \mu m$ and $a_0 = \pm 12.5\ \mu m$ – four initial modulations). First of all, we obtained a very high quality image and we are able to visualize easily the interface Cu-CH_2 in addition to the shock front in the CH_2 together with the unperturbed CH_2. Moreover, we can recognize the four initial modulations which have became four 'arches'. In addition, the shock front, as well as the interface high/low-density material are bowed. This is due to 2D effects and the shock does not remain plane. On the other hand, the growth of the RTI seems to be stopped by the arrival of a second shock, probably launched by the collision between the shield which explodes (into the cavity) and the base of the target. This interpretation has been confirmed by numerical simulations (see next paragraph). Figure 1c shows a blow-up of the interface at $t = 30$ ns in the case $\lambda = 30\ \mu m$ and $a_0 = \pm 3\ \mu m$. The instability enters the nonlinear regime since the amplitude at that time (40 μm) is of the same order as λ.

3. Modeling

One of the main interests of such a laser experiment is to compare the results with those of simulations performed with astrophysical hydrodynamical codes. Careful comparisons with realistic laboratory conditions allow upgrades of the codes and more reliable astrophysical simulations. Two codes developed initially at CEA-Saclay for stellar dynamics have been modified to compute the evolution of common laser experiments. With the 1D Lagrangian code, *CHARM* (fully implicit with multigroup radiation transfer and multi-materials equations of state) the interaction laser/target has been calculated. Then, a 2D Eulerian code, *HYDREL* (multi-materials equations of state, average atom opacities – More's method – and SESAME equations of state), we have computed the evolution of the RTI after the end of the light pulse. First of all, we have calculated (code *CHARM*) the 1D evolution (40 ns duration) of a planar slab composed of 4 different materials: the inner gold shield of the cavity, the cavity vacuum and the target (60 μm of copper and 600 μm of plastic). A first shock is launched into the copper slab and reaches, at 4 ns, the Cu/CH_2 interface which gets strongly decelerated and, therefore, is subject to both Richtmyer-Meshkov instabilities (RMI) and RTI. At the same time, the gold shield explodes and collides with the front end of the copper slab. A second shock is launched in the target, much stronger than the first one. This second shock quickly catches up and accelerates the interface (roughly at 15 nanoseconds). The flow becomes, therefore, purely R-M unstable. Next, we computed the 2D evolution of the target with *HYDREL*, using the results obtained by *CHARM* as initial/boundary conditions. Starting at $t = 4$ ns, when the first shock hits the copper/plastic interface, we simulate the time sequence plotted in Figure 2. As the first shock hits the copper/plastic interface, one clearly sees that the RMI, followed by the RTI, produce a strong increase in the amplitude of the perturbation. As the second shock hits the interface, the modulation gets reversed again, due to the secondary RMI.

4. Conclusion

In this paper, we have presented a laser experiment making possible the study of a new area which can be called 'Laboratory Astrophysics'. Several domains ranging from nuclear physics (reaction rates), dense and/or hot plasmas, atomic physics (opacities, equation of states), to magnetized plasmas ... can be, now, examined at the laboratory. In our opinion, the use of high energy lasers will be a very promizing and challenging way to study the futures of Space Science in the 21st century.

References

Arnett, D.: 1996, *Supernovae and Nucleosyntesis*, Princeton University Press, Princeton.

Baclet, Ph., Benuzzi-Mounaix, A., Bouquet, S., Cherfils, C., Chièze, J.-P., Muchielli, F., Munsch, P., Polès, L., Reverdin, Ch., Teyssier, R., Thais, F., Thébault, J.-P. and Troussel, Ph.: 1999, *First International Conference on Inertial Fusion Science and Applications (IFSA'99)*, Sept. 12–17, Bordeaux, France.

Kumagai, S., Shigeyama, T., Nomoto, K., Itoh, M., Nishimura, J. and Tsuruta, S.: 1989, *Astrophys. J.* **345**, 412–422.

Remington, B.A., Arnett, D., Drake, R.P. and Takabe, H.: 1999, *Science* **284**, 1488–1493.

Remington, B.A., Kane, J., Drake, R.P., Glendinning, S.G., Estabrock, K., London, R., Castor, J., Wallace, R.J., Arnett, D., Liang, E., McCray, R., Rubenchick, A. and Fryxell, B.: 1997, *Phys. Plasma* **4**, 1994–2003.

III

BASICS OF WINDS

PHYSIQUE DES VENTS

A SIMULATION METHOD FOR SEMICOLLISIONAL PLASMAS

FILIPPO PANTELLINI and SIMONE LANDI
Observatoire de Paris, France

Abstract. We present a one dimensional kinetic simulation model which can be used to simulate the stationary state of a semicollisional plasma. Results of a simple simulation are presented and compared to Fokker-Planck calculations. The model is particularly well suited for the diluted solar atmosphere.

1. Introduction

It is sometimes impossible to conveniently simulate a plasma using either fluid (i.e. fully collisional) or collisionless model. The solar atmosphere beyond the upper chromosphere and out to heliocentric distances of the order of an astronomical unit is an example of such a plasma. The method we present in this paper is suited for stationary configurations where the collisional mean free path for a thermal particle is not too small compared to the typical gradients of macroscopic quantities (e.g. temperature, density, etc.) and can be seen as alternative to Fokker-Planck and Monte Carlo simulations. Most basic aspects of the model have been discussed elsewhere (Pantellini, 2000) for the case of identical particles undergoing hard-sphere type collisions and shall not be repeated here. Only the extension of the model to the case of charged particles undergoing Coulomb type collisions is briefly discussed here.

2. The Model

A qualitative sketch of the model is shown in Figure 1. The model is based on the numerical integration of the one dimensional motion (along the z axis) of N protons and N electrons plunged in a z aligned gravitational field g and an external self-consistent charge neutralizing constant electric field E which is determined experimentally by trying a number of different values. Particles are vanishing small, have non zero mass and three dimensional velocities. Whenever the world lines of any two particles encounter they may make an elastic collision according to a prescribed velocity dependent probability distribution P. The choice of P specifies the character of the interaction. A probability distribution $P \propto u^{-4}$, where u is the relative velocity of the two particles, mimics a Coulomb type collision whereas a

Astrophysics and Space Science **277:** 149–152, 2001.

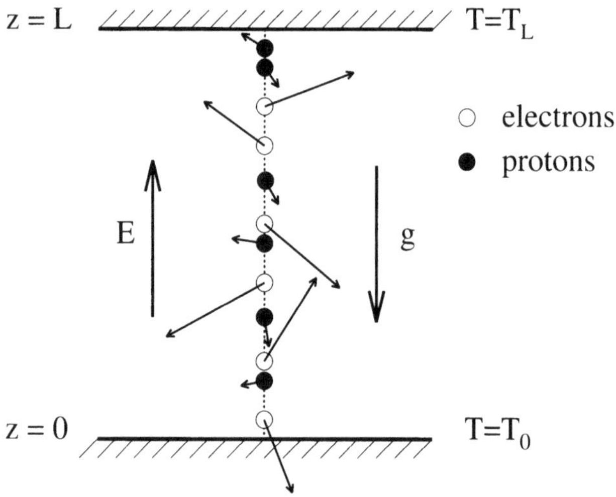

Figure 1. Schematic illustration of the model for a proton-electron plasma. Note that the particles' velocity is 3D even though the model is spatially 1D.

velocity independent P mimics a hard sphere type collision. Given the low velocity divergency of the probability distribution for Coulomb collisions we set $P(u) = 1$ for relative velocities $u < u_0$ where u_0 is typically of the order, or smaller, than half the typical relative velocities between the colliding particles. A particle hitting one of the boundaries at $z = 0$ or $z = L$ is instantly reinjected into simulation domain following a not yet specified velocity distribution function.

3. An Example: The Thermoelectric Field

We simulate a plasma in a $g = 0$ gravitational field using 80 electrons and 80 protons. A proton to electron mass ratio of 100 has been used to speed-up the simulation. Particles reaching the boundaries are reinjected into the simulation domain according to Maxwellian velocity distribution functions with temperatures $T(0) = T_0$ and $T(L) = T_L$, respectively. For moderate temperature gradients a constant electric field is generally good enough to ensure local charge neutrality everywhere in the system. Now, if the plasma is sufficiently collisional such that the electron temperature $T_e(z)$ and the proton temperature $T_p(z)$ are equal we expect the electric field to be related to the temperature gradient via (e.g. Chapter 7.9 in Golant *et al.*, 1980)

$$eE = -\alpha \frac{\partial}{\partial z}(kT_e) \tag{1}$$

where k is the Boltzmann constant and e the proton charge. This field is called the thermoelectric field as it is solely due to the temperature gradient in the plasma slab. It is reminiscent of the Seebeck effect in metals (first described by Seebeck in the early 1820s) which makes the familiar thermocouple thermometer work. The field is ultimately due to the unbalanced frictional force exerted by the heavy protons on the electrons flowing from the hot to the cold boundary and vice versa so that the appearance of the numerical constant α is primarily an effect of proton-electron collisions. Figure 2 shows electron number density n_e (normalized to the average

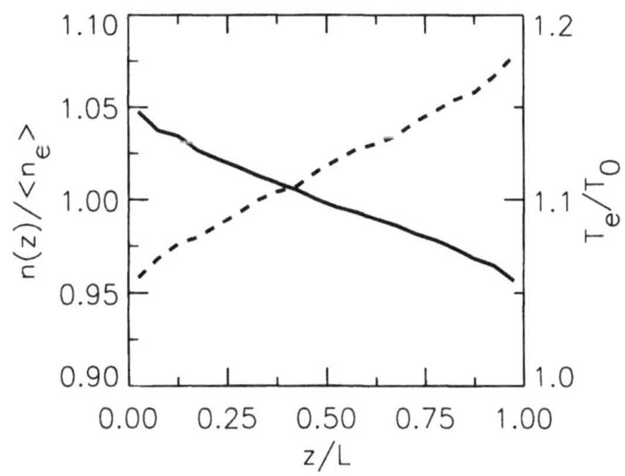

Figure 2. Simulation of a plasma slab with imposed temperature gradient. The electron temperature (dashed profile) and the electron density have been obtained by periodically sampling the position and the velocity of all electrons in the system.

electrons density $\langle n_e \rangle$) and electron temperature T_e (in units of T_0) profiles for a simulation based on our model. The reason for the normalized temperature profile to differ from the theoretical profile $T_e(z) = 1 + 0.2\ z/L$ one would expect based on the imposed temperatures at the boundaries is due to the fact that the typical collisional mean free path is finite. As expected both the proton and electron pressures are equal and constant so that the only contribution to the charge neutralizing electric field is the thermoelectric field which we find to be $E = -0.0895$ (in units of kT_0/eL). On the other hand the temperature gradient, which can be measured on the figure, is $\partial T_e/\partial z = 0.116/L$. Using Equation (1) we find $\alpha_{sim} = 0.77$ which compares quite well with the Fokker-Planck result $\alpha_{FP} = 0.71$ found by Spitzer and Härm (1953).

Unlike the thermoelectric field, which depends on the electron-proton collisions, the electron heat flux is mainly supported by electron-electron collisions. Accurate Fokker-Planck calculations lead to the following expression for the total electron heat flux (e.g. Chapter 7.10 in Golant *et al.*, 1980)

$$q^e_{FP} = -3.16 \times 10^{-3} \frac{n_e k T_e}{\nu_{ep} m_e} \frac{\partial k T_e}{\partial z} \quad \text{(SI units)} \tag{2}$$

where ν_{ep} is called the transport frequency of the elastic electron-proton collisions which is of the order of the number of collisions per time unit of an electron with the protons in the system. Replacing the collision frequency measured in the simulation and the temperature gradient of Figure 2 into Equation (2) one obtains a heat flux $q^e_{FP} = -0.075$ (in units of $\langle n_e\rangle m_e(kT_0/m_e)^{3/2}$) which, again, compares quite well with the heat flux $q^e_{sim} = -0.087$ measured in the simulation. This value differs substantially from hard sphere type collisional heat flux, which for the observed temperature gradient can be estimated (cf. most books on statistical physics) to be $q^e_{HS} = -0.027$. Even though a collisionless plasma is unable to support a temperature gradient it can support a strong heat flux. The latter is due to the fact that the upward and downward traveling particles have different temperatures given by the injection temperature T_0 and T_L, respectively. In that case the temperature is constant $T_e(z) = \sqrt{T_0 T_L}$ but the heat flux is roughly twice the Fokker-Planck value, i.e. $q^e_{NC} = -0.167$. We conclude by noting that the Fokker-Planck results Equations (1) and (2) are valid in the limit of small values of the thermal Knudsen number $K_T = (\partial T_e/\partial z)\lambda_{ee}/T_e$ where λ_{ee} is the average collisional mean free path for an electron-electron collision. In the simulation of Figure 2 we find $\lambda_{ee} = 0.19\,L$ and $K_T \approx 0.02$ which justifies comparing our simulation with results from Fokker-Planck calculations. We note that increasing the number of particles by roughly a factor 10 leads to $K_T = O(10^{-3})$ which is a typical value in the solar corona or the solar transition region. Simulations in this regime of the Knudsen number, including gravitational effects, will be published elsewhere (Landi and Pantellini, 2001).

References

Golant, V., Zhilinsky, A. and Sakharov, I.: 1980, *Fundamentals of Plasma Physics*, John Wiley & Sons, New York.

Landi, S. and Pantellini, F.G.E.: 2001, On the temperature profile and the heat flux in the solar corona: Kinetic simulations, *A&A* (in press).

Pantellini, F.G.E.: 2000, A simple numerical model to simulate a gas in a constant gravitational field, *Am. J. Phys.* **68**(1), 61–68.

Spitzer, L. and Härm, R.: 1953, Transport Phenomena in a Completely Ionized Gas, *Phys. Rev.* **89** (May), 372–380.

ON THE MAGNETIC ENERGY AVALANCHE IN THE SOLAR ATMOSPHERE

M.P. RYUTOVA
Lawrence Livermore National Laboratory, Institute of Geophysics and Planetary Physics, Livermore, CA 94550
E-mail: ryutova@beowulf.ucllnl.org

T.D. TARBELL
Lockheed Martin Solar and Astrophysics Laboratory, 3251 Hanover Street, Palo Alto, CA 94304
E-mail: tarbell@lmsal.com

Abstract. We discuss a mechanism of energy production associated with the acoustic and MHD shocks produced by colliding and reconnecting flux tubes and present the observational results showing a connection between the dynamic changes in the photosphere and the high velocity and heating events in the transition region. We suggest that these processes may provide a constant energy supply for the origin of the fast wind and help to advance studies of coronal dynamics.

1. Introduction

The mechanisms that heat the corona are still unknown. All the existing models of the coronal heating ignore the existence of the transition region (TR) and its energy reservoir which, we know, consists of bright radiative transients and a wide variety of steady and unsteady flows with velocities ranging from 10 to 400 km s^{-1}. The physics of the corona cannot be understood without these unsteady phenomena occurring in its very 'bottom'. It is even less clear why the corona should be the base of the solar wind, and not already a part of it. We believe that the energy source lies far below these heights, and that the chromosphere and transition region flows, which are the first to consume the energy coming from below, form the base of the fast wind.

The idea that the energy for the fast solar wind may come from a chromosphere network has been invoked by Axford and McKenzie (1992). The observations also indicate a connection between the solar wind outflow and the chromospheric magnetic network (Hassler *et al.*, 1999). A new view on the origin of the solar wind (Woo, 1996; Habbal *et al.*, 1997) suggests that the fast solar wind originates not only from coronal holes but from the entire surface of the quiet sun. This requires some universal mechanism, free from restrictions posed by the coronal hole open magnetic structures. Such a mechanism may be associated with the post-reconnection shocks produced by interaction of small scale magnetic elements in the photosphere (Tarbell *et al.*, 1999).

Astrophysics and Space Science **277:** 153–156, 2001.

In the present paper we show that simultaneous measurements taken by the TRACE and SOHO instruments in different lines support the governing role of small-scale magnetic interactions in heating and jet formation observed in the chromosphere and TR.

2. Magnetic Energy Avalanche

The key elements of energy transport by the post-reconnection acoustic shocks (Tarbell *et al.*, 1999) are as follows. Brought together by convective motions, magnetic flux tubes collide and reconnect. Straightening the reconnection products create a 'sling-shot' effect that generates acoustic waves (in non-magnetic environment), that quickly become shocks in the sharply stratified low atmosphere. Having a complex wavy surface shocks experience self focusing which leads to hydrodynamic cumulation, i.e. concentration of the energy into a small volume. Shock fronts may experience either head-on collision, or form a figure of revolution, or converge at some angle. Respectively, cumulative energy may be either converted entirely into heat or into strong jets, or be distributed between the two. For example, if reconnection occurs in the photosphere ($\rho_0 \simeq 2.8 \times 10^{-7}$ g cm^{-3}, $B = 1000\ G$), at the height $h = 1100$ km the shock collision produces (in the heating regime) the temperature $T \simeq 1.2 \times 10^5$ K; respectively, in the jetting regime cumulative jet will have a velocity $v_{jet} \simeq 124$ km s^{-1}. In the photosphere/chromosphere region small

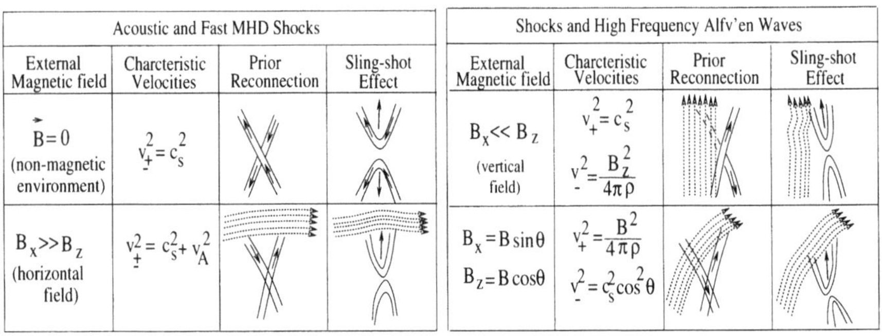

Figure 1. Reconnection of magnetic flux concentrations in non-magnetic and magnetic environments. The second columns in both panels show the characteristic velocities, the two last columns depict the flux tubes prior the reconnection and the post-reconnection sling-shot effects.

scale network magnetic elements form the arcades of loops of different sizes. In these arcades reconnection of flux tubes may occur in various magnetic configurations of the environmental plasma. The most typical cases are shown in Figure 1. The left panel shows two cases: (1) collision of flux tubes in non-magnetic environment, that leads to generation of the acoustic shocks, and (2) reconnection near the horizontal magnetic field in which case the fast MHD shocks are generated. The right panel shows the reconnection of flux tubes near the vertical magnetic structure (the upper set), where one branch provides formation of the acoustic shocks, and the other is responsible for generation of high frequency Alfvén waves. In all these cases it is an acoustic or fast MHD component that leads to the formation of high velocity jets and microflares in the TR. The Alfvén waves have quite restricted parameters for shock formation, but their frequency, determined by the reconnection rate, may be high enough ($\omega \simeq 0.1 - 2.5\ s^{-1}$) to carry the energy through the TR. The overall picture of high velocity and heating events in the TR carries

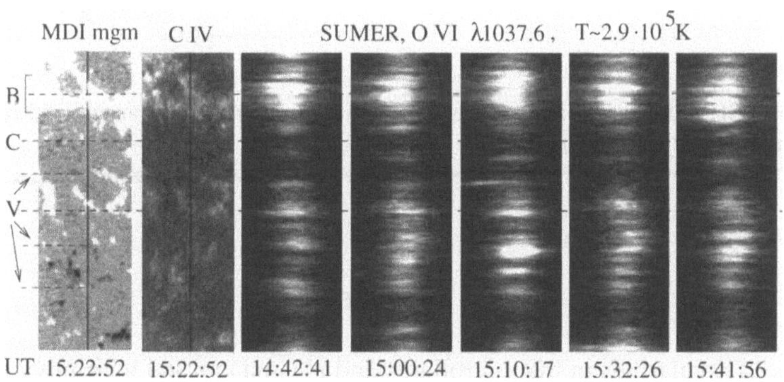

Figure 2. Imprint of the photospheric magnetic field in the transition region (C IV image and O VI spectra). The solid line is the SUMER slit position. A shift to the right in the SUMER spectra indicates upflow (and v.v.). Field of view $70'' \times 200''$.

an obvious imprint of the photospheric magnetic field. Results from simultaneous measurements taken by TRACE in chromospheric and TR (C IV) images, high resolution magnetograms taken by MDI, and spectra of chromospheric (C II) and the TR lines (O VI) obtained with the SUMER instrument on SOHO show that (i) enhanced emission in the C IV line ($T \sim 10^5$ K) is usually co-spatial with the magnetic pattern in the photosphere (cf. two left panels in Figure 2); (ii) microflares are often accompanied by plasma jets; (iii) plasma flows are always seen whenever the slit crosses strong magnetic flux tubes. Figure 2 shows 5 snapshots of SUMER spectra in the O VI line extracted from a 2.2 h time sequence taken with $\sim$ 13 s cadence and co-aligned with the MDI magnetogram and C IV image of the studied region. By 'B', 'C' and 'V' we mark the regions of magnetic field, cell interior and vertices. Jets and microflares in the TR overlying strong magnetic field site ('B') were observed continuously during the entire observing period. Frequent multiple flows are seen above the regions where the slit crosses supergranular vertices ('V'). Examples of the O VI intensity curves for a 2 h period are shown in Figure 3 (left panel). The right panel shows three positions along the slit corresponding to the intensity curves plotted on the left and spectra at t = 14:39:45; y = 170 is a cell interior producing no activity in the TR. These data clearly show the governing role of the photospheric fields in the ubiquitous TR flows.

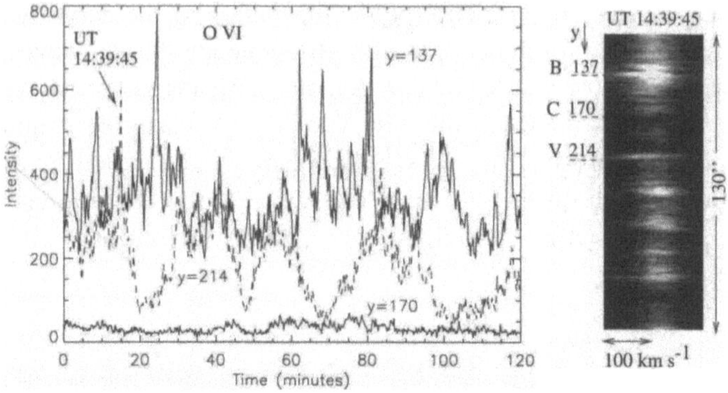

Figure 3. O VI SUMER spectra: left panel is an intensity variation in time; upper solid curve is a response of the TR to dynamic changes in the strong magnetic flux concentrations; lower solid curve corresponds to the convection cell interior, and the dashed curve to vertex. Right panel is a SUMER image at t = 14:39:45.

3. Summary

The observed regularities in the microflares and plasma jets and their obvious connection with the small scale interactions of the magnetic fields which we are now able to see more clearly than before with the SOHO and TRACE missions strongly support a ubiquitous nature of the energy production by the network magnetic field. This field totally replaces itself in about 40 hours. High rate of the emergence of new fluxes and diversity of their parameters result in the cascade of shock waves, providing thus relatively 'steady' energy input into higher atmosphere. The resulting energy flux estimated on the basis of observed time scale (40 h) is enough to explain the observed UV/EUV radiation of the order of $5 - 10 \times 10^5$ erg cm^{-2} s^{-1}.

References

Axford, W.I. and McKenzie, J.F.: 1992, in: *Solar Wind Seven*, Pergamon Press, Oxford, p. 1.

Habbal, S.R., Woo, R., Fineschi, S., O'Neal, R., Kohl, J., Noci, G. and Korendyke, C.: 1997, *Astrophys. J. Lett.* **489**, 103.

Hassler, D.M., *et al.*: 1999, *Science* **283**, 810.

Tarbell, T., Ryutova, M., Covington, J. and Fludra, A.: 1999, *Astrophys. J.* **514**, L47.

Woo, R.: *Nature* **379**, 321.

HYDRODYNAMICS OF THE SOLAR WIND EXPANSION

Why is the solar wind supersonic?

M. VELLI
Dipartimento di Astronomia e Scienze dello Spazio, Università di Firenze

Abstract. A pedagogical introduction to the classical hydrodynamic steady state solutions for flows in a spherically symmetric atmosphere is presented along with a detailed analysis of the stability of such flows and a discussion of the Parker/Bondi phase diagram of solutions in the Mach number - stellar radius plane. This leads naturally to the scenario presented by Velli (1994) for wind-accretion transitions, which helps to explain results from numerical simulations in many contexts and gives a more comprehensive answer to the question of why the solar wind and other steady state flows in spherical symetry must quite generally become supersonic.

1. Solar Wind Theory Origins

Although knowledge of a solar influence at the earth's orbit dates back from Lord Carrington's observations that aurorae often occurred several hours after white light solar flares in the second half of the last century, the first direct indication of a continuous outflow of fast particles from the sun came from Biermann's investigation in the fifties of the shape of the cometary ion tails, from which he deduced an average speed of around 475 km sec^{-1} for this flow. In 1957 Chapman showed how a static conductive corona starting at 10^6 K at the sun should maintain a high density out to far distances (in fact, after an initial decrease, the density should increase outward again!), and in 1958 Parker argued, on the basis of the unreasonably high pressures which static solutions yielded at large distances, that 'probably it is not possible for the solar corona, or, indeed, perhaps the atmosphere of any star, to be in complete hydrostatic equilibrium out to large distances'. He then proceeded to show how a viable solution yielding negligible pressures at infinity consisted in a flow accelerating continuously and becoming supersonic at large distance.

Consider hydrostatic balance for a spherically symmetric atmosphere with gravity

$$\frac{\partial p}{\partial r} = -m_p n \frac{g}{r^2}, \tag{1}$$

where $g = GM_\odot/R_0$ and $M_\odot$ is the solar mass, while R_0 is the solar radius, G is Newton's constant, and distances have been normalized with the solar radius. Also

Astrophysics and Space Science **277:** 157–167, 2001.

m_p is the proton mass and the mass density is $\rho = m_p n$. Recalling that $p = n\kappa T$ we may integrate to find that

$$n\kappa T = n_0 \kappa T_0 \exp\Big(- \int_1^r dr \, \frac{m_p}{\kappa T} \frac{g}{r^2} \Big) \tag{2}$$

so that a static spherically symmetric extended atmosphere with a temperature profile decreasing with distance less rapidly than $1/r$ requires a finite pressure at infinity to be confined, the same being true if the atmosphere 'evaporates' a subsonic flow, or breeze.

The subsequent Parker-Chamberlain debate on supersonic/subsonic evaporation was made academic, though of great theoretical interest, by the in situ measurement of the steady, supersonic wind by the Luna 2 (1959) and Explorer (1961) spacecraft (see,e.g., Hundhausen, 1972). However, Mestel (quoted in Roberts and Soward, 1972) first remarked that 'were the temperature at the base of the solar corona 10^5 K rather than the generally accepted $1 - 2 \times 10^6$ K, the total pressure far from the sun would suffice to suppress the solar wind entirely'. Indeed, the pressure of the interstellar medium, or ISM, $p_{ISM} \simeq 1.24 \; 10^{-12}$ dyne cm^{-2} would suffice to confine a 4 10^5 K static corona with base density 10^9 cm^{-3}. So although correct, the argument for a supersonic wind does not appear to be as cogent on the basis of pressure arguments only. In reality, the dependence of spherically symmetric, non-rotating atmospheres with flows on changes of the external conditions is somewhat more subtle, as described by Velli (1994). Here this problem will be rexamined both analytically as well as through a numerical experiment. For the sake of simplicity, we will consider only isothermal flows, i.e. flows for which the temperature may be considered constant out to great radial distances,.

2. Stationary Isothermal Flows: Breezes, Winds, Accretion

The equations of motion for one-dimensional, spherically symmetric, stationary isothermal flow neglecting self-gravity may be written in the form

$$\frac{\partial}{\partial r}(\rho v r^2) = 0, \quad p = c^2 \rho \tag{3}$$

$$v \frac{\partial v}{\partial r} = -\frac{1}{\rho}\frac{\partial p}{\partial r} - \frac{g}{r^2} \tag{4}$$

where v is the velocity, c the constant sound speed. For a static atmosphere, the pressure profile is given by $p = p_0 \exp(-g/c^2 + g/rc^2)$ which, as discussed above, implies a non-vanishing asymptotic value for the pressure at large distances $p_\infty^s = p_0 e^{-g/c^2}$. In terms of the mach number $M = v/c$ the flow equations may be written (a prime denoting radial derivatives throughout this section)

$$\Big(M - \frac{1}{M} \Big) M' = \frac{2}{r} - \frac{g}{r^2 c^2} \tag{5}$$

which may be integrated and expressed in two equivalent ways

$$\frac{1}{2}(M^2 - M_0^2) - \log\left(\frac{M}{M_0}\right) = 2\log r + \frac{g}{rc^2} - \frac{g}{c^2}. \tag{6}$$

$$M^2/2 + \log p - \frac{g}{rc^2} = M_0^2/2 + \log p_0 - \frac{g}{c^2}, \tag{7}$$

where M_0 is the base Mach number. The second form is essentially the conservation of energy flux, where for an isothermal atmosphere (polytropic index $\gamma = 1$, the enthalpy is expressed as $\log p$ instead of $\gamma p/(\gamma - 1)$. By definition, in this case, attention is being limited to positive energy flows. A general and very interesting discussion of the behaviour of flows with a more realistic energy equation in the cases of positive energy as well as vanishing total energy, and various intermediate limits, may be found in Roberts and Soward (1972).

Equation (5) has a singular point at the sonic point, $r = g/2c^2$, $M = 1$. Solutions to the above equations may be represented in the (M, r) phase plane, or Parker-Bondi diagram, illustrated in Figure 1, which, following the symmetry of Equation (5) is symmetric in the sign of M. The diagram is divided into 4 (8, considering positive and negative M) areas (labeled I–IV) by the two critical (upward and downward transonic) solutions which cross at the sonic point $r = g/2c^2$, $M = 1$. Single valued continuous flow profiles $M(r)$ which are subsonic for all r, the breezes, lie below both transonic curves (region I). Among flows which are subsonic at the atmospheric base the accelerating transonic has the special property that density and pressure tend to zero at large distances: because of the small but finite values of the pressure of the ambient 'external' medium, a terminal shock transition, connecting to the lower branch of the double valued solutions filling region II will in general be present (see e.g. Holzer and Axford, 1970 where a complete discussion of the above equations in the case of general polytropic flows, may be found).

The jump conditions across such a shock are found from conservation of mass and momentum across the shock, which read (superscripts −,+ denote the solution immediately upstream and downstream from the shock respectively)

$$\rho^- M^- = \rho^+ M^+, \tag{8}$$

$$\rho^+ (M^+)^2 + \rho^+ = \rho^- (M^-)^2 + \rho^-, \tag{9}$$

from which one finds immediately $M^+ M^- = 1$. This gives a way to graphically construct shock transitions: it is sufficient to plot the curves corresponding to $1/M$ for the transonics (dashed lines in Figure 1), and connect the transonic with the double valued curve in region II where the dashed line intersects them: such solution is given by curve W.

The downward transonic in itself is not a possible solution for outflows, since a continuous transition from supersonic to subsonic flows is unstable. This solution was the one discovered by Bondi for spherically symmetric accretion (Bondi, 1952). It plays the same role the Parker wind solution does, but for accretion flows:

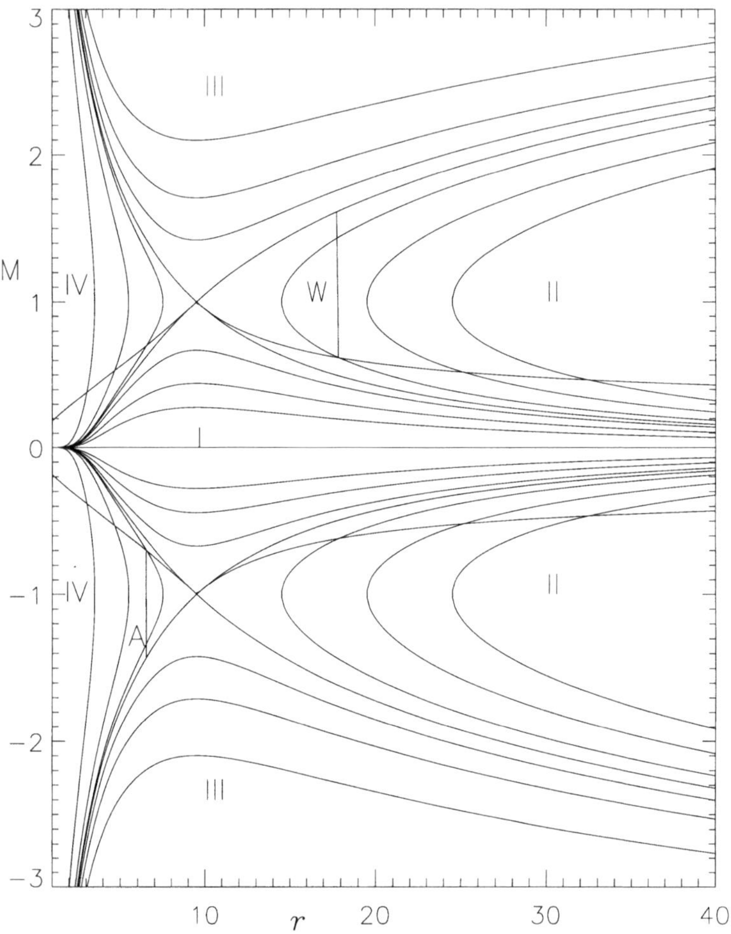

Figure 1. The (M,r) phase plane. The dashed line intersection with double valued curves defines the shock position for winds (region II, curve W) or accretion flows (region IV, curve A).

in negative M solutions, th equivalent construction leads to accretion shocks in the flow, as shown in McCrea (1956), this time in the region labeled IV. One such transition is shown in Figure 1 (curve A).

For given base values of the pressure, the position of the shock is uniquely determined by the pressure of the ISM, and the distance from the critical point to the shock decreases as the pressure increases. Physically, this is obvious, but mathematically it may be shown by noticing that conservation of mass across the shock allows the writing of the asymptotic pressure in terms of the upstream Mach number M^- as

$$p_\infty = p_0 M^- \exp\left({M_*}^2 - 2\frac{g}{c^2} - {M^-}^2 + \frac{1}{{M^-}^2} \right)/2, \tag{10}$$

where M_* is the base Mach number of the upward transonic. p_∞ is a monotonically decreasing function of M^-, which is itself, obviously, a monotonically *increasing* function of the shock position r_s, so that increasing p_∞ decreases r_s. When p_∞ reaches the critical value $p^c_\infty = p_0 \exp({M_*}^2/2 - g/c^2)$, the shock distance $r_s = r_c = g/2c^2$ and the discontinuity in the flow velocity reduces to a discontinuity in the derivative of the profile. This is the fastest possible, or critical, breeze, made up of the section of upward transonic below r_c and the section of downward transonic beyond r_c. For the breeze solutions, with a base Mach number M_0 such that $M_* > M_0 \geq 0$, the asymptotic pressure is easily calculated to be

$$p_\infty = p_0 \exp\left(M_0^2/2 - g/c^2\right) \geq p^s_\infty . \tag{11}$$

It follows that the pressure required to confine a breeze *increases* with increasing base Mach number and is greater, by the exponential of a very small number, than that of a static atmosphere. The limiting value of p_∞ is again p^c_∞. Therefore, for a given base pressure and asymptotic pressures $p^c_\infty > p_\infty \geq p^s_\infty$ *two* possible stationary outflow solutions exist, a supersonic shocked wind and a subsonic breeze.

3. Stability of the Flow

When such conditions occur, it is frequently the case that one of the solutions is found to be unstable, i.e. that any small perturbation may lead the flow to evolve away from stationarity. In this case, it is the breeze solutions that are unstable: to prove this we introduce small perturbations (sound waves) and linearize the equations of motion around the stationary state. We will apply boundary conditions which allow the configuration to evolve from one to the other of the stationary solutions we have found: the perturbing modes, a superposition of sound waves, must leave the pressure (and density) unperturbed at the atmospheric base and infinity, i.e. they will be standing waves. It is convenient to introduce characteristic variables $y^\pm = \delta v/c \pm \delta p/\rho c^2$, where δp is the pressure perturbation and δv is the radial velocity fluctuation. In these variables an outward (inward) propagating sound wave has the property that $y^- = 0$ ($y^+ = 0$). Assuming a time dependence $y^\pm = y^\pm(r) \exp(-i\omega + \gamma)t$ the linearized equations become

$$\left(M \pm 1\right) y^{\pm\prime} - i(\omega + i\gamma) y^\pm + \frac{1}{2}\left(y^\pm + y^\mp\right) \frac{M'}{M} \left(M \mp 1\right) = 0. \tag{12}$$

In the presence of a nonuniform but stable flow, Equation (12) describes wave propagation and reflection, and a conserved quantity, the wave action flux, exists (in a static medium, the wave energy flux is conserved; when there are mass motions, this is replaced by the wave action flux: see, e.g. Kadomtsev, 1983 §4 for a discussion of wave energy and Velli, 1993 and references therein for a discussion of wave-action conservation). When $\gamma \neq 0$ the wave action evolution equation becomes

$$\left[\frac{(M+1)^2}{M}|y^+|^2 - \frac{(M-1)^2}{M}|y^-|^2\right]' + \tag{13}$$

$$+2\frac{\gamma}{M}\left[(M+1)|y^+|^2 - (M-1)|y^-|^2\right] = 0,$$

the first square bracket being proportional to the wave action. Notice that for $|M| < 1$ the term in the second square bracket is positive definite. Integrating this equation between 1 and r and imposing the boundary condition that the pressure perturbation vanish (i.e. $y^+ = y^-$) both at the solar surface and at great distances, we find the following estimate for γ:

$$\gamma = \frac{2\left(|y^+|_0^2 - |y^+|_r^2\right)}{\int_1^r dr\, M^{-1}\left[(M+1)|y^+|^2 - (M-1)|y^-|^2\right]}, \tag{14}$$

where $|y^+|_0^2, |y^+|_r^2$ are the fluctuation amplitudes at the atmospheric base and r respectively. It follows then that if the perturbation amplitude is non vanishing at the base but tends to 0 at great distances, the flow is unstable. Now for $\omega = 0$ and large r Equations (12) have, as may be easily verified, leading order asymptotic solutions for breeze velocity profiles,

$$y^+ \sim \frac{e^{\mp\gamma r}}{r}\left(1 \pm \frac{1}{2\gamma r}\right), \qquad y^- \sim \pm\frac{e^{\mp\gamma r}}{2\gamma r^2} \tag{15}$$

so the boundary conditions are satisfied either by the first solution, if γ is positive, or the second, if γ is negative. But in both cases the amplitudes tend to zero at great distances, and the numerator and denominator of Equation (14) are always positive, so that the only possibility is that γ must also be positive, and therefore breeze solutions must be unstable.

Numerical solutions show that this is indeed the case: in Figure 2a we plot the growth rate of the instability as a function of base Mach number. The growth rate is largest for high values of the base Mach numbers but both the static atmosphere ($M_0 = 0$) and the critical breeze ($M_0 = M_*$) are marginally stable. In the latter case the perturbation equations become singular at the sonic point, because the phase speed of the inward propagating wave vanishes there: an additional regularity condition must be imposed in the stationary equations, effectively isolating the region below the sonic point from the region beyond it. The presence of this additional boundary condition is the mathematical reason behind the stability of flows with a continuous subsonic/supersonic transition.

The reason breezes must be unstable lies in the unfavourable stratification, a consequence of mass continuity and the asymptotic velocity decrease, which cause

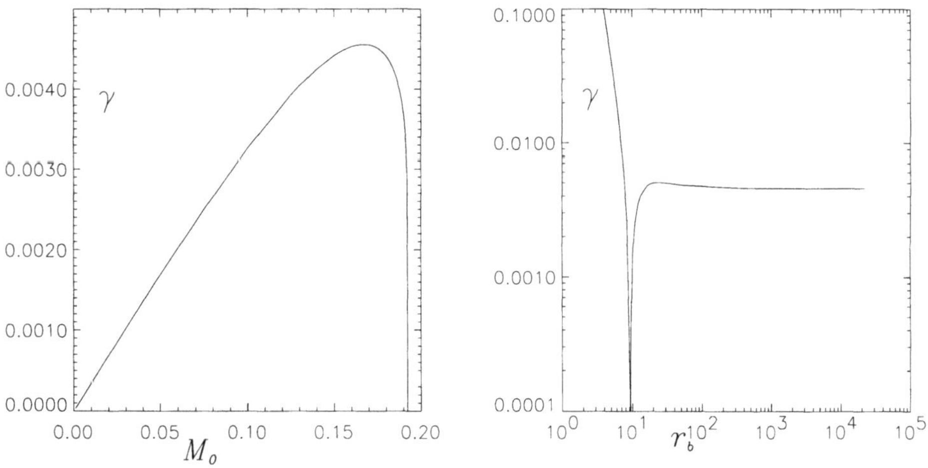

Figure 2. Left: Growth rate γ of breeze instability as a function of base Mach number ($g/c^2 = 5.0$). Marginal stability is obtained for $M_0 = 0$, $M_0 = M_* \simeq 0.192$. Right: Maximal growth rate for breezes (continuous line) and subsonic accretion (dotted line) as a function of the position of the outer boundary r_b. The two lines join in $r_b = r_s$, where $\gamma = 0$.

the asymptotic pressure to increase with increasing base mach number (Equation (11)). For imagine a static atmosphere in equilibrium, and let the pressure at infinity increase: an *inflow*, not an outflow, would be the expected result. On the other hand, that accretion breezes are stable follows immediately from the analysis presented above; the denominator in Equation (14) changes sign, so the only consistent way to satisfy the boundary conditions is to choose the second solution in Equation (15), implying a negative value for γ. In fact, the stationary equations are symmetrical in M, while the perturbation equations are invariant under a change in sign of both M *and* γ.

4. A Hysteresis Cycle

Given that breezes are unstable, even in the pressure range $p^c_\infty > p_\infty \geq p^s_\infty$ the only stable outflow is a supersonic wind with a terminal shock. But what happens if the pressure difference between the coronal base and the distant medium varies? As p_∞ increases, the shock moves inward, decreasing in amplitude as the critical point is approached. However, when the critical breeze is reached, there is no neighbouring outflow solution capable of sustaining a higher pressure at infinity. The only possibility for the flow is to collapse into its symmetrical ($M \rightarrow -M$) critical breeze accretion profile, which is also marginally stable. As the pressure is increased further, an accretion shock is formed *below* the sonic point, connecting the symmetrical of the downward transonic to one of the double valued curves in region IV (as shown by the curve labelled A in Figure 1). For $p_\infty > p^c_\infty$ there is a unique shocked accretion flow (McCrea, 1956), and the shock position moves inward from the critical point as the pressure is increased beyond p^c_∞ (if the pressure is too high, the shock may collapse onto the star). Consider now what happens if, starting from a shocked accretion flow, the pressure at the surface increases, or alternatively the pressure of the ISM decreases. The shock moves outwards, but this time, as p_∞ decreases below p^c_∞, the flow can evolve with continuity into subsonic accretion. As p_∞ decreases further, the accretion-breeze velocities decrease, but when p_∞ decreases below p^s_∞, the flow must accelerate again into a supersonic shocked wind.

This scenario, predicted in Velli (1994), seems to have been overlooked previously. Indeed, some numerical simulations of spherically symmetric flows were interpreted as showing how the transitions in the Parker/Bondi diagram were continuous (Korevaar, 1989). In that paper however, the author did indeed seem to notice a very strong dependence of the flow on the boundary conditions, as well as on their location. It is easy to understand the origin of the sensitivity. The stratification produced by breezes, though globally unstable, is not locally unstable everywhere; for example, below the critical point the pressure in breezes decreases with height more rapidly than in the static case. Inspection of Equation (7) actually shows that this is true out to the radius r_s where the Mach number of the flow has decreased to the same level as the base Mach number, which may be calculated by imposing $M = M_0$ in the adimensional version of Equation (7), i.e. $2\log r_s + g/r_s - g = 0$. This equation is independent of base mach number, which also means that at this height the pressure is the same for all breezes, while below this radius, the pressure at a given height is a monotonically *decreasing* function of base Mach number. As the the boundary conditions are imposed at closer and closer distances r_b the growth rate of the instability is reduced, and marginal stability is obtained when $r_b = r_s$. Imposing boundary conditions below this radius stabilizes the breezes, but consequently destabilizes subsonic accretion, as is shown in Figure 2b, where the maximal growth rate for breezes (continuous line) and accretion (dashed line) as a function of r_b is plotted. For large values of g this marginal stability radius depends exponentially on g as $r_s \simeq \exp(g/2) - g/2$.

When the boundary is at $r_b < r_s$ the equilibrium flows still present an hysteresis-type cycle in terms of the pressure (enthalpy) jump between r_b and the coronal base ($r = 1$), but in a reversed order with respect to that previously described: supersonic accretion is blown into supersonic winds as the base pressure is increased beyond a critical value, while an outflow breeze phase exists before the collapse to accretion as the pressure at the outer boundary is increased beyond the value appropriate to a static atmosphere. It is this parameter regime which was unknowingly simulated in Korevaar (1989). This explanation of the sensitive dependence on boundary conditions also resolves the sometimes contradictory stability statements made in the litterature (contradictions remarked in Theuns and David, 1992).

Del Zanna *et al.* (1998) carried out a numerical simulation to test the full hysteresis cycle predicted by Velli (1994). The numerical domain was large enough so that the outer pressure boundary condition was imposed in the regime of unstable

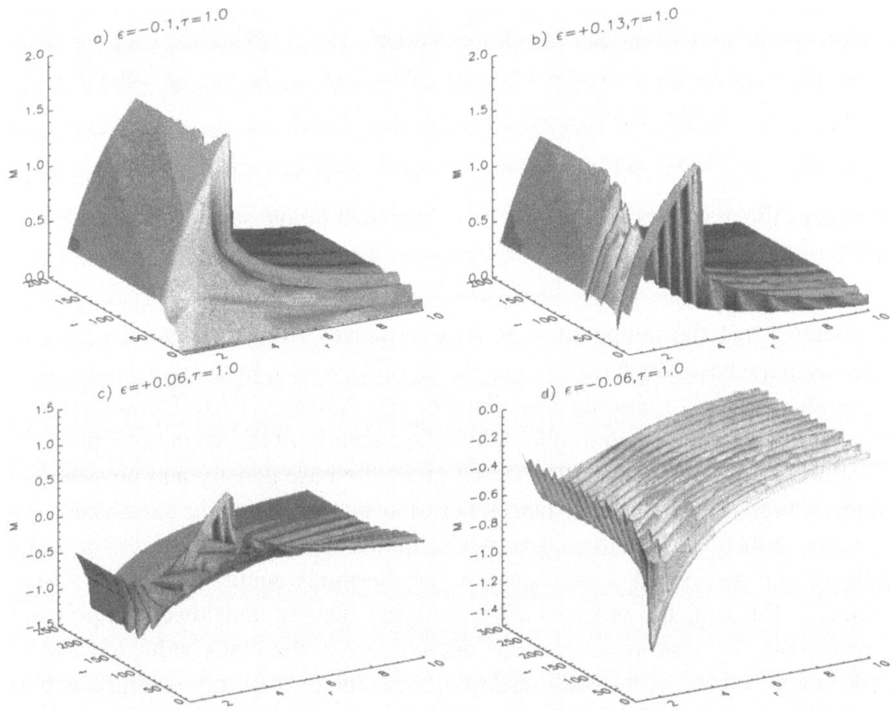

Figure 3. Hysteresis cycle in the Parker/Bondi diagram: a) from the static atmosphere to shocked wind b) the pressure p_∞ is increased to the critical value p_∞^C c) a further small increase in pressure p_∞ causes the wind to collapse into shocked accretion d) return to the conditions in b) yields a stable accretion breeze.

outflow breezes. The results are summarized in Figure 3 (a,b,c,d). First, (Figure 3a) the static atmosphere was taken as initial condition and the pressure at the outer boundary was lowered by a small amount. The solution develops into a shocked wind. Then, (Figure 3b), the pressure was increased, pushing the shock inward close to the critical point. Subsequently, (Figure 3c), the pressure was further increased slightly, collapsing the wind to supersonic shocked accretion. Finally, the outer pressure was decreased again to the value it had before the shocked wind collapsed, but this time a stable accretion breeze resulted. The reader is referred to Del Zanna *et al.* (1998) for details of the experiment and to Del Zanna and Londrillo (2000) for the numerical technique.

5. Discussion

In generalizing to polytropic or other more realistic equations for the energy, some attention is necessary since the density and pressure may fall to 0 at a finite distance, and transonic flows do not exist for all polytropic indices ($\gamma \leq 3/2$ below the sonic point is a necessary condition Parker, 1963). With these caveats, the discussion of the isothermal case is easily generalized. The energy equation now becomes (as the sound speed varies, c_0 is its base value)

$$v^2/2 + c^2/(\gamma - 1) - g/r = v_0^2/2 + c_0^2/(\gamma - 1) - g.$$

For breezes the asymptotic behavior $v \sim 1/r^2$ still holds, so that in fact we may write

$$c_\infty^2/(\gamma - 1) = v_0^2/2 + c_0^2/(\gamma - 1) - g$$

which shows that the temperature at great distances from the central object increases with the base Mach number, up to the value which, for a given base density and pressure, gives a transonic flow (Holzer and Axford, 1970). Conservation of energy across the shock then implies that independently of the asymptotic pressure, c_∞ is always the same. It is still true that given the base density and pressure, for a range of pressures at great distances between that of the static atmosphere and that of the critical breeze there are two solutions, an unstable breeze (or stable accretion) and a shocked wind, but now the thermodynamic state of the distant medium is different, the breeze having a higher density and lower temperature. Analytic work for more realistic cases, including e.g. thermal conduction and radiative energy losses, or an additional source of energy and momentum has been limited to boundary conditions in which both the temperature and the pressure tend to vanish at great distances, i.e., the case of expansion in a void (see, e.g., Roberts and Soward, 1972). In this case only transonic flows have a positive net energy flux, while because both kinetic and internal energy tend to zero breezes have a zero net energy flux. The equations being highly nonlinear, it is not clear whether in the positive energy case uniqueness of solutions exists, and this difficulty is compounded by also considering shocks in the flow. Also, thermodynamic instabilities related to the heating and cooling of the plasma (e.g. thermal instabilities,) might become important. Nonetheless, simulations such as those of Korevaar (1989) seem to show that the same hysteresis-cycle scenario holds even in situations where a realistic energy equation is integrated and the polytropic approximation is relaxed: the transition from accretion to wind and back is necessarily of an abrupt nature, and very special, and hence rare conditions are necessary to establish steady state quasi-static or subsonic flows: in other words, when it rains it pours.

Acknowledgements

I would like to thank R. Grappin, A. Mangeney, N. Meyer, J. Leorat, F. Pantellini, D. Hubert, C. Lacombe, for many interesting discussions. More in general, I would like to thank DESPA and the Observatory of Paris for having provided an intense and stimulating atmosphere during my post-doc years. Finally, I would like to thank J.-L. Steinberg, with whom conversations on all topics continue to be most enriching.

References

Bondi, H.: 1952, *Mon. Not. R. Astron. Soc.* **112**, 195.

Del Zanna, L., Velli, M. and Londrillo, P.: 1998, Dynamical response of a stellar atmosphere to pressure perturbations: numerical simulations, *Astron. Astrophys.* **L13**.

Holzer, T.E. and Axford, W.I.: 1970, Theory of stellar winds and related flows, *Annu. Rev. Astron. Astrophys.* **8**, 30.

Hundhausen, A.J.: 1972, *Coronal Expansion and Solar Wind*, Springer Verlag, Berlin.

Kadomtsev, B.: 1983, *Phenomènes Collectifs dans le Plasma*, MIR, Moscow.

Korevaar, P.: 1989, Time-dependent corona models: coronae with accretion, *Astron. Astrophys.* **226**, 209.

Londrillo, P. and Del Zanna, L.: 2000, High order upwind schemes for multidimensional magnetohydrodynamics, *Astrophys. J.* **530**, 508.

McCrea, W.: 1956, Shock Waves in Steady Radial Motion Under Gravity, *Astrophys. J.* **124**, 461.

Parker, E.N.: 1958. *Astrophys. J.* **128**, 664.

Parker, E.: 1963, *Interplanetary Dynamical Processes*, Interscience, New York.

Roberts, P.H. and Soward, A.M.: 1972, Stellar Winds and Breezes, *Proc. Roy. Soc. London* **328** A, 185.

Theuns, T. and David, M.: 1992, Spherically symmetric polytropic flow, *Astrophys. J.* **384**, 587.

Velli, M.: 1993, On the propagation of Alfvén waves in radially stratified atmospheres and winds, *Astron. Astrophys.* **270**, 304.

Velli, M.: 1994, From supersonic winds to accretion: comments on the stability of stellar winds and related flows, *Astrophys. J.* **43**, L55.

KINETIC MODELS OF SOLAR AND POLAR WINDS

JOSEPH LEMAIRE and VIVIANE PIERRARD
Institut d'Aéronomie Spatiale de Belgique, Brussels, Belgium

Abstract. The study of the solar corona has been strewn with great discoveries, surprises and controversies. The major steps since van de Hulst's (1953) and Chapman's (1957) early hydrostatic models of the extended corona, until the most recent generations of kinetic models of the coronal expansion and of the supersonic solar wind flows, are presented. These models are compared to in-situ observations. Progress in polar wind models went through a somewhat similar evolution that is outlined also. The advantages and limitations of the successive brands of solar wind and polar wind models are considered.

1. Introduction

In this paper, a brief historical review of the theoretical models proposed to explain the main observational features of the solar and the polar wind, is given. We show how the solar wind and polar wind models have followed rather parallel development. This review is non exhaustive due to the limited space but just recounts the main steps in the discoveries and model development. In the often quoted review paper of Lemaire and Scherer (1973), it was shown that solar and polar wind models have followed similar initial evolution. Starting with hydrodynamic models of increasing complexity, exospheric kinetic models have completed the description of the phenomenon in the collisionless regions. Models based on the solution of the Fokker-Planck equation were recently developed to study the wind in any collisional regimes.

Although current theoretical models of the radial expansion of the solar corona have improved step by step to provide a more or less comprehensive description of the solar wind, the heating mechanism of the solar corona remains controversial. Indeed, from a thermodynamical point of view, it is difficult to conceive that the corona is so much warmer (10^6 K) than the lower layers of the chromosphere and photosphere. These high coronal temperatures were eventually confirmed by Edlèn, who identified in 1942 coronal red and green lines in the emission spectrum of highly ionized iron ions.

The large flow velocities observed in the fast solar wind are still a matter of debate. Indeed, in all current models, the velocity of the particles is related to their temperatures in the corona. But the fast solar wind originates from coronal holes, where, at least the electron temperature is lower than in the equatorial region of the corona. This leads most people to believe that an additional acceleration

Astrophysics and Space Science **277:** 169–180, 2001.

mechanism is responsible for the fast speed solar wind. Heating by wave-particle interactions has been proposed so far and recently a promising alternative: the velocity filtration process.

2. Models of the Solar Corona

2.1. First Hydrostatic Models

The first models considered that the solar corona is in isothermal hydrostatic equilibrium (van de Hulst, 1953). From the determination of the scale height of the electron number densities observed during solar eclipses, Alfvén (1941) deduced that the coronal temperature has to be higher than 10^6 K, assuming the solar corona was in hydrostatic equilibrium.

In 1957, Chapman developed also an hydrostatic model of the solar corona extending beyond 1 AU. Assuming that heat is mainly transported by conduction, he was able to predict the temperature profiles at large radial distances of the corona. Due to the $T^{5/2}$ temperature dependence of the thermal conductivity of the hydrogen plasma, he found that the temperature should then decrease with r, the radial distance, as

$$T = T_0(r_0/r)^{2/7}.$$

But in such a model, the temperature gradient becomes superadiabatic at 34 R_s and the number density reaches a minimum value at 123 R_s. Such an hydrostatic model of solar atmosphere is indeed convectively unstable. Furthermore, Lemaire (1969) showed that turbulent convection alla Bohm-Vitense (1953) is not much more efficient than thermal conduction to evacuate the heat deposited at the base of the corona. He came to the conclusion that it is only by steady state radial expansion or CME like explosions that this excess heat can be most efficiently evacuated out of the corona into interplanetary space. This result constitutes an additional physical argument in support of a radial expansion of the solar corona. It complements therefore the conclusion obtained much earlier by Parker (1958) that a continuous radial expansion of the corona is required instead of hydrostatic equilibrium, because the latter models all predicted too high kinetic pressure at infinity. Consequently, hydrostatic equilibrium had to be abandoned, not only because of mechanical constraints (too large hydrostatic pressure at infinity compared to the pressure observed in the interstellar medium), but also because of thermal constraints (convection being the most efficient mechanism to transport heat away from the base of the corona).

2.2. First Hydrodynamic Models

Parker (1958) developed first an isothermal hydrodynamic model based on the classical Euler hydrodynamic equations assuming the pressure tensor is isotropic presumably not due to the effect of Coulomb collisions but as a consequence of many postulated wave-particle interactions. Although not justified nor supported by any observational evidence in 1958, this assumption has been widely accepted and became an unquestionable paradigm for almost a quarter of century in space plasma physics. Parker (1958) selected the critical solution of the momentum equation (and its Euler hydrodynamic approximation) to obtain a sufficiently small isotropic pressure at large heliocentric distances. In this model, the bulk velocity of the solar wind increased from subsonic values in the low corona to a supersonic speed at 1 AU, as it was confirmed by MARINER-2 plasma measurements (Neugebauer and Snyder, 1966). Bulk velocities around 400 km/s at 1 AU, corresponding to the slow speed solar wind, were obtained. But to reach bulk velocities > 700 km s^{-1} at 1 AU comparable to those observed in the high speed solar wind, temperatures higher than those expected to exist in the coronal holes (around 2×10^6 K) would be needed in Parker's one fluid hydrodynamic models. Since this first brand of steady state hydrodynamic models (Euler approximation), many more elaborated kinds of hydrodynamic codes have been produced (Navier Stokes, 13 or 20 moments approximations, multifluid models, ...). In particular, wave-particle interactions were introduced to transfer momentum and accelerate the fast solar wind or to add extra heat to the corona. But there is not yet clear evidence that there is enough power in the postulated wave spectrum to accelerate the solar wind plasma up to 600–900 km s^{-1}.

The classical hydrodynamic equations used in the fluid models are derived under the assumption that the velocity distribution function of the particles is steady state and close to an isotropic displaced Maxwellian distribution (at zero order approximation); furthermore, to justify the $T^{5/2}$ dependence of the thermal conductivity in the energy transport equation, it has to be assumed implicitly that λ, the Coulomb collision mean free path of the ions and electrons, is small compared to the density scale height $H = -(d \ln n/dr)^{-1}$. Nevertheless, Hundhausen (1968) calculated that $\lambda > H$ above 7 solar radii, and he deduced that above this level, called exobase, a kinetic approach should be in order. Moreover, he noted that the solar wind protons and ions should be collisionless approximately beyond 15 solar radii in order to explain the significant anisotropy of the observed proton and ions velocity distribution functions.

Brasseur and Lemaire (1977) indicated how hydrodynamic models from the exobase can be fitted accross this surface to the exospheric models outlined in the next section.

2.3. First kinetic exospheric models

Chamberlain (1960) developed the first kinetic model of the solar wind. Kinetic models are obtained by solving the equations of evolution of the particles velocity distribution functions (VDFs). Chamberlain developed an exospheric model based on a simple solution of the Vlasov equation, considering that the plasma has a truncated displaced Maxwellian VDF at a reference altitude corresponding to the exobase where $\lambda = H$. Above this altitude, the plasma is assumed completely collisionless; it is confined by the gravitational and polarization electric field. Only particles with a velocity exceeding the critical escape speed contribute to a net evaporation flux. Assuming a Pannekoek-Rosseland electric potential distribution (Pannekoek, 1922; Rosseland, 1924), he calculated the density, bulk velocity and average temperature of the ballistic and escaping particules as a function of the heliospheric radial distance. In this solar breeze model, the bulk velocity at 1 AU is only 20 km s^{-1} and vanishes at the infinity as in hydrostatic models. Since the supersonic speed predicted by Parker's hydrodynamic models was confirmed by in-situ solar wind observations, Chamberlain's model was disregarded as an academic curiosity by the MHD community.

Furthermore, this community gained the false believe that the calculated density, bulk velocity and temperature distributions of this first exospheric model of the corona as well as the later ones outlined below, did not satisfy the moment equations from which the various hydrodynamical approximations of the transport equations are derived. Of course, this is a wrong impression and damaging misinterpretation, since all moments of any appropriate VDF necessarily satisfy the whole hierarchy of moments equations. This has been carefully checked to test the validity of all our numerical codes calculating exospheric density, bulk speed, temperature and energy flux distributions.

2.4. Maxwellian kinetic exospheric model

Nevertheless, the interest for kinetic exospheric description revived when Jockers (1970) and Lemaire and Scherer (1971) noted that the Pannekoek-Rosseland electric potential distribution postulated in the solar breeze as well as in polar breeze models, is not applicable when there is a net flux of escaping particles, i.e., when the plasma distribution is not precisely in hydrostatic equilibrium. Indeed, for such an electric potential distribution, the flux of escaping electrons is $(m_p/m_e)^{1/2} = 43$ times larger than the escaping flux of protons. Lemaire and Scherer (1972) calculated the electric potential difference between the exobase level and infinity that is necessary to warrant the equality of outward fluxes for the electrons and protons, i.e., zero net electric current. Local quasi-neutrality was used to determine the distribution of the potential at all altitudes above the exobase. With a truncated maxwellian at the exobase, they found that the zero electric current condition requires an electrostatic potential drop of 690 V, much larger than the 150 V corresponding to the Pannekoek-Rosseland potential. They showed that

TABLE I

Comparison between measurements and models of the solar wind for the number density, bulk velocity, parallel and perpendicular temperatures, temperature anisotropies, energy flux and heat conduction flux. During quiet solar wind conditions, the observations are taken from Hundhausen (1968) at 1 AU and theoretical results are obtained with the Lemaire and Scherer's kinetic exospheric model for the slow speed solar wind (Lemaire and Scherer, 1971); the conditions imposed at the exobase $r_0 = 6.6$ R_s are: $n_e(r_0) = n_p(r_0) = 3.1 \times 10^{10}$ m^{-3}, $T_e(r_0) = 1.52 \times 10^6$ K and $T_p(r_0) = 9.84 \times 10^5$ K. For the high speed solar wind, the observations are made by Helios-1/2 (Maksimovic, 1995); the theoretical results are obtained with the Lorentzian kinetic exospheric model with $\kappa = 2$ and $r_0 = 6.4$ R_s, $n_e(r_0) = n_p(r_0) = 3.2 \times 10^{10}$ m^{-3}, $T_e(r_0) = 1.5 \times 10^6$ K and $T_p(r_0) = 10^6$ K (Maksimovic *et al.*, 1997b)

	Slow wind Observations (Hundhausen)	Exospheric Maxwellian model (LS)	Fast wind Observations (Helios-1/2)	Exospheric Lorentzian model (MPL)
Bulk velocity (km s^{-1})	320	320	667	667
Number density (cm^{-3})	5.4	7.18	3	2.7
Protons temperature (K)	4.8×10^4	4.8×10^4	2.8×10^5	1.22×10^4
Electrons temperature (K)	1.1×10^5	1.17×10^5	1.3×10^5	1.34×10^6
Anisotr. protons $T_{p\parallel}/T_{p\perp}$	3.4	164	1.2	46
Anisotr. electrons $T_{e\parallel}/T_{e\perp}$	1.2	3.05	1.2	4.4
Energy flux (erg cm^{-2} sec^{-1})	0.24	0.20		
Heat cond. flux (erg cm^{-2} sec^{-1})	1×10^{-2}	5.1×10^{-2}		

this larger electrostatic potential difference accelerates the protons to supersonic velocities and enhances the critical escape speed of the electrons, limiting therefore their evaporation flux (see Maksimovic *et al.* (2000b) in the present issue for a more detailed discussion). Considering realistic conditions at the exobase taken at 6.6 R_s for the protons and the electrons, Lemaire and Scherer's kinetic model predicts radial profiles for the number density, bulk velocity, temperature and heat flux of the solar wind electrons and protons. Speeds around 300 km s^{-1} are obtained at 1 AU in their exospheric model. Table I shows the values obtained with a maxwellian model at 1 AU compared with observations collected by Hundhausen (1968) during quiet solar wind conditions.

Satisfactory agreement is obtained between the results of the model and the average slow solar wind observations, except for temperature anisotropies which are too large in the theoretical model. The excessive temperature anisotropies are the consequence of the simplifying assumption that the particles are completely collisionless. In the discussion of their paper, Lemaire and Scherer (1972) argue that pitch angle scattering by Coulomb collisions would be able to reduce significantly the too large temperature anisotropies without changing however the average energies and mean temperatures of the electrons and protons. Moreover, Chen *et al.* (1972) and Pierrard *et al.* (2001) have shown that the introduction of the spiral solar magnetic field (instead of the radial B-field assumed in Lemaire and Scherer's models) significantly reduces the proton anisotropy.

Although the second generation of exospheric models with a more appropriate electrostatic field distribution happens to be rather adequate to reproduce the observed slow solar wind properties (see Table I and Figure 12 in Lemaire and Scherer, 1973), nevertheless, the exospheric models described above are unable to account for the fast solar wind, which is characterized by velocities higher than 500 km/s at 1 AU. To reach such velocities, exobase temperatures larger than 2×10^6 K would have to be postulated in these exospheric models, as it was the case also for Parker's hydrodynamic models. Moreover, it was found that fast speed streams originate from coronal holes at high heliospheric latitudes where the electron coronal temperature is lower than in the equatorial source region of the slow wind. The slow and fast solar winds seem to be two different flow regimes driven by different physical mechanisms. The same conclusion applies for the high latitude polar wind and the slower upward ionization flows at more equatorial latitudes in the topside ionosphere of the Earth.

2.5. Lorentzian Exospheric Model

In 1992, Scudder (a, b) suggested a new physical mechanism to explain the high temperature observed in the corona without invoking dissipation of energy or momentum by wave-particle interactions in the inner corona. He called this mechanism 'velocity filtration effect'. It implies the assumption that the velocity distribution function of the electrons has an enhanced population of suprathermal electron at the base of the corona. Lorentzian distributions are characterized by enhanced suprathermal tails with a phase space density decreasing as a power law of v^2 instead of $e^{-\beta v^2}$ when $v \to \infty$. The slope of the tail is determined by the value of an index kappa; when $\kappa \to \infty$, one recovers the isotropic Maxwellian VDF.

In the fast solar wind, the observed electron distributions are characterized by enhanced high velocity tails. Their energy spectrum is harder than in the slow solar wind. They can be fitted by Lorentzian (or kappa) functions with a kappa index between $\kappa = 2$ and 5 (Maksimovic *et al.*, 1997a). The characteristic suprathermal electron tails are known as the halo population.

Following Scudder's work, Pierrard and Lemaire (1996) developed a generalized Lorentzian ion-exosphere model, which was adapted for the solar wind (Maksimovic *et al.*, 1997b, 2000b). This new family of exospheric models shows that suprathermal tails increase the electric potential difference between the exobase and infinity. Thus, larger bulk velocities are obtained at 1 AU by reducing the value of kappa without unreasonably large coronal temperatures and without additional heating process in the inner or outer region of the corona (see Table I). The evaporation of the electrons is very sensitive to the value of the index κ. Nevertheless, the temperatures obtained with this model remain too low for the protons.

Predictions of asymptotic electron temperature at large radial distances have been deduced recently for exospheric models (Meyer-Vernet and Issautier, 1998). These authors established that the electron temperature radial profile is equal to the sum of a term $\propto r^{-4/3}$ plus a constant, with both terms being of the same order of magnitude at 1 AU. Interestingly enough, this characteristic asymptotic behavior is independent of the velocity distributions assumed at the exobase deep in the corona. This characteristic electron temperature profile precisely corresponds to the observations of $T_e(r)$ as determined from the plasma wave experiment onboard of Ulysses (Issautier *et al.*, 1998; Maksimovic *et al.*, 2000a). We do not think that any hydrodynamic model of the solar wind has been able to predict this observed characteristic temperature distribution for the solar wind electrons.

2.6. Fokker-Planck Kinetic Models

Analytic exospheric models are useful but they assume that the plasma of the solar wind is completely collisionless above the exobase. Nevertheless, Coulomb collisions still have some effects above this level. Since the Coulomb cross section is inversely proportional to the fourth power of the relative velocity between the colliding particles, the exobase for a given particle decreases with its energy. Therefore, no unique exobase corresponding to the mean thermal velocity of particles should possibly be used. On the other hand, the assumption that the plasma is collision-dominated, as implicitly assumed in fluid models, is also difficult to justify since the mean free path of the particles becomes larger than the scale height above 7 R_S. Furthermore, particles with velocities larger than the mean thermal velocity become collisionless even at lower altitudes.

At this step, we tentatively ignore the effects of wave-particle interactions. The gyroradii of the solar wind electrons and protons are much smaller than the scale lengths of presumed DC magnetic irregularities in the interplanetary medium. Therefore, these irregularities in the IMF are not likely to thermalize nor scatter the pitch angles of the solar wind electrons and protons. Of course, this is not the case for the much more energetic solar and galactic cosmic ray particles whose pitch angles are more easily scattered by the irregularities in the interplanetary magnetic field.

The classical hydrodynamic description and the kinetic exospheric approach are complementary, they should be applied in collision-dominated region and in collisionless region respectively. But the solar wind is neither a purely collision-dominated medium, nor a purely collisionless one. Therefore, to study the solar wind from very low radial distances where the plasma is dominated by collisions to very large radial distances where the solar wind becomes almost collisionless, it is required to solve the Fokker-Planck equation. In this equation, the effects of external forces (gravitational, electric and Lorentz forces) are taken into account, as well as the Coulomb collisions between the particles.

Lie-Svendsen *et al.* (1997) developed a kinetic model of the solar wind that is a solution of the Fokker-Planck equation. In order to simulate the density distribution of the background particles, they adopted a radial density profile taken from an hydrodynamic solar wind model. Two boundary conditions were imposed, one at the base of the corona in the collision-dominated region, and one at large radial distances well above the exobase. They solved this equation using a finite difference numerical method. Pierrard *et al.* (1999) developed a spectral method to solve the Fokker-Planck equation. To avoid singular (unphysical) solutions, they imposed regularity conditions to the electron VDF at the boundaries of the integration domain. Typical electron velocity distribution functions observed at 1 AU by the instrument 3DP on WIND are used as boundary conditions to determine the velocity distribution function of the electrons at 4 solar radii in the corona. Pierrard *et al.* (1999, 2000) found that suprathermal tails have to be present in the velocity distribution function of the electrons already at low altitude in the corona in order to match the observed distributions at 1 AU. But deeper into the solar corona, the relative number density of these halo electrons forming these non-Maxwellian tails becomes negligibly small compared to that of the core electrons.

3. Models of the Polar Winds

3.1. First Hydrodynamic Models

The polar wind is the escape of thermal ions out of the terrestrial polar ionosphere along the open magnetotail field lines. It is a phenomenon similar to the solar wind and the theoretical studies of the polar wind have followed quite the same historical evolution.

Nicolet (1961) first pointed out that the thermal escape of neutral helium cannot support the large production of He^4. As the photoionization rate of He^4 is nearly equal to its production rate, Nicolet argued that terrestrial helium should escape in its ionized state. Banks and Holzer (1968) developed an hydrodynamic model of the outward-flowing plasma escaping from the Earth, called polar wind at the suggestion of Axford (1968). The model is based on the integration of the hydrodynamic transport equations (Euler's hydrodynamic approximation) for an isothermal temperature distribution in the topside polar ionosphere. Oxygen and helium ions are produced by photoionization and hydrogen ions by charge exchange between O^+ and H. Among the solutions of their transport equations, Banks and Holzer selected the critical solution which provides supersonic bulk speed at high altitude, since this is the only solution for which kinetic pressure tends to zero at large distances.

3.2. First exospheric kinetic models

A controversy started similar to that between Parker, the father of hydrodynamic solar wind models, and Chamberlain, challenging these kind of models in favor of kinetic ones. Indeed, the Coulomb collision frequency throughout the topside polar ionosphere is 10 times too weak for the hydrodynamic approximation to be valid above the exobase altitude. Dessler and Cloutier (1969) proposed an exospheric model in which the collisions between particles are neglected above the exobase located at an altitude of 1000 km. They also assumed a Pannekoek-Rosseland electic field and obtained a kinetic model called the polar breeze model. Lemaire and Scherer (1970) calculated a more appropriate electrostatic potential distribution satisfying everywhere the quasi-neutrality and the zero electric current conditions. The thermal electrons and oxygen ions are decelerated by the sum of the electric force and gravitational force; but the protons are accelerated outward by the large electric force and they reach supersonic velocities of 20 km s^{-1} at large radial distances.

Since lorentzian distributions are generally observed in space plasmas, Pierrard and Lemaire (1996) developed a lorentzian exospheric model. This new kind of model proved to be very useful to account for larger supersonic polar wind bulk velocities, but also to account for the positive gradients observed in the plasmaspheric temperatures. It has been applied also to determine a generalized current-voltage relation in magnetospheric and auroral flux tubes (Pierrard, 1996).

3.3. Fokker-Planck kinetic models

The transformation of the velocity distribution function in the transition region between the collision-dominated regime at low altitudes and the collisionless regime at high altitudes has been studied by solving the Fokker-Planck equation (Lie-Svendsen and Rees, 1996; Pierrard and Lemaire, 1998). In these more comprehensive kinetic models, the hydrogen ions are accelerated upward to supersonic velocities through a background of oxygen ions, like in the earlier exospheric polar wind models. The proton VDF is almost isotropic and maxwellian at low altitudes and becomes highly anisotropic at high altitudes. In the transition region, a double hump velocity distribution function is obtained and the shape of the VDF for upward moving suprathermal protons becomes like a kidney bean at very high altitudes.

These results support the overall characteristics of VDF assumed in exospheric models where subthermal protons do not contribute to the net flux of particles nor to the outflow of energy. Only the suprathermal protons contribute to these fluxes in the polar wind. These mathematical solutions of the Fokker-Planck equation are in good agreement with results obtained by Direct Monte Carlo Simulations (Barakat *et al.*, 1995). On the contrary, the solution of Grad's moment equations obtained by Lie-Svendsen and Rees (1996) corresponds to VDFs which are close to displaced

Maxwellians all the way through the transition region into the collisionless part of the polar wind.

4. Conclusions

Solar wind and polar wind theories and model approximations have followed quite the similar development over a period of 30 years. Hydrodynamic models and exospheric models are appropriate for different collisional regimes. From kinetic models, we have learned that the acceleration of the protons forming these winds is due to an internal electric field. The resulting electric potential distribution is induced between the exobase level and infinity by thermoelectric effects, to warrant the zero current condition and the quasi-neutrality of the plasma. In hydrodynamic models, the zero pressure condition at infinity is generally used to determine the choice of the critical solution which then becomes supersonic beyond a critical point of singularity of a particular approximation of the transport equations. This explosive type of solution is also that for which the energy deposited at the base of the corona is most efficiently evacuated to the outer space.

In order to close the system of moment equations in the Chapman-Enskog's or Grad's theory of non-uniform gases, the VDF is assumed to be a displaced Maxwellian at zero order approximation with first order corrections depending linearly on the pressure tensor components and on the heat flux, as in classical hydrodynamics where collisions between particles are dominant. These assumptions fail however to be valid when the Knudsen number (i.e. the smallness parameter) becomes of the order or larger than unity. When the plasma is dominated by collisions, the Knudsen number is small. It is then reasonable to assume that the velocity distribution function of the particles is close to a displaced Maxwellian. But when the plasma becomes collisionless, the velocity distribution function of the particles can become quite different from a displaced Maxwellian. This is indeed confirmed by solar wind observations for over 30 years; therefore, it is necessary to use a proper kinetic theory. The polynomial expansion of the VDF developed by Pierrard (1997) or the more restricted expansions proposed by Leblanc and Hubert (1997) can indeed fit almost any of the complex spectral and pitch angle distributions of electrons observed at 1 AU in the solar wind. The Grad's 20 moments approximations, although more flexible than the Chapman-Enskog expansion, cannot be expected to fit the most complex VDF observed in the interplanetary medium where Coulomb collisions are not frequent enough to maintain any almost isotropic or displaced Maxwellian distribution.

Exospheric models yield too high temperature anisotropies. Collisional kinetic models based on the Fokker-Planck equation are applicable in the collision-dominated regions as well as in the collisionless regions. The addition of non-local Coulomb collisions or other pitch angle scattering and energization mechanisms might eventually be needed to fit the solutions of generalized Fokker-Planck equations to the observed VDFs, but unless we have comprehensive statistical models of the wave distributions in the corona, interplanetary medium and topside polar ionosphere, we better investigate first how much can be achieved with Coulomb collisions alone, and stay away from wild guesses and hand waving argument.

Acknowledgements

V. Pierrard acknowledges the FNRS for the grant of 'chargée de recherches' (A4/5-DM-CR96). The authors thank Dr M. Maksimovic for his remarks.

References

Alfvén, H.: 1941, *Ark. F. Math., Astr. o. fysik* **27A**, 25.
Axford, W.I.: 1968, *J. Geophys. Res.* **73**, 6855.
Banks, P.M. and Holzer, T.E.: 1968, *J. Geophys. Res.* **73**, 6846.
Barakat, A.R., Barghouthi, I.A. and Schunk, R.W.: 1995, *Geophys. Res. Let.* **22**, 1857.
Brasseur, G. and Lemaire, J.: 1977, *Planet. Space Sci.* **25**, 201.
Chamberlain, J.W.: 1960, *Astrophys. J.* **131**, 47.
Chapman, S.: 1957, *Smithson. Contrib. Astrophys.* **2**, 1.
Chen, W.M., Lai, C.S., Lin, H.E. and Lin, W.C.: 1972, *J. Geophys. Res.* **77**, 1.
Dessler, A.J. and Cloutier, P.A.: 1969, *J. Geophys. Res.* **74**, 3730.
Hundhausen, A.J.: 1968, *Space Sci. Rev.* **8**, 690.
Issautier, K., Meyer-Vernet, N., Moncuquet, M. and Hoang, S.: 1998, *J. Geophys. Res.* **103**, 1969.
Jockers, K.: 1970, *Astron. Astrophys.* **6**, 219.
Leblanc, F. and Hubert, D.: 1997, *Astrophys. J.* **483**, 464.
Lemaire, J.: 1969, PhD thesis, Univ. de Liège, Physics PhD.
Lemaire, J.: 1971, *paper presented at 1st European Earth and Planetary Physics Colloquium*, Reading, England.
Lemaire, J. and Scherer, M.: 1970, *Planet. Space Sci.* **18**, 103.
Lemaire, J. and Scherer, M.: 1971, *Phys. Fluids* **14**, 8, 1683.
Lemaire, J. and Scherer, M.: 1972, *Bull. Cl. Sci. Acad. Royale Belg.* **58**, 1112.
Lemaire, J. and Scherer, M.: 1973, *Rev. Geophys. Space Phys.* **2**, 427.
Lie-Svendsen, O., Hansteen, V.H. and Leer, E.: 1997, *J. Geophys. Res.* **102**, 4701.
Lie-Svendsen, O. and Rees, M.H.: 1996, *J. Geophys. Res.* **101**, A2, 2415.
Maksimovic, M.: 1995, *PhD thesis*, University of Paris VII-Denis Diderot, Physics.
Maksimovic, M., Gary, S.P. and Skoug, R.M.: 2000a, *J. Geophys. Res.*, in press.
Maksimovic, M., Pierrard, V. and Lemaire, J.: 1997b, *Astron. Astrophys.* **324**, 725.
Maksimovic, M., Pierrard, V. and Lemaire, J.: 2000b, *Astrophys. Space Sci.*, this issue.
Maksimovic, M., Pierrard, V. and Riley, P.: 1997a, *Geophys. Res. Lett.* **24**, 9, 1151.
Meyer-Vernet, N. and Issautier, K.: 1998, *J. Geophys. Res.* **103**, 29705.
Neugebauer, M. and Snyder, C.W.: 1966, *J. Geophys. Res.* **71**, 4469.
Nicolet, M.: 1961, *J. Geophys. Res.* **66**, 2263.
Pannekoek, A.: 1922, *Bull. Astron. Inst. Meth.* **1**, 107.
Parker, E.N.: 1958, *Astrophys. J.* **128**, 664.
Pierrard, V.: 1996, *J. Geophys. Res.* **101**, 2669.
Pierrard, V.: 1997, *PhD Thesis*, UCL, 1997. Physics PhD.

Pierrard, V., Issautier, K., Meyer-Vernet, N. and Lemaire, J.: 2001, *Geophys. Res. Letters* **28**, 223.
Pierrard, V. and Lemaire, J.: 1996, *J. Geophys. Res.* **101**, 7923.
Pierrard, V. and Lemaire, J.: 1998, *J. Geophys. Res.* **103**, 11701.
Pierrard, V., Maksimovic, M. and Lemaire, J.: 1999, *J. Geophys. Res.* **104**, 17021.
Pierrard, V., Maksimovic, M. and Lemaire, J.: 2000, *Astrophys. Space Sci.*, this volume.
Rosseland, S.: 1924, *Mon. Not. R. Astron. Soc.* **84**, 720.
Scudder, J.D.: 1992a, *Astrophys. J.* **398**, 299.
Scudder, J.D.: 1992b, *Astrophys. J.* **398**, 319.
van de Hulst, H.C.: 1953, in: G.P. Kuiper (ed.), *The Sun* **307**, University of Chicago Press, Chicago.
Vitense, E.: 1953, *Zeitsch. Astrphysik* **32**, 135.

ON THE EXOSPHERIC APPROACH FOR THE SOLAR WIND ACCELERATION

MILAN MAKSIMOVIC[1], VIVIANE PIERRARD[2] and JOSEPH LEMAIRE[2]
[1] *DESPA, Observatoire de Paris-Meudon, France*
[2] *Institut d'Aéronomie Spatiale de Belgique, Brussels, Belgium*

Abstract. We present the basics of the exospheric models of the solar wind acceleration. In these models the plasma is assumed fully collisionless above a typical altitude in the corona. The solar wind is accelerated by the interplanetary electrostatic potential which is needed to warrant the equality of the proton and electron fluxes. These models suggest that the fast wind emanating from the polar regions could be due to the presence of non-thermal electron distributions in the corona.

1. Introduction

An important assumption inherent to the hydrodynamic solar wind models (Parker, 1958) is that the Coulomb collisional mean free path λ of the particles is much smaller than the density scale height H, i.e. that the plasma is dominated by collisions. Therefore the hydrodynamic approach implies that the particles velocity distribution functions (VDFs) are rather close to a Maxwellian. The observed solar wind electron VDFs depart from nearly isotropic maxwellian (Feldman *et al.*, 1975; Pierrard *et al.*, in this volume), indicating the limited validity of this hypothesis.

The problem of using the hydrodynamic approach for the solar wind is that the coronal particles become rapidly collisionless with increasing height. Therefore, it is not unreasonable to assume that the plasma is completely collisionless above a given altitude called the *exobase*. This particular altitude is actually defined as the location where λ for a thermal particle is equal to H. The exobase is typically located between 2 and 10 $r_\odot$, depending on the density and temperature profiles adopted to model the corona.

In the present paper, we simply illustrate the basics of the exospheric approach and its consequences on the solar wind acceleration. For instance, the magnetic moment conservation is not taken into account here. For the complete and rigorous development of solar wind exospheric theories, the reader can refer to (Lemaire and Scherer, 1971; Maksimovic, Pierrard and Lemaire, 1997; Lemaire and Pierrard, in this volume). We recommend also the reading of the very interesting didactic paper by Meyer-Vernet (1999) that clearly illustrates, in a different perspective, the exospheric acceleration of the solar wind.

Astrophysics and Space Science **277:** 181–187, 2001.

2. The Interplanetary Electrostatic Potential

The appropriate calculation of the interplanetary electrostatic potential $\phi_E(r)$ is a key point in the exospheric models. Let us describe briefly how to determine its value $\phi_{E0} = \phi_E(r_0)$ at a typical exobase $r_0 = 5r_\odot$, assuming it is equal to zero at infinity. We assume that the electron and proton temperatures are equal at the exobase , $T_{e0} = T_{p0} = 10^6$ K, and that the plasma density at infinity is equal to zero.

Above the exobase, the collisionless plasma is confined by the gravitational and electrostatic fields. Only particles with velocities exceeding the local escape speed v^* contribute to the evaporating flux. The critical escape velocity, v^*, can be determined from the conservation of the total energy of the particles which is given by:

$$\frac{mv^2(r)}{2} + m\phi_g(r) + Ze\phi_E(r) = \frac{mv^2(r_0)}{2} + m\phi_g(r_0) + Ze\phi_E(r_0) \tag{1}$$

where m is the mass of the particle, Z its charge and $\phi_g(r) = -GM_\odot/r$ the interplanetary gravitational potential.

Since $\phi_E(\infty) = 0$, the escape velocity at the exobase is determined by:

$$v_0^* = v^*(r_0) = \left(-2\phi_g(r_0) - \frac{2Ze\phi_E(r_0)}{m}\right)^{1/2} \tag{2}$$

For electrons, gravity can be neglected at all altitudes (this can be checked at the end of the calculation). Thus, the expression of their escape velocity reduces to $v_{e0}^* \simeq \sqrt{2e\phi_{E0}/m_e}$. The escape flux of any species at the exobase $F_0 = F(r_0)$ is found by integrating the velocity distribution function at the exobase $f_0(v)$ for all the particles with $v > v_0^*$. Assuming that no particles are coming from infinity, i.e. that the electron VDF is a truncated Maxwellian (no particles with $v < -v_0^*$, i.e. $f(v < -v_0^*) = 0$) with a thermal speed $\omega_{e0} = \sqrt{2k_BT_{e0}/m_e}$, one finds

$$F_{e0} = \frac{n_{e0}}{2\sqrt{\pi}}\omega_{e0}(1 + U_{e0})e^{-U_{e0}} \tag{3}$$

where $U_{e0} = v_{e0}^{*2}/\omega_{e0}^2$ and n_{e0} is the electron number density at the exobase.

In Chamberlain's exospheric breeze model, the Pannekoek-Rosseland (*PR*) electric potential $\phi_{E0}^{PR} = -(m_p - m_e)\phi_{g0}/(2e)$ was used. This potential is obtained by imposing quasi-neutrality to a plasma in hydrostatic equilibrium. However, this is incorrect since the plasma is not in hydrostatic equilibrium. Using this expression for $\phi_E(r_0)$ in Equation (2), it comes that $v_{e0}^*/v_{p0}^* = \sqrt{m_p/m_e}$ and $U_{e0} = U_{p0}$. Applying this in Equation (3) yields $F_{e0}^*/F_{p0}^* = v_{e0}^*/v_{p0}^* = \sqrt{m_p/m_e} \simeq 42$. Thus, with a PR potential and $T_{e0} = T_{p0}$, the Sun would charge positively since the flux of electrons would be 42 times larger than that of the positive charges!

It has been first recognized by Jockers (1970) and Lemaire and Scherer (1971) that the appropriate value of ϕ_{E0} is the one which warrants the equality of the escape fluxes for the electrons and protons, i.e., that ensures a zero net electric current, in addition to charge quasi-neutrality. It can be easily demonstrated that the correct value of ϕ_{E0} is larger than ϕ_{E0}^{PR}. Its value is such that the gravity term in (2) confining the protons is smaller than the electric potential one which pushes them outwards. Thus, with such ϕ_{E0}, all the protons are in a total (gravitational + electric) potential which is repulsive, and they all escape since $v_{p0}^* = 0$. Therefore, $U_{p0} = 0$ and consequently the net escape flux of protons is $F_{p0} = n_{p0}\omega_{p0}/\sqrt{\pi}$, where n_{p0}, the proton number density at the exobase, is equal to n_{e0}. Note that compared to the expression of the electron escape flux, a factor $1/2$ is missing in the latter expression of F_{p0}. This factor is due to a density normalisation at the exobase (see Meyer-Vernet, 1999 for more details). The appropriate expression of ϕ_{E0} is then obtained by equalizing the proton and electron fluxes. It is the solution of the algebraic equation:

$$(1 + U_{e0}) \exp(-U_{e0}) = 2\sqrt{m_e/m_p}, \text{ with } U_{e0} = \frac{e\phi_{E0}}{k_B T_{e0}} \tag{4}$$

For $m_e/m_p \simeq 5.4 \times 10^{-4}$, this yields $U_{e0} \simeq 5$ and $e\phi_{E0} \simeq 5k_B T_{e0}$. With a one million Kelvin corona we obtain $\phi_{E0} \simeq 490$ V, while $\phi_{E0}^{PR} \simeq 200$ V.

Note that for lower exobase altitudes, and therefore larger gravitational potentials, the above conditions are no longer valid. In such a case, ballistic protons, for which the total energy potential is attractive, would appear and the solution of the problem would be more complex.

3. The Evolution of the Velocity Distribution Functions

Figure 1 shows cuts, in the radial direction, of the logarithm of both VDF for electrons (left hand side) and protons (right hand side). $\log(f)$ is plotted as a function of $\tilde{E}$, which is defined as the kinetic energy $E = mv^2/2$ of the particles times the sign of their velocity (positive for anti-sunward and negative for sunward directions). With this convention, a Maxwellian distribution maps as two straight lines. Two different class of solutions are displayed on Figure 1: (a) both electron and proton VDFs at r_0 are Maxwellian (solid lines) and (b) the proton VDF is still Maxwellian while the electron VDF is a non-thermal Lorentzian distribution (dashed lines) defined by Equation (5). The VDFs are displayed for three different altitudes: the exobase r_0, one AU and infinity. Note that the scales on both axis are arbitrary, this plot being shown for illustration purpose only.

3.1. The Velocity Distribution Functions at the Exobase r_0

The two bottom panels represent the electron and proton VDFs at r_0. Let us first examine the shapes of the VDFs in case (a), when both VDFs are Maxwellian. This case is represented by solid lines on Figure 1. For the electrons, the VDF is

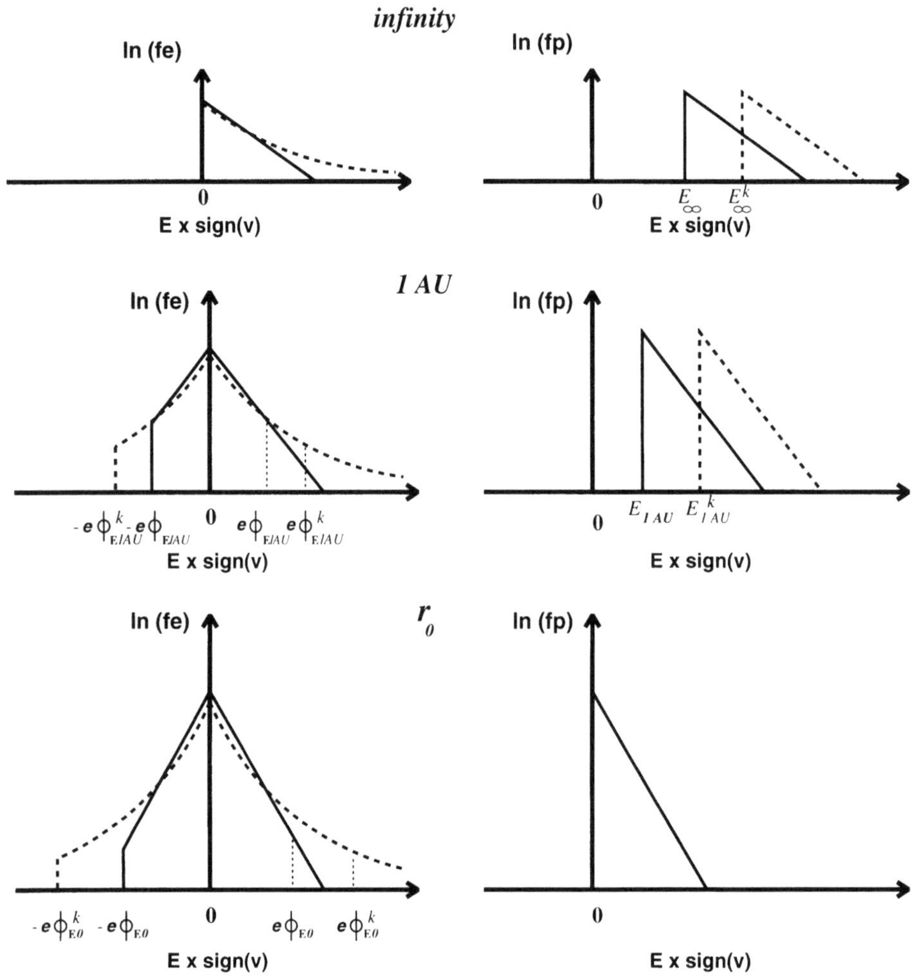

Figure 1. Radial evolution of the velocity distribution functions.

composed of two distinct populations. First the electrons with energies larger than $e\phi_{E0}$, i.e. with velocities larger than v_{e0}^*. These are the particles contributing to the total electron flux F_{e0}. These escaping particles do never come back to r_0, because we ignore collisions. Therefore, if no particles are coming from infinity, the electron VDF has no particles having $\tilde{E} < -\phi_{E0}$. The second population of electrons for which $-e\phi_{E0} < \tilde{E} < \phi_{E0}$ corresponds to particles on ballistic trajectories. These particles leave the exobase with $\tilde{E} > 0$, cannot reach $r = \infty$ since they do not have enough energy to overcome the electrostatic potential barrier. They come back to the exobase with $\tilde{E} < 0$.

Since $v_{p0}^* = 0$, there is only one population for the protons. All the protons are in a repulsive potential, they are all on escaping trajectories. Therefore, the VDF is empty for $\tilde{E} < 0$.

3.2. The Velocity Distribution Functions at 1 AU

Given the distribution functions $f_e(r_0)$ and $f_p(r_0)$ at the exobase, the distributions $f_e(r)$ and $f_p(r)$ at any distance in the exosphere are obtained by applying Liouville's theorem (Lemaire and Scherer, 1971). The moments of $f_e(r)$ and $f_p(r)$ can then be calculated analytically. These moments depend only on the radial profile of the electrostatic potential $\phi_E(r)$, which is calculated by imposing local quasi-neutrality $n_e[\phi_E(r)] = n_p[\phi_E(r)]$ (in addition to zero current flux). With the coronal conditions defined previously, we find an electrostatic potential at 1 AU which has a typical value of $\phi_{E1AU} \simeq 9$ Volts.

Since the electrons are attracted towards the Sun by the interplanetary electric field, only those particles whose energy is larger than the potential barrier $e(\phi_{E0} - \phi_{E1AU})$ are present at 1 AU. All escaping electrons at 1 AU are coming from r_0. The minimum value of their kinetic energy distribution is now reduced to ϕ_{E1AU} because of the potential barrier. But not all of the ballistic electrons from r_0 are present at 1 AU: those with an energy larger than the potential barrier $e(\phi_{E0} - \phi_{E1AU})$, which overcome 1 AU ($0 < \tilde{E} < e\phi_{E1AU}$), do not reach $r = \infty$ since they are ballistic, and fall back to the exobase ($-e\phi_{E1AU} < \tilde{E} < 0$). As at r_0, the electron VDF is truncated for $\tilde{E} < -e\phi_{E1AU}$ which would correspond to trajectories coming from infinity. Finally, note that in order to actually reach the quasi-neutrality in the exosphere a third electron population has to be taken into account: the trapped electrons. This third population appears naturally in the rigorous exospheric developments by taking into account the magnetic moment conservation (Lemaire and Scherer, 1971).

As for the electrons, all escaping protons are coming from r_0. But the minimum value of their energy distribution is larger since they are in a repulsive potential. The minimum value E_{1AU} of the energy that a proton can have at 1 AU can easily be determined from Equation (1). Neglecting the gravitational potential at 1 AU, we find $E_{1AU} = e(\phi_{E0} - \phi_{E1AU}) - GM_{\odot}m_p/r_0$. For the coronal conditions adopted above, this corresponds to a minimum velocity $v_{1AU} = (2E_{1AU}/m_p)^{1/2} \simeq 123$ km s^{-1}, which is supersonic. Finally, note that the electron and proton bulk speeds are equal since by definition the electron and proton fluxes and densities are equal at any distance in the exosphere.

3.3. The Velocity Distribution Functions at ∞

At $r = \infty$, everything is much simpler because only escaping particles are present for both electrons and protons. Since we have set $\phi_E(\infty) = 0$, the potential barrier that the escaping electrons have to overcome is ϕ_{E0}. Therefore, the minimum value of their energy distribution is zero.

The proton distribution is very similar to the VDF at 1 AU. The difference is the minimum value of the energy distribution E_∞.

4. The Effect of Non-Thermal Electron VDFs in the Corona

What happens now when the electron VDF is non-thermal in the corona, i.e., when it exhibits a suprathermal tail. This point is very important since recent SOHO observations indicate that this could indeed be the case (Esser and Edgar, 2000). Maksimovic (1995) and Maksimovic, Pierrard and Lemaire (1997) have studied in detail this possibility. These authors have assumed generalized Lorentzian electrons VDFs in the form:

$$f_{e\kappa} = \frac{n_e}{(\pi\kappa\omega_\kappa^2)^{3/2}} \frac{\Gamma(\kappa+1)}{\Gamma(\kappa-1/2)} \left[1 + \frac{v^2}{\kappa\omega_\kappa^2}\right]^{-(\kappa+1)} \tag{5}$$

where the thermal speed is defined by $\omega_\kappa = \left(\frac{2\kappa-3}{\kappa}\frac{k_B T_e}{m_e}\right)^{1/2}$. For $v^2 \gg \kappa\omega_\kappa^2$ this distribution decreases as a power law, which is much slower than the exponential decrease of a Maxwellian. Note that the Lorentzian VDF (5) tends to a Maxwellian distribution when $\kappa \to \infty$.

Let us examine the case (b) for which the VDFs on Figure 1 are represented by dashed lines. The proton VDF is still assumed Maxwellian with $T_{p0} = 10^6$ K, while a Lorentzian VDF with a typical value $\kappa = 3$ will be adopted for the electrons, keeping the same exobase electron temperature $T_{e0} = 10^6$ K. From Figure 1, it can be seen that the electron suprathermal tail is significantly enhanced (and therefore the electron escaping flux is drastically increased) compared to the Maxwellian case. As a consequence, the electric potential at the exobase ϕ_{E0} determined by the new zero net current condition must be larger than 400 Volts, the value obtained for the corresponding maxwellian VDF. ϕ_{E0}^κ is now obtained by balancing the proton flux $F_{p0} = n_{p0}\omega_{p0}/\sqrt{\pi}$ and the electron escape flux, which in the case of Lorentzian VDFs is given by:

$$F_{e0}^\kappa = \frac{n_{e0}\omega_{\kappa 0}}{2\sqrt{\pi}} \frac{\Gamma(\kappa+1)}{\Gamma(\kappa-1/2)} \frac{\kappa^{\kappa-1/2}}{\kappa-1} (1+U_{e0})(\kappa+U_{e0})^{-\kappa}, \tag{6}$$

$$\text{where } U_{e0} = \frac{v_0^{*2}}{\omega_{\kappa 0}^2} = \frac{\kappa}{\kappa-3/2} \frac{e\phi_{E0}^\kappa}{k_B T_{e0}} \tag{7}$$

With $\kappa = 3$ and the coronal temperatures as before, we obtain $\phi_{E0}^\kappa \simeq 1230$ V, which is significantly larger than the exobase potential found for the Maxwellian case. From the quasi-neutrality condition at 1 AU we obtain $\phi_{E1AU}^\kappa \simeq 40$ V.

Examining now the VDFs displayed by dashed lines in Figure 1, one notes that their radial evolution is roughly the same as in the Maxwellian case. However, the higher value of ϕ_{E0}^κ increases the terminal bulk speed by increasing the minimum energy that a proton can have at 1 AU or at $r = \infty$. We obtain $E^{\kappa}_{1AU} = e(\phi^{\kappa}_{E0} - \phi^{\kappa}_{E1AU}) - GM_{\odot}m_p/r_0 \simeq 790$ eV which corresponds to a minimum velocity at 1 AU of ≈ 390 km s^{-1}! This shows that the presence of non-thermal electron distributions in the corona considerably increases the proton flow speed at large radial distances.

Note that the acceleration effect of non-Maxwellian VDFs has been first suggested by Scudder (1992). In his model Scudder proposed that non-Maxwellian VDFs in the solar atmosphere may lead to an apparent heating of the corona through kinetic 'velocity filtration' of the distributions. Applying then the simple Parker isothermal model, Scudder showed that this apparent heating, which is proportional to the importance of the VDFs suprathermal tails, leads in turn to an increase of the solar wind bulk speed.

5. Final Remarks

The solar wind exospheric models suggest that the fast wind emanating from the polar regions might be due to the presence of non-thermal electron distributions in the corona. Although this is not the ultimate explanation for the existence of the fast wind, it contributes to reduce the gap between observations and models.

However, much progress remains to be done in kinetic modeling of the solar wind to reach a satisfactory fit between theory and observations. Besides wave-particle interactions whose importance is difficult to determine, other effects need to be incorporated in the future, like spiral interplanetary magnetic field lines (Pierrard *et al.*, 2001) or (and) non-local Coulomb collisions.

References

Esser, R. and Edgar, R.J.: 2000, *Astrophys. J.* **532**, L71–L74.
Feldman, W.C., *et al.*: 1975, *J. Geophys. Res.* **80**, 4181–4196.
Jockers, K.: 1970, *Astron. Astrophys.* **6**, 219–239.
Maksimovic, M.: 1995, *Thèse de doctorat*, Université Paris VII – Denis Diderot.
Maksimovic, M., Pierrard, V. and Lemaire, J.: 1997, *Astron. Astrophys.* **324**, 725–734.
Meyer-Vernet, N.: 1999, *Eur. J. Phys.* **20**, 167–176.
Lemaire, J. and Scherer, M.: 1971, *J. Geophys. Res.* **76**, 7479–7490.
Lemaire, J. and Pierrard, V.: in this issue.
Parker, E.N.: 1958, Dynamics of the interplanetary gas and magnetic fields, *Astrophys. J.* **128**, 664–676.
Pierrard, V., Issautier, K., Meyer-Vernet, N. and Lemaire, J.: 2001, *Geophys. Res. Lett.* **28**, 223.
Pierrard, V., Maksimovic, M. and Lemaire, J.: 2001, in this volume.
Scudder, J.D.: *Astrophys. J.* **398**, 319–349.

ELECTRON TEMPERATURE IN THE SOLAR WIND FROM A KINETIC COLLISIONLESS MODEL: APPLICATION TO HIGH-LATITUDE ULYSSES OBSERVATIONS

KARINE ISSAUTIER and NICOLE MEYER-VERNET
Observatoire de Paris, DESPA, CNRS UMR 8632, Meudon, France

VIVIANE PIERRARD and JOSEPH LEMAIRE
IASB/BIRA, Avenue Circulaire, Brussels 1180, Belgium

Abstract. We use a kinetic collisionless model of the solar wind to calculate the radial variation of the electron temperature and obtain analytical expressions at large radial distances. In order to be compared with Ulysses observations, the model, which initially assumed a radial magnetic field, has been generalized to a spiral magnetic field. We present a preliminary comparison with Ulysses observations in the fast solar wind at high heliospheric latitudes.

1. Introduction

The transport of thermal energy by electrons and the large flow speeds observed in the solar wind are far from being understood. Due to their small mass, electrons play a key role in the transport of energy and measuring the electron temperature gradient may be an important step to understand the solar wind. However, the electron temperature gradients measured in the solar wind during four decades show a complete lack of mutual agreement (Issautier *et al.*, 1999). This is not surprising since (i) the data suffered from large instrumental uncertainties and (ii) the observations were made near the ecliptic plane where different types of winds had to be sorted out, thus introducing bias due to selection. The Ulysses mission, exploring for the first time the high solar latitudes, gives us the opportunity to analyse several months of data in the stationary high-speed wind.

2. Observations

For this study, we use the method of quasi-thermal noise spectroscopy applied to the observations made by the radio receiver (URAP) on board Ulysses to obtain accurate in situ plasma diagnostics (see Issautier *et al.*, in this volume). The Ulysses fast pole-to-pole transit allowed us to obtain accurate radial profiles of the electron density and thermal temperature out of the ecliptic, at a time near solar activity minimum. The data obtained between 40° and 80° southern latitudes constitute a

Astrophysics and Space Science **277:** 189–193, 2001.

large set of stationary high-speed wind since Ulysses was continuously immersed in this kind of flow, for several months at a distance between 1.5 and 2.3 AU from the Sun. During this whole period the measured density has been seen to decrease as r^{-2}. Given that such a density profile corresponds to the density profile in a spherically expanding solar wind at constant velocity (and since at such large distances from the Sun the radial velocity of the wind can effectively be assumed to be roughly constant) we concluded that the latitudinal variations of the solar wind had to be negligible during that period. With this assumption we found that the thermal electron temperature varies as $r^{-0.64}$ between 1.5 and 2.3 AU, which is midway between adiabatic $r^{-4/3}$ and isothermal r^0 (Issautier *et al.*, 1998).

3. Kinetic Collisionless Model

How can one explain such a profile ? A simple way to understand it emphasizes the role of the ambipolar electric potential between the Sun and infinity. Indeed, in the absence of this potential, the Sun would eject much more electrons than protons because of the difference in their thermal speeds. At equilibrium, an electrostatic field is set up, which binds the electrons sufficiently strongly to make their outward flux be balanced by the proton flux. As a result, there are two electron populations: the escaping electrons which have enough energy to overcome the potential barrier, and those which are bound in the potential well. In exospheric models, the former have a strongly anisotropic velocity distribution whose parallel temperature remains roughly constant in a radial magnetic field. On the other hand, the bound electrons are isotropically distributed, and their temperature varies adiabatically with distance. Meyer-Vernet and Issautier (1998) showed that both contributions to the temperature have the same order of magnitude near 1 AU, yielding a profile midway between isothermal and adiabatic.

The exospheric approach, where the plasma is assumed collisionless beyond the distance where the mean free path becomes larger than the density scale height (3–10 solar radii) is expected to be justified to calculate the electron temperature. Indeed, electron/proton collisions do not affect the total electron temperature due to the large ion-to-electron mass ratio. Imposing no net charge and flux, the electrostatic potential is calculated self-consistently and the electron distribution at any distance is calculated from Liouville's theorem (with conservation of energy and magnetic moment) given its shape at the exobase. Assuming a radial magnetic field, we find (Meyer-Vernet and Issautier, 1998) that the ambipolar potential varies as $r^{-4/3}$ and that the electron temperature profile is given by

$$T_e \propto 0.4 \left(\frac{m_p}{m_e}\right)^{5/6} \left(\frac{r_0}{r}\right)^{4/3} + 1$$

where r_0 is the exobase radius. The two terms in the sum, coming from the isotropically distributed non-escaping electrons and from the escaping electrons respect-

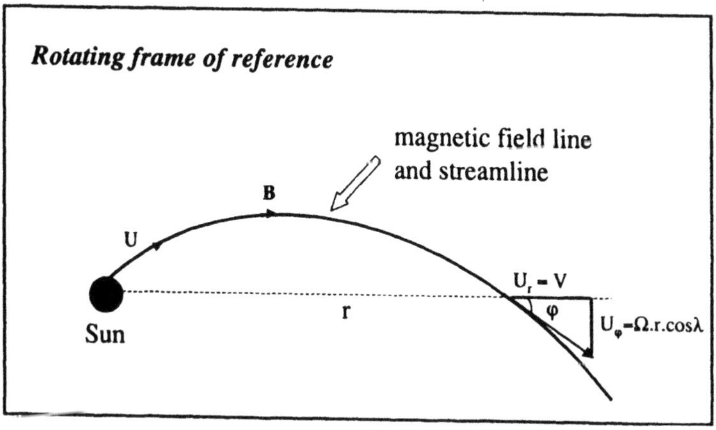

Figure 1. Illustration of a flow streamline or magnetic field line in a frame of reference rotating with the Sun. The expansion speed, V, is constant and the nonradial velocity component U_ϕ depends on the angular velocity Ω, the latitude λ, and the radial distance r.

ively, turn out to be of the same order of magnitude at 1 AU. This result, which does not depend on the assumed electron velocity distribution at the exobase, agrees with the common observation that the temperature profile flattens at large distances. The exospheric model also yields an analytic expression for the wind terminal velocity. Unlike the temperature profile, the terminal velocity depends on the choice of the electron distribution at the exobase. It has also been analytically shown that the wind velocity increases significantly when the electron distribution in the corona has a suprathermal tail (Meyer-Vernet, 1999). This is because a small increase in the number of suprathermal particles dramatically enhances the electron flux; to keep a balance in the electron and proton fluxes, the ambipolar potential must increase, thereby accelerating the protons and increasing the terminal speed. These results are in good agreement with those obtained numerically by Maksimovic *et al.* (1997).

In order to compare these results with Ulysses data in the fast polar wind, the model must be improved to take into account the spiral structure of the solar magnetic field. We assume the magnetic field at distance r to be given by

$$B(r) = B(r_0)\left(\frac{r_0}{r}\right)^2\left(1+\frac{\Omega^2 r^2\cos\lambda^2}{V^2}\right)^{1/2}$$

where $\Omega \simeq 2.865 \times 10^{-6}$ rad s^{-1} is the angular velocity of the Sun, λ is the heliographic latitude, V is the solar wind speed and $B(r_0)$ is the radial component of the magnetic field at the exobase. In a rotating frame of reference as shown in Figure 1, the spiral magnetic field lines coincide with the flow streamlines (Parker, 1958). In this frame, we still have two invariants of motion: the total energy of the particles,

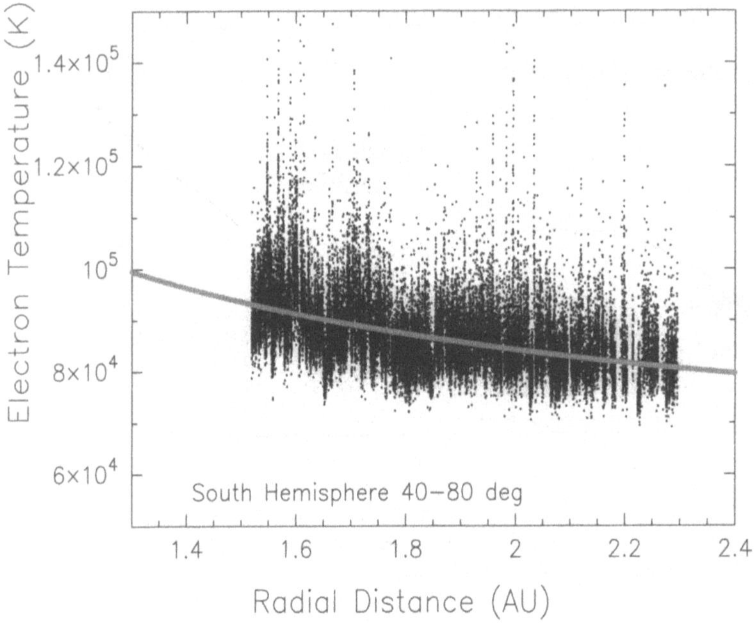

Figure 2. Comparison of the kinetic model generalized to a spiral magnetic field and the Ulysses observations acquired in the fast solar wind with the URAP receiver using quasi-thermal noise analysis.

which now contains a centrifugal energy term, and the magnetic moment. Using this model, we find (Pierrard *et al.*, 2001) that (i) the total electron density still decreases as r^{-2} at large distances, in agreement with Ulysses observations, and the electric potential varies roughly as in a radial magnetic field. Even the terminal speed is only marginally affected by the spiral structure of the magnetic field. The main difference occurs for the electron temperature profile at low latitudes, which is found to decrease less rapidly with distance than in the radial magnetic field case. It is worth noting that these effects may be important in stars rotating faster than the Sun.

4. Preliminary Comparison and Discussion

We give a preliminary comparison between the above model and the Ulysses observations acquired in the polar region of the Sun's southern hemisphere. Figure 2 shows the measured electron temperature (dots) versus the radial distance r, between 1.5 and 2.3 AU, for a latitude range between $\lambda = 40°$ and $\lambda = 80°$. As the distance increases, the profile tends to flatten. The solid line is the result of the fit between the exospheric model presented here and the data. From this fit, we are able to derive the exobase radius and the electron temperature at the base of the corona; we find that $r_0 \simeq 3$ solar radii, which is in agreement with the exospheric model assumption and $T_0 \simeq 2 \times 10^6$ K, which is in the upper range of the observed values. Note that the small range of involved radial distances limits the accuracy of these results, and a comparison involving a larger range of radial distances is in progress, using particle analysers measurements on Ulysses.

Comparing to the analytical expression of the temperature obtained in the radial case, the term due to the escaping electrons is affected by the spiral field, since it is proportional to the ratio of the escaping to the total electron density. The density of these escaping electrons varies as the inverse of the flux tube section, which decreases more slowly than the total density in the spiral case, so that their contribution to the temperature increases with distance. On the other hand, the non escaping electron term still varies as $r^{-4/3}$ with distance, as in the radial case. Detailed expressions of the electron temperature will be given in a forthcoming paper.

References

Issautier, K., *et al.*: 1998, *J. Geophys. Res.* **103,** 1969.

Issautier, K., *et al.*: 1999, *Solar Wind Nine, AIP Conference Proceeding* **471**, 581.

Maksimovic, M., Pierrard, V. and Lemaire, J.: 1997, *Astron. Astrophys.* **324**, 725.

Meyer-Vernet, N.: 1999, *Eur. J. Phys.* **20**, 167.

Meyer-Vernet, N. and Issautier, K.: 1998, *J. Geophys. Res.* **103**, 29,705.

Parker, E.: 1958, *Astrophys. J.* **128**, 664.

Pierrard, V., Issautier, K., Meyer-Vernet, N. and Lemaire, J.: 2001, *Geophys. Res. Lett.* **28**, 223.

CORE, HALO AND STRAHL ELECTRONS IN THE SOLAR WIND

VIVIANE PIERRARD[1], MILAN MAKSIMOVIC[2] and JOSEPH LEMAIRE[1]
[1] *Institut d'Aéronomie Spatiale de Belgique, Brussels, Belgium*
[2] *DESPA, Observatoire de Paris-Meudon, France*

Abstract. Electron velocity distribution functions (VDF) observed in the low speed solar wind flow are generally characterized by 'core' and 'halo' electrons. In the high speed solar wind, a third population of 'strahl' electrons is generally observed. New collisional models based on the solution of the Fokker-Planck equation can be used to determine the importance of the different electron populations as a function of the radial distance. Typical electron velocity distribution functions observed at 1 AU from the Sun are used as boundary conditions for the high speed solar wind and for the low speed solar wind. Taking into account the effects of external forces and Coulomb collisions with a background plasma, suprathermal tails are found to be present in the electron VDF at low altitudes in the corona when they exist at large radial distances.

1. Introduction

It is now established that the observed velocity distribution functions (VDFs) of the electrons in the solar wind exhibit three different components: a thermal core and a suprathermal halo which are both present at all pitch angles, and a sharply field aligned 'strahl', which is observed mostly in the fast solar wind (Feldman *et al.*, 1975; Pilipp *et al.*, 1987; Hammond *et al.*, 1996).

What are the origins of such velocity distribution functions? The classical hydrodynamic approach is inadequate to answer this question since it assumes that the particles VDFs are very close to a Maxwellian. A kinetic approach can be expected to answer this question.

In a collisionless approximation (Lemaire and Scherer, 1971; Schulz and Eviatar, 1972), the core corresponds to the relatively cool population which is trapped within the ambipolar thermo-electric potential well of the heliosphere (ϕ_E), whereas the halo and strahl are composed of electrons energetic enough to escape this potential barrier. The strongly anisotropic strahl component can be understood as the result of focussing along the magnetic field due to conservation of the magnetic moment. However, the isotropy of the halo component is not well understood. Indeed in the collisionless exospheric models, the electron VDF is truncated for local velocities larger than $\sqrt{2e\phi_E/m_e}$ and there should be a halo component in the parallel anti-sunward direction for velocities larger than the local escape velocity.

Kinetic models for the heat conduction in the solar wind acceleration regions and attempts to describe the origin of typical 1 AU electron VDFs have been pro-

Astrophysics and Space Science **277:** 195–200, 2001.

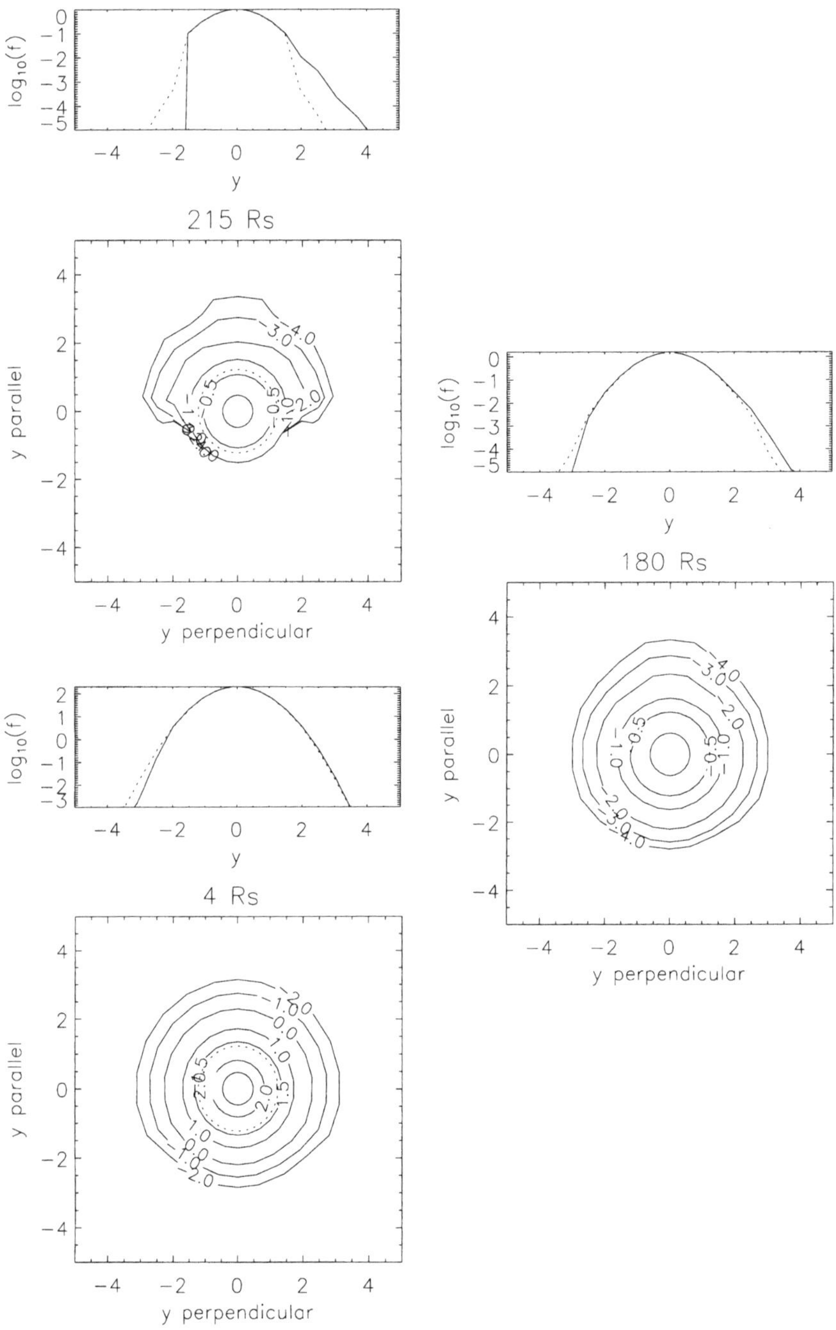

Figure 1. Velocity distribution function of electrons at three different radial distances in a collisional model with two boundary conditions. The parallel and perpendicular velocities y are normalized to the thermal speed of the electrons. The dotted circle corresponds to the thermal speed of the electrons.

2.3. Boundary condition at 1 AU and results

As already mentioned, isotropic suprathermal tails are observed at large radial distances. In P99, we have developed a model where a VDF measured at 1 AU is used as a unique boundary condition for the Fokker-Planck equation. The VDF obtained at low altitude (4 solar radii) is determined in order to match the VDF of electrons observed at 1 AU. It is important to note here that, when an observed VDF is used as boundary condition at 1 AU, it is assumed that the region beyond 1 AU is still collisional and that this collisionality is responsible for the isotropic halo component in the anti-sunward direction.

The VDF obtained at 4 solar radii by solving the Fokker-Planck equation, and taking into account the external forces and Coulomb collisions, demonstrates that suprathermal tails already exist in the corona at 4 R_S. The suprathermal tails of the electron VDF are less asymmetric deep in the corona than at 1 AU. The index κ is found to be between 4 and 10 in the corona, depending if we model the slow or fast solar wind. Finally, note that there are additional indications that suprathermal tails exist in the corona (for instance Ko *et al.*, 1996).

3. Conclusions

To study the solar wind from very low altitudes, where the plasma is dominated by collisions, to very large radial distances, where the solar wind becomes almost collisionless, it is necessary to solve the Fokker-Planck equation. We have developed a kinetic model based on the solution of this equation. The strategy is to study the diffusion of a population of test electrons undergoing Coulomb collisions with a background solar wind plasma (Pierrard *et al.*, 1999). Our results show that, in order to match the observed velocity distributions at 1 AU, suprathermal tails must already exist in the corona at 4 solar radii, but they are much less developed at these altitudes than at larger distances.

Acknowledgements

V. Pierrard thanks the FNRS for her grant of 'chargée de recherches' (A4/5-DM-CR96) and the director of BIRA-IASB for his hospitality.

References

Feldman, W.C., Asbridge, J.R., Bame, S.J., Montgomery, M.D. and Gary, S.P.: 1975, *J. Geophys. Res.* **80**, 4181.
Hammond, M.C., *et al.*: 1996, *Astron. Astrophys.* **316**, 350.

Hinton, F.L.: 1983, in: A.A. Galeev and R.N. Sudan (eds.), *Basic Plasma Physics I and II*, North-Holland Publishing Company, Amsterdam.
Hubert, D. and Leblanc, F.: 2001, *Astrophys. Space Sci.*, this volume.
Ko, Y.-K., Fisk, L.A., Gloeckler, G. and Geiss, J.: 1996, *Geophys. Res. Lett.* **20**, 2785.
Lemaire, J. and Scherer, M.: 1971, *Phys. Fluids* **14**, 8, 1683.
Lie-Svendsen, O., Hansteen, V.H. and Leer, E.: 1997, *J. Geophys. Res.* **102**, A3, 4701.
Lie-Svendsen, O. and Rees, M.H.: 1996, *J. Geophys. Res.* **101**, 2415.
Pantellini, F. and Landi, S.: 2001, *Astrophys. Space Sci.*, this volume.
Pierrard, V.: 1997, PhD thesis, UCL, *Aeron. Acta A*, 401, Physics.
Pierrard, V. and Lemaire, J.: 1998, *J. Geophys. Res.* **103**, 11701.
Pierrard, V., Maksimovic, M. and Lemaire, J.: 1999, *J. Geophys. Res.* **104**, 17021.
Pilipp, W.G., Miggenrieder, H., Montgomery, M.D., Mulhauser, K.-H., Rosenbauer, H. and Schwenn, R.: 1987, *J. Geophys. Res.* **92**, 1075.
Schulz, M. and Eviatar, A.: 1972, *Cosmic Electrodynamics* **2**, 402.
Scudder, J.D.: 1992a, *Astrophys. J.* **398**, 299.
Scudder, J.D.: 1992b, *Astrophys. J.* **398**, 319.
Scudder, J.D. and Olbert, S.: 1979a, *J. Geophys. Res.* **84**, 2755.
Scudder, J.D. and Olbert, S.: 1979b, *J. Geophys. Res.* **84**, 6603.
Spitzer, L.: 1956, *Physics of fully ionized gases*, Interscience, New York, 105 pp.
Treumann, R.A.: 1997, *Geophys. Res. Lett.* **14**, 1727.

A FOKKER-PLANCK APPROACH FOR THE EXPANSION OF THE FAST SOLAR WIND: WHY AND HOW?

F. LEBLANC
University of Virginia

D. HUBERT
Département de Recherche Spatiale, Observatoire de Paris

Abstract. We present recent observations of the plasma parameters in coronal holes at the origin of the fast solar wind and in the interplanetary medium. A model based on the heat conductivity law in a dilute plasma shows the coherency of the electron and proton temperature observations from coronal holes to the interplanetary medium. These new observations are severe constraints for any model of the expansion of the fast solar wind. We discuss why and how non-equilibrium multispecies Fokker-Planck approach must be developed and present a generalized Grad's solution.

1. Introduction

Since the prediction of a supersonic flow in the interplanetary space by Parker (1958) in the late fifties and the first *in situ* observations at the rising of the space age, the interest in the solar wind formation has been growing. The heliosphere ploughed by spacecraft appears as a natural laboratory which challenges many ground facilities. Awareness has appeared that the corona heating and the solar wind expansion must be tackled together. One of the recurrent questions along these decades is to know whether astrophysicists have in their hands the right tools to undertake the two aspects of this same problem. The answer is related to the degree of completeness of the prediction needed, compared to progress in the field of plasma transport theory, particularly in the context of laser-produced plasma (Luciani *et al.*, 1985).

The fast solar wind, whose steady state has been confirmed by the Ulysses observations at high heliospheric latitude, is our concern. We recall in section 2, the main constraints obtained recently on the SOHO and Ulysses missions, which complete the invaluable Helios observations between Mercury's and the Earth's orbits. In section 3, we study the evolution of fundamental parameters of the fast solar wind from coronal holes to few AU in the high latitude heliosphere. We then discuss why the Fokker-Planck approach should be used and we sketch our original approach. A conclusion suggests future perspectives.

Astrophysics and Space Science **277:** 201–204, 2001.

2. The Expansion of the Fast Solar Wind

Important temperature measurements have been made during solar minimum activity in coronal holes, sources of the fast solar wind. David *et al.* (1998) carried out observations on SOHO on May 1996 and found that the electron temperature reaches a maximum of less than 1 MK at 1.20 $R_\odot$ (Sun radius) and decreases gradually. Esser *et al.* (1999) showed that in coronal holes the proton temperature slowly increases without exceeding 3MK from 1.35 to 2.7 $R_\odot$. Measurements indicate that the terminal fast solar wind speed is almost reached at 15 $R_\odot$, much closer to the Sun than expected (Grall *et al.*, 1996; Esser *et al.*, 1998).

In situ measurements from Mercury's to the Earth's orbits (Schwenn and March, 1991), and recently over the solar poles (Feldman *et al.*, 1996; Issautier *et al.*, 1998) show that in the fast wind the proton temperature is larger than the electron temperature and that the alpha particle temperature is about 4 times the proton temperature. The proton temperature anisotropy should be important and $T_{p\perp} > T_{p\parallel}$ at distance lower than 0.5 AU, with $T_{p\perp}$ the perpendicular temperature to the interplanetary magnetic field (IMF) (Schwenn and Marsch, 1991). The radial evolution of species temperature provided by Helios 2 and Ulysses is non isothermal and non adiabatic (Schwenn and Marsch, 1991; Pilipp *et al.*, 1990; Issautier *et al.*, 1998). The species velocity distribution functions are all characterized by important heat flux and suprathermal tails elongated in the outward IMF direction.

3. How to Model the Fast Solar Wind Expansion?

A simple calculation shows that at 1.2 $R_\odot$ the coronal plasma is collisional and that we can apply the classical conductivity law (Spitzer and Härm, 1959) which provides an electron temperature evolution $T_{e1} \propto r^{-2/7}$. The core electron temperature law deduced from *in situ* data in the fast solar wind on Helios and Ulysses (Pilipp *et al.*, 1990; Issautier *et al.*, 1988) is $T_{e2} \propto r^{-0.62}$. These two laws both predict a temperature at 8.6 $R_\odot$ equal to $T_e = 5.7 \times 10^5$K. Although no measurement of the electron temperature has been made between 2 $R_\odot$ and Mercury's orbit, it is expected that in this gap the electron temperature decreases faster than $T_{e1} \propto r^{-2/7}$ (Hu *et al.*, 1997). However, in this paper, in order to be conservative we consider the law T_{e1} for $1.2 \leq r \leq 8.6\ R_\odot$ and the law T_{e2} for $r > 8.6R_\odot$; let us underline that this choice overestimates the mean free path in the corona. Similar considerations lead us to estimate a nearly isothermal radial evolution of the proton temperature in the first ten solar radii, which tends at larger distances progressively to $T_p \propto r^{-0.80}$ (Schwenn and Marsch, 1991).

Table I provides the evolutions between 2$R_\odot$ and 2 AU of the electron Knudsen number K_n based on the density scale height and of the ratio of the mean free path to the electron temperature scale height K_T. For this calculation we have considered the two laws T_{e1} and T_{e2} and also the electron-electron collisions which are nearly as efficient as the electron-proton collisions to transversally diffuse the electrons. Table I shows that the electron Knudsen number K_n never reaches the value 1

TABLE I

The heliospheric distance r is in solar radius, then in astronomical unit

r	2 $R_\odot$	5 $R_\odot$	10 $R_\odot$	0.3 AU	1 AU	2 AU
K_n	0.09	0.4	0.6	0.5	0.4	0.3
K_T	0.007	0.08	0.10	0.15	0.12	0.09

defining the exobase position (Lemaire and Scherer, 1973). K_n reaches a maximum value around 10 $R_\odot$, then decreases gradually to 0.3 at 2 AU; K_T evolves as K_n but is always lower. Table I confirms that any fluid model, for which K_n must be $\ll 1$, as well as the Chapman-Enskog approach, for which $K_T < 10^{-2}$, cannot be applied at distances larger than 3 $R_\odot$ from the coronal base (Shizgal, 1977). Actually fluid approaches which model the heat flux by an ad-hoc law have been shown to be unadapted in many laboratory experiments (Luciani *et al.*, 1983). As regards the validity of the exospheric model, let us stress that, i) K_n is always < 1 from 2$R_\odot$ to 2 AU, ii) the electron temperature is less than 1 MK in coronal holes and iii) the proton temperature is always 3-4 times higher than the electron temperature in the lower corona. These values in the source region of the fast wind, represent strong constraints for the exospheric models in predicting a fast terminal velocity at about 15 $R_\odot$, despite interesting recent developments (Maksimovic *et al.*, 1997).

New approaches are necessary and should be established from the Fokker-Planck equation, the only kinetic equation adapted to describe the expansion from a collisional region to a rarefied region. This approach should take into account the main processes in the solar wind: the anisotropic wave-proton and wave-alpha interactions, the particle-particle coulomb energy transfers, the parallel-perpendicular energy transfer in anisotropic plasma for the different species and the heat conductivity along the expansion. Grad's method to solve the Fokker-Planck equation has the great advantage to provide simultaneously two types of descriptions: at the level of the transport equations and at the level of the velocity distribution function.

A model which generalizes the Grad's method has been developed by Leblanc and Hubert (1997, 1998). The originality of this model with respect to the so called bi-Maxwellian model (Demars and Schunk, 1979) consists of the non-Maxwellian zeroth order distribution which generates far from equilibrium states and new truncation properties. This model fits without generating negative values the measured distributions with important heat flux value (Leblanc and Hubert, 1997). Collisional transfer properties have been established for multispecies with important relative velocities, large temperature anisotropy and typical high heat flux as observed for instance for the solar wind protons (Leblanc *et al.*, 2000a). We have shown that in the corona the region of energy transfer between electrons and protons should be larger than previously thought. A recent application of this model to the terrestrial polar wind illustrates the potentialities of this approach with respect to the bi-Maxwellian model; and that interesting properties of the proton heat conductivity are induced by the new truncation (Leblanc *et al.*, 2000b).

4. Conclusion

We have presented the up to date plasma measurements from the origin of the fast solar wind to the interplanetary medium at few AU from the Sun. The fast solar wind appears to reach its terminal velocity at a distance less than 15 $R_\odot$ above the limb, much faster than predicted in the first models of the solar wind. From these data, a conservative model based on our knowledge of the heat conductivity in dilute laboratory plasma implies strong limits to fluid and exospheric approaches.

A generalized Grad's solution has been developed and provides self consistently two levels of description. Most of the characteristics of the solar wind observations at the macroscopic and microscopic levels can be compared. As a consequence it provides the capability to tackle the wave-proton and wave-helium interactions which control the distribution shapes along the expansion. A first application of this model to the terrestrial polar wind expansion, has shown improvements with respect to similar approaches based on the original Grad's method.

References

David, C. *et al.*: 1998, *Astron. Astrophys.* **336**, L90.
Demars, H.G. and Schunk, R.W.: 1979, *J. Phys. D* **12**, 1051.
Esser, R. *et al.*: 1998, *Astrophys. J.* **498**, 448.
Esser, R. *et al.*: 1999, *Astrophys. J.* **510**, L63.
Feldman, W.C. *et al.*: 1996, *Astron. Astrophys.* **316**, 355.
Grall, R.R. *et al.*: 1996, *Nature* **379**, 429.
Hu, Y.Q. *et al.*: 1997, *J. Geophys. Res.* **102**, 14,661.
Issautier, K. *et al.*: 1998, *J. Geophys. Res.* **103**, 1969.
Leblanc, F. and Hubert, D.: 1997, *Astrophys. J.* **483**, 464.
Leblanc, F. and Hubert, D.: 1998, *Astrophys. J.* **501**, 375.
Leblanc, F. *et al.*: 2000a, *Astrophys. J.* **530**, 478.
Leblanc, F. *et al.*: 2000b, *J. Geophys. Res.* **105**, 2551.
Lemaire, J. and Scherer, M.: 1973, *Rev. Geophys. Space Phys.* **11**, 427.
Luciani, J.F. *et al.*: 1985, *Phys. Fluids* **28**, 835.
Maksimovic, M. *et al.*: 1997, *Astron. Astrophys.* **324**, 725.
Parker, E.N.: 1958, *Astrophys. J.* **128**, 664.
Pilipp, W.G. *et al.*: 1990, *J. Geophys. Res.* **95**, 6305.
Schwenn, R., and Marsch, E.: 1991, in: *Physics of the Inner Heliosphere* **2**, Springer-Verlag.
Shizgal, B.: 1977, *Planet. Space Sc.* **25**, 203.
Spitzer, L. and Härm, R.: 1953, *Phys. Res.* **89**, 977.

HELIOPAUSE AND ASTEROPAUSES

R. LALLEMENT
Service d' Aéronomie du CNRS, Verrières-le-Buisson, France

Abstract. In this paper I will use recent results about the heliosphere and the so-called 'H walls' around nearby star 'asteropauses' (the equivalent of the heliopause for other stars) to show new connections between heliospheric physics and other fields in space physics.

1. The Solar Environment at Different Scales

Figures 1 and 1bis illustrate the structure of the gaseous medium around the Sun at six different scales, from the galactic disk down to the planetary system scale. This will be useful for the Heliosphere-Interstellar Medium (ISM) connections. The Sun is in an outer region of the galactic disk, where the average thickness of the dense disk is about 300 pc. However, the Sun is in a peculiar region, a kind of 'hole' in the dense disk, i.e. a large volume devoid of dense gas, whose size is of the order of 200 pc. This so-called Local Bubble (LB) is thought to be filled with very tenuous gas at about one million degrees. The LB has been found to be elongated along an axis tilted by about 20 degrees from the polar axis, and perpendicular to the 'local' galactic disk, which is itself tilted by 20 deg with respect to the main galactic disk (Sfeir *et al.*, 1999). Inside the hot gas are embedded a few condensations, and also, here and there, low-density warm clouds (called diffuse clouds). There is an important group of such clouds in the solar vicinity: the so-called Local Fluff. The typical size of such clouds is one parsec, i.e. the average distance between stars.

The Local Interstellar Cloud (LIC) is simply one of these unconspicuous masses of tenuous gas. Its temperature is 7000K and its velocity relative to the Sun is 26 km/s. The Sun blows inside the LIC a cavity filled by solar plasma, elongated along the Sun-LIC relative motion vector. Neutral particles and high energy cosmic rays enter the heliosphere while ionized species are diverted around the heliopause. The size of this heliosphere is governed by the ambient LIC plasma density, magnetic presure and pressure of the reflected low-energy cosmic rays. Before reaching equilibrium at the heliopause, the solar wind (and probably also the interstellar flow) are decelerated by a shock (for the solar wind: the so-called termination shock: Parker, 1963). Inside the heliosphere, neutral atoms get ionized close to the Sun by solar radiation and charge-exchange with solar wind particles (Axford *et al.*, 1963), and the newly formed pick-up ions are convected outwards by the solar wind electromagnetic field. Atoms from the neutral flow which survive ionization backscatter the solar radiation, which makes them detectable. Up to now

Astrophysics and Space Science **277:** 205–217, 2001.

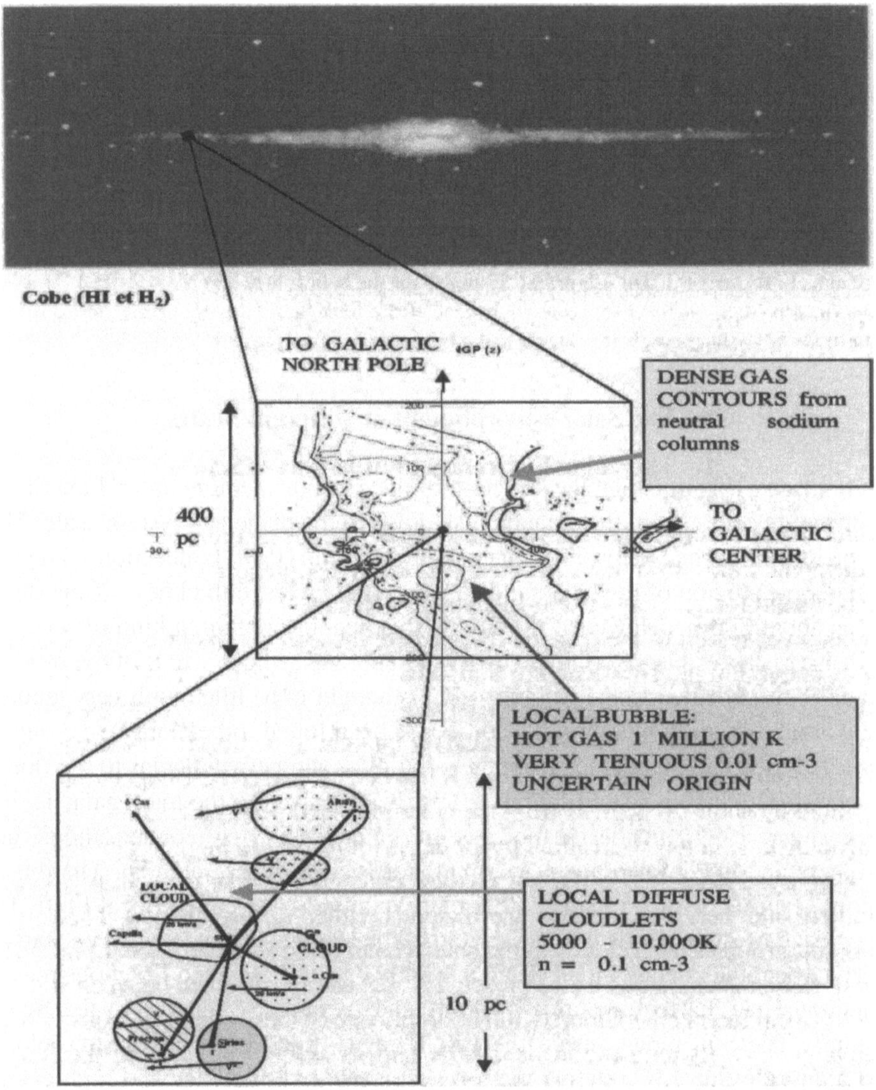

Figure 1. Environment of the Sun at decreasing spatial scales, from its location in the Local Arm (1-top), in the low density region called the Local Bubble (LB), and in a group in small diffuse clouds with sizes of the order of a parsec (1-bottom) embedded in the LB.

the H and He glows have been intensively studied with their resonance lines at 121.6 and 58.4 nm.

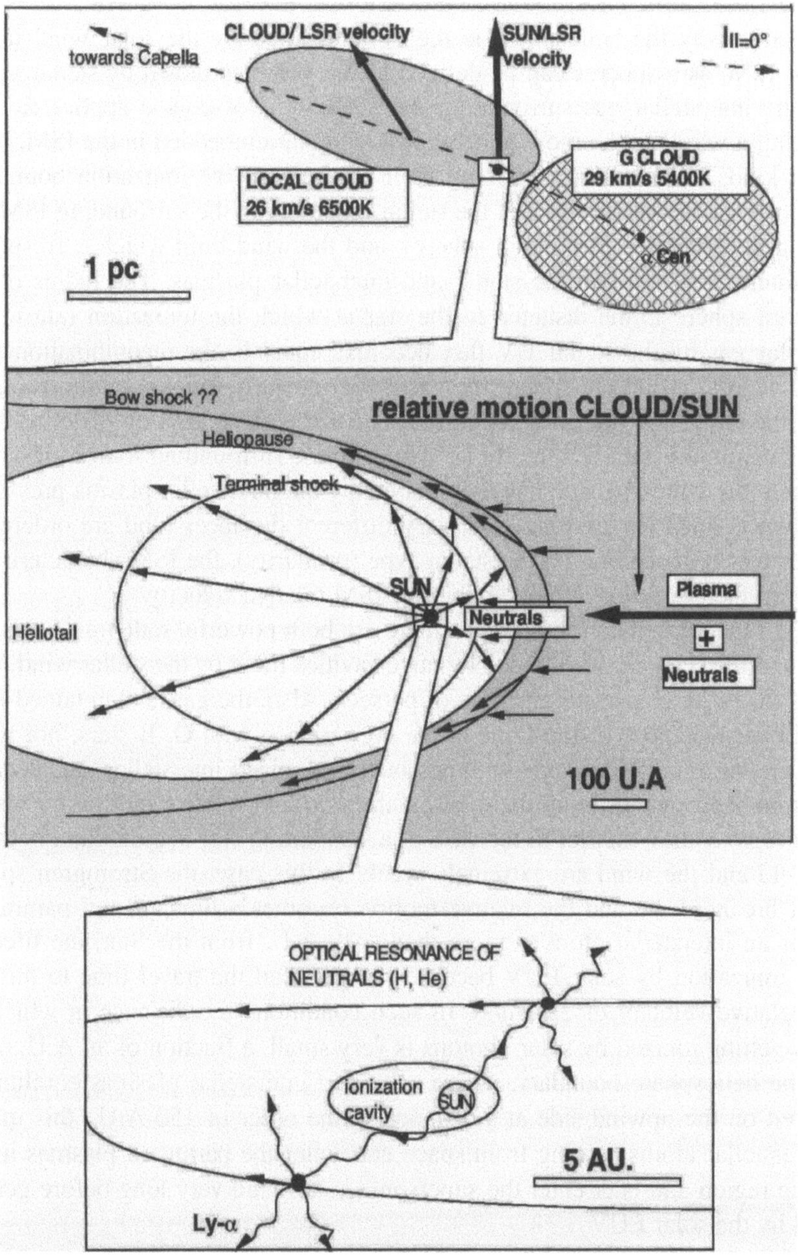

Figure 1bis. The Sun in the Local Interstellar Cloud (LIC). The solar wind blows from inside the LIC the heliosphere (middle) in which interstellar neutrals enter and scatter the solar EUV which makes them visible (bot).

2. Astrospheres in General and the Particular Case of Our Heliosphere

In the same way the heliosphere is the volume filled by the solar wind in the ambient ISM, astrospheres can be defined as the volumes blown by stellar winds inside the interstellar gas surrounding the stars. This of course applies to stars possessing a wind! In the more general case of a star embedded in the ISM, there are two kinds of boundaries defining their interaction: the ionization boundary, which corresponds to the effect of the stellar radiation on the surrounding ISM gas (defining the so-called Stromgren sphere), and the wind limit which corresponds to the interaction between the stellar and interstellar plasmas. The radius of the Stromgren sphere is the distance to the star at which the ionization rate of the interstellar gas by the stellar UV flux becomes equal to the recombination rate. In other words, inside the sphere the gas is significantly ionized by the star, and outside the ionization becomes negligible. The astrosphere limit on the other hand is the discontinuity between the stellar wind and the surrounding ionized ISM, i.e. it is where the wind pressure has dropped below the interstellar plasma presssure. These two boundaries are located at very different distances (and are ordered in different ways) depending on the stellar type (primarily), the ISM characteristics, and a third factor often overlooked, the star/ISM relative velocity.

For hot O and B stars for example, there are both powerful radiation fields and winds, and the stars are surrounded by huge cavities filled by the stellar wind, with sizes of the order of parsecs and tens of parsecs. Also, the gas is maintained fully ionized over distances of the same order. At variance with O, B stars, hot white dwarfs are surrounded by huge volumes of fully ionized interstellar gas, because there is no stellar wind. In all these cases, the star/ISM relative motion makes the ionized regions dissymetric. In the case of a standard G star like the Sun, both the EUV field and the wind are extremely weak. In this case, the Stromgren sphere concept breaks down and the relative motion becomes a fundamental parameter. E.g., for an interstellar atom at more than $\approx$30 A.U. from the Sun, the lifetime against ionization by solar EUV becomes longer than the travel time to the star (for a relative velocity of 25 km/s). In such conditions the distance at which an atom is getting ionized by solar photons is very small, a fraction of an A.U. only! Since the heliosphere boundary, where solar and interstellar plasmas equilibrate, is located on the upwind side at a distance of the order of 150 A.U., this means that interstellar atoms coming from space encounter the perturbed plasmas in the interface region and later enter the supersonic solar wind very long before getting ionized by the solar EUV.

Moreover, in the inner heliosphere and supersonic solar wind, photoionization is still so inefficient that it is exceeded by charge-exchange with the solar wind protons. This has a very important (and kind of fortunate) consequence: since the interstellar gas is not photoionized in the interstellar space during its approach to the Sun, all information on the interstellar medium immediately outside the heliosphere, such as elemental abundances or ionization degrees, which can be derived from heliospheric measurements, applies to the unperturbed ISM, i.e., the matter exactly as it is in the interstellar cloud, before any modification by the Sun! In other words, the local ISM outside the heliosphere is not perturbed by the tool used to measure it *in situ*, i.e. the heliosphere!

3. Heliosphere-ISM Connections

3.1. Interstellar Neutrals: From Interstellar Space Thermal Particles to Pickup Ions and Anomalous Cosmic Rays

After neutral atoms of interstellar origin have reached the inner heliosphere and experienced ionization by solar EUV photons and charge-exchange with solar protons, the new pickup ions (PUI's) are characterized by a 'ring' velocity distribution and during their journey in the expanding solar wind they become preferential sources for acceleration in the interplanetary shocks (Fisk *et al.*, 1974) and finally at the solar wind termination shock, giving rise to the so-called anomalous cosmic rays (Pesses *et al.*, 1981). A series of comparisons can thus be done between the consecutive states of the particles: – a) the ISM 'quiet' initial state (neutrals) – b) the outer heliosphere state which may be different from the previous one due to the charge-exchange with the ions in the heliospheric interface, essentially the decelerated ISM protons – c) the pickup ion state after ionization of the neutrals in the Sun vicinity and convection in the ambient solar wind, d) the energetic form after acceleration of the pickup ions in the solar wind CIR's and GMIR's, e) the final state: the anomalous cosmic rays after acceleration at the solar wind termination shock . All this chain is currently under study, thanks mainly to measurements of the Voyager, Ulysses and SOHO instruments. At all stages one can learn about the physical processes, the heliosphere structure, and the ISM composition. This chain itself shows the multi-disciplinary nature of the analysis of all these measurements.

3.2. Identification of the LIC from both Heliospheric and Astronomical Observations

Some neutral species, as helium, do not interact with the ions in the heliosphere interface on distance scales as small as a few hundred A.U. because the charge-exchange cross-section is too low. As a consequence, those neutrals keep unchanged their characteristics from interstellar space once they have entered the heliosphere. This is why helium measurements inside the heliosphere and the subsequent analysis of the helium flow properties allow the extrapolation back to the neutral helium density in the LIC in the Sun vicinity. In other words, interstellar neutrals like helium which are not coupled to the plasma, provide in a very direct way an '*in situ*' measurement of the surrounding interstellar medium.

Analyses of the interstellar absorption lines in the nearby star spectra have allowed, through the Doppler triangulation method, to identify two masses of gases on both sides of the Sun and to measure their relative velocities and their temperatures (Figure 1bis, top) (Lallement *et al.*, 1989; Lallement and Bertin, 1992). One of these masses of gas or clouds was found to have exactly the same temperature (7000 K) and velocity (26 km/s) as the flow of helium inside the heliosphere (Witte *et al.*, 1993). This allowed us to identify which one from the two clouds contains the Sun and is the Local Cloud (Bertin *et al.*, 1993). This was one of the first examples of heliosphere/ISM connections.

3.3. Determination of the Interstellar Plasma and Neutral Densities from Both Heliospheric and Astronomical Observations

Another illustration of the link between interstellar and heliospheric physics is the recent determination of the interstellar neutral and plasma densities outside the heliosphere. The basic steps are as follows:

a) Recent measurements of helium atoms and He^+ and He^{++} PUI's have provided number densities for the helium flow inside the heliosphere, which, although corresponding to three totally independent diagnostics, are in rather good agreement to give $n(He^0)=n(HeI) \approx 0.014$ cm^{-3} (Von Steiger *et al.*, 1996). Since interstellar helium keeps its properties at the interface, this number applies to the surrounding local ISM.

b) Helium is homogenously ionized in the local interstellar clouds at about 40% (Dupuis *et al.*, 1995; Wolff *et al.*, 1999). This result comes from analyses of white dwarfs spectra recorded by the Extreme Ultraviolet Explorer (EUVE), which provide column-densities of HI, HeI, HeII towards about 20 nearby stars in all directions. One can reasonably assume that this applies also to the Sun region, and correcting for the 40% ionization of He one derives from a) that the total helium density in the surrounding ISM is n(He)=n(HeI)+n(HeII)= 0.023 cm^{-3}.

c) From the cosmological abundance ratio H/He $\approx$ 10, this means that n(H)= n(HI)+n(HII)= 0.23 cm^{-3} in the LIC in the Sun vicinity.

d) Contrary to neutral helium, neutral hydrogen is substantially perturbed when flowing through the interface, due to the interaction with the interstellar protons through the charge-exchange reaction $H+H^+ \rightarrow > H^+ + H$. Neutral hydrogen flow properties are modified all through the interface and the gas is heated and decelerated. Moreover, a fraction of the neutral atoms is excluded from the heliosphere. The changes in bulk velocity V, temperature T and density n along the Sun-wind axis are shown in Figure 2. These results come from a self-consistent neutral-plasma model of the entire heliosphere, in the case of a supersonic wind and negligible interstellar magnetic field (Baranov and Malama, 1993; Izmodenov *et al.*, 1999). The differences ΔV, ΔT, Δn betwen the outside properties on one hand (i.e. in the LIC, derived from helium observations or interstellar lines absorption measurements) and the flow properties just inside the termination shock on the other hand, are an increasing function of the coupling, and thus of the interstellar plasma density n(e-). One can make thus use of the three diagnostics separately to derive n(e-). The SOHO/SWAN measurements of the velocity distributions (from backscattered solar radiation by neutral H) and the corresponding modeling of the flow in the inner heliosphere provide V and T (Quémerais *et al.*, 1999; Costa *et al.*, 1999). They provide also the density n, but with a larger uncertainty than the measurements of n from the PUI's (Gloeckler *et al.*, 1993). A preliminary analysis shows that the data favor an interstellar plasma density n(e-)= 0.04 cm^{-3} for the three diagnostics (Lallement, 1999). This would mean that the ISM in the Sun region is characterized by 20% ionization of H. Interestingly, this number density for the electrons is close to the density corresponding to the plasma frequency of 1.8 kHz, the cutoff of the lower band of the radio emissions detected by the Voyager and believed to originate in the heliospheric interface (Kurth *et al.*, 1984). This suggests that the 2 kHz band is generated outside the heliopause, as originally considered by Gurnett and Kurth (1993).

But the plasma density, in turn, gives in principle the main parameter for the confinement of the heliosphere. For an electron density of only 0.04 cm^{-3}, the 'supersonic' models (those assuming that the SUN/ISM velocity is superfastmagnetosonic) predict a closest distance to the termination shock of the order of 90 A.U., and to the heliopause of 185 A.U. This last distance of 185 A.U. is above the 110–160 A.U. distance range predicted by Gurnett *et al.* (1993) from the time-delay between the strong solar wind discontinuities and the radio emissions believed to be associated to their interactions with the heliopause. An additional pressure other than the plasma pressure is necessary to reconcile both diagnostics. It could be a magnetic pressure linked to the interstellar magnetic field. However, MHD models are required in this case to better represent the heliospheric structure. In this respect, one of the keys for a realistic representation of the outer heliosphere will be a full understanding of the physical processes at work in the outer heliosphere which produce 2–3 kHz radio emissions.

3.4. The Local Bubble-Heliosphere Connection

Another illustration of the link between heliospheric and interstellar studies is the discovery of a strong inconsistency in the pressures of the hot gas which fills the local bubble (LB) on one hand, and the pressure in the LIC on the other. Diffuse clouds embedded in the LB (Figure 1) should be in pressure equilibrium with the surrounding gas, since the LB is an old (about 1 million years) bubble and the clouds are not seen in the process of contraction or expansion. Surprisingly, the soft X-ray data imply a hot gas pressure $nT \approx 14{,}000$ Kcm^{-3}, while the numbers above for the LIC imply $nT \approx 2{,}000$ Kcm^{-3}. For a number of reasons (e.g. Lallement, 1996), it is very unlikely that the difference is explained by a strong magnetic field

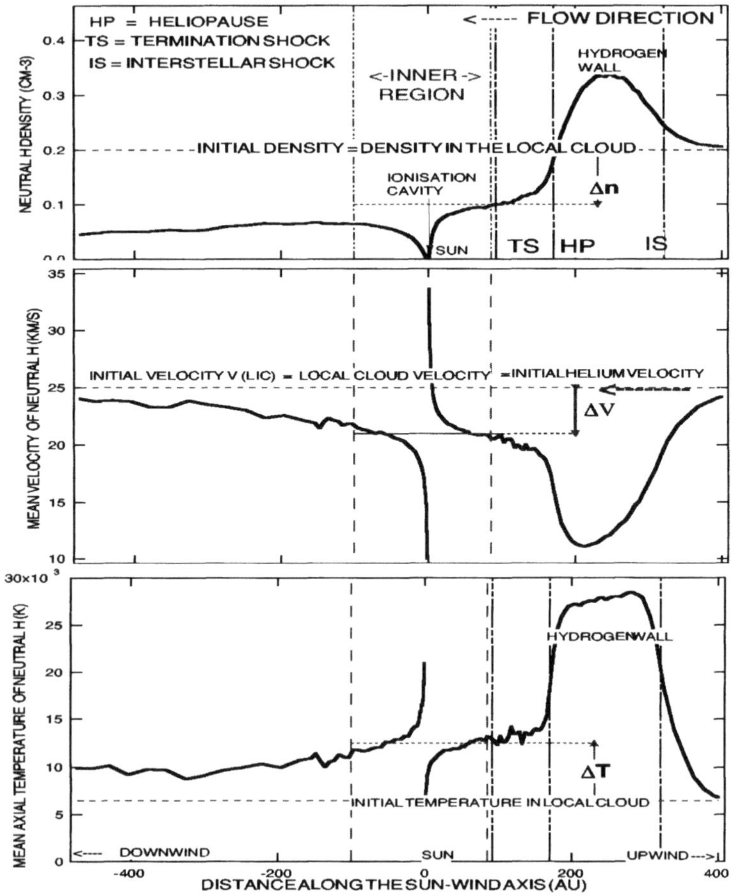

Figure 2. Neutral hydrogen characteristics along the Sun-wind axis according to Baranov and Malama (1993) and Izmodenov *et al.* (1999a). The density, average velocity and apparent temperature are shown as a function of distance to the Sun for the 'average' neutral flow. It is made of two components, primary atoms which have not suffered any charge-exchange with the plasma and secondary atoms produced by charge-exchange. When entering in the supersonic solar wind (left of the TS transition in red), the parameters N, V and T have changed from their initial values in the unpertubed interstellar medium (right side). The H wall, conspicuous in this 2-shock model computed for a plasma density of 0.07 cm^{-3}, is seen between the bow shock and the heliopause.

in the LIC. Also, a difference in the cosmic ray pressure is excluded. Today, the reason for such a pressure imbalance is still unknown.

4. Heliosphere and Cosmology

4.1. The Heliosphere as a Tool for Measurements of the ISM Elemental and Isotopic Composition

Many pickup ion species have now been detected with the SWICS instrument on board Ulysses (e.g. Gloeckler *et al.*, 1993). Among them are H^+, He^+, He^{++}, N^+, O^+, and Ne^+ (e.g Geiss *et al.*, 1994). From the velocity distributions and the fluxes it is possible to derive the properties of the initial fluxes of the corresponding neutral 'parents', after correction for the ionization and dynamical processes close to the Sun, and the perturbations and filtrations at the heliospheric interface. The relative strengths of the fluxes are the relative abundances in the LIC.

One of the most interesting measurements of this type has been the 3He/4He isotopic ratio, found to be $2.25 \pm 0.65\ 10^{-4}$ from the relative fluxes of $^3He^+$ and $^4He^+$ (Gloeckler and Geiss, 1996). In this last case, because all the processes which transform neutrals in the LIC to pickup ions very likely affect isotopes of the same element in a very similar way, this is directly the isotopic ratio in the Local Interstellar Cloud. This ratio is related to the Big Bang nucleosynthesis and galactic chemical evolution and places constraints on the production of 3He in the low mass stars.

In the same way, future measurements should allow the determination of $^2H/^1H$, $^{18}O/^{16}O$, and $^{22}Ne/^{20}Ne$.

4.2. The Heliosphere as an Additional Lyman-Alpha Absorber along the Line-of-Sights to Astronomical Objects

There is another link between cosmological abundances and heliospheric properties, under the form of a 'contamination' of the D/H measurements by the heliospheric boundary region. Absorption by interstellar H and D at Lyman-alpha in the spectra of hot nearby stars has been used for years to derive the deuterium relative abundance in the local interstellar clouds. However, due to charge-exchange, there is a volume of compressed and decelerated hot (up to 100,000 K) neutral hydrogen around the heliopause: the so-called 'H wall'. Hot hydrogen is a strong absorber at the transitions of the Lyman series, especially at Ly-alpha. Indeed, the heliospheric H wall is detected in absorption in the spectra of nearby stars (alpha Cen: Linsky and Wood, 1996; Sirius: Izmodenov *et al.*, 1999). Though the optical thickness of the wall is small compared to the total thickness to the star, it adds absorption because it is Doppler shifted and wide. This means that in order to get correct determinations of D/H, the heliospheric absorption must be taken into account, whatever the star direction.

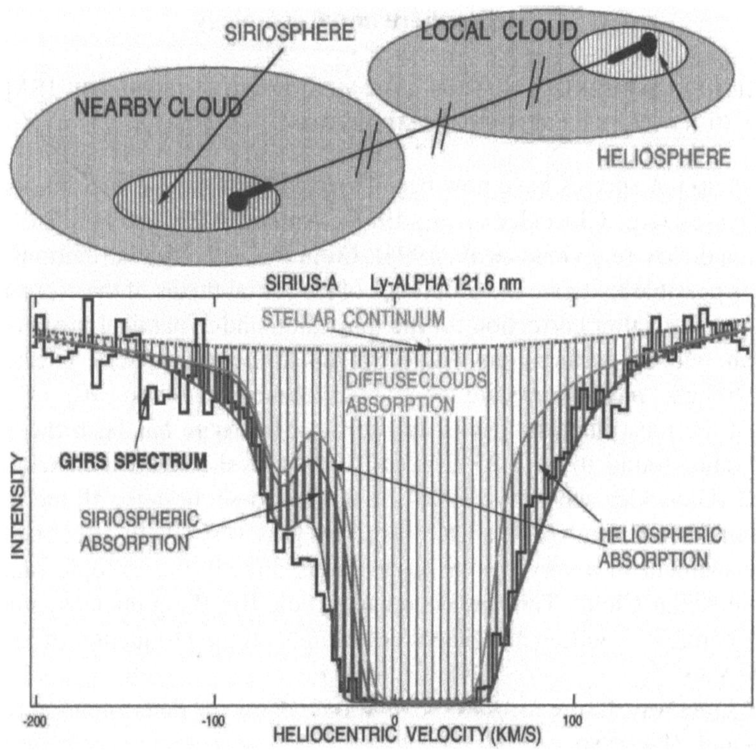

Figure 3. Absorption at Ly-alpha by hot neutral hydrogen from the heliosphere and very likely from the astrosphere around Sirius add to absorption lines by the interstellar gas in the Local Interstellar Cloud and its neighbour. Despite the very small size of the astrospheres (100–1000 A.U., not to scale in the top schematic view) with respect to interstellar distances, the lines are detectable.

5. Astrospheres of Nearby Stars: Revealed by Their EUV Absorptions at Ly-Alpha

If a star emitting a wind is traveling through a partially ionized interstellar cloud, there should be an ‘H wall’ of the same nature as the heliospheric ‘H wall’ around the heliopause. Such ‘H walls’ have been inferred from their absorptions at Ly-alpha in the spectra of some nearby stars. The first detections have been for the fast moving stars ϵ Ind and λ And (Wood *et al.*, 1996). As a matter of fact, H wall temperatures increase with the relative velocity of the star with respect to the ambient interstellar medium and such hot H walls are more easily detected. Subsequently, H walls have been inferred around other stars (Piskunov *et al.*, 1997) and the detected absorptions around alpha Cen and Sirius have been shown to be of the expected magnitude and velocity (Izmodenov *et al.*, 1999; Lallement, 1999).

Such detections can be very useful in bringing constraints on stellar winds and interstellar medium structure simultaneously.

6. Heliosphere and Earth

Any change in the size and structure of our heliosphere is followed by changes in the spectra and fluxes of galactic cosmic rays reaching the inner heliosphere and the Earth. The heliosphere can be modified in two ways: as a consequence of solar wind changes and as a consequence of the temporal evolution of the circumsolar interstellar medium. The first aspect is already well understood, but new developments, as e.g. a possible 'cosmic rays' origin of the upper atmosphere phenomena of 'sprites' type, deserve further studies. As for the second aspect, improvements in the knowledge of our interstellar medium environment can be used to understand (for past events) or predict the consequences on Earth of the response of our heliosphere to the crossing of interstellar clouds of various types, or to galactic events on time scales from 10^3 to 10^6 years (e.g. Zank and Frisch, 1999).

7. Conclusions

The study of the solar wind- ISM interaction, and by extension of stellar winds-ISM interactions, is connected in various ways to other research activities. For this reason, a better understanding of the physical processes at work in the heliospheric gas mixture would also benefit to these other fields. At present, most of the advances have been done with deep space probes, in conjunction with satellites devoted to various projects. There are however unsolved questions, as the strength and the role of the interstellar magnetic field, the influence of the energetic particles and the pickup ions on the interface structure, etc..., and, on the interstellar side, some of the crucial isotopic ratios in the ISM can not be measured from inside the heliosphere since the elements do not exist under the neutral form in the ISM and do not enter the interface. A fast (20 A.U./yr) 'interstellar probe' catching up the two Voyager and reaching 100–150 A.U. would solve these questions and open many new perspectives, while helping in understanding past, present and future of our solar wind 'cocoon', the heliosphere.

References

Axford, W.I., Dessler, A.J. and Gottlieb, B.: 1963, *Astrophys. J.* **137**, 1268.

Baranov, V.B. and Malama, Y.G.: 1993, The model of the solar wind interaction with the local interstellar medium: Numerical solution of the self-consistent problem, *J. Geophys. Res.* **98**, 15157.

Costa, J., Lallement, R., Quémerais, E., Bertaux, J.L., Kyrola, E. and Schmidt, W.: 1999, Heliospheric Interstellar H temperature from SOHO/SWAN H cell data, *A&A* **349**, 660.

Dupuis, J., Vennes, S., Bowyer, S., Pradhan, A.K. and Thejll, P.: 1995, Hot white dwarfs in the local interstellar medium: Hydrogen and helium interstellar column densities and stellar effective temperatures from extreme-ultraviolet explorer spectroscopy, *Astrophys. J.* **455**, 574.

Fisk, L.A., Kozlovsky, B. and Ramaty, R.: 1974, An interpretation of the observed oxygen and nitrogen enhancements in low-energy cosmic-rays, *Astrophys. J.* **190**, L35.

Geiss, J., Gloeckler, G., Mall, U., von Steiger R., Galvin, A.B. and Ogilvie, K.W.: 1994, Interstellar oxygen, nitrogen and neon in the heliosphere, *Astron. Astrophys.* **282**, 924.

Gloeckler, G., Geiss, J., Balsiger, H., Fisk, L.A., Galvin, A.B., Ipavich, F.M., Ogilvie, K. W., von Steiger, R. and Wilken, B.: 1993, Detection of interstellar pick-up hydrogen in the solar system, *Science* **261**, 70.

Gloeckler and Geiss: 1996, Abundance of 3He in the local interstellar cloud, *Nature* **381**, 210.

Gurnett, D.A., Kurth, W.S., Allendorf, S.C. and Poynter, R.L.: 1993, Radio emission from the heliopause triggered by an interplanetary shock, *Science* **262**, 199.

Izmodenov, V.V., Lallement, R., Malama, Y.: 1999c, Heliospheric and astrospheric neutral hydrogen absorption towards Sirius: No need for interstellar hot gas, *Astron. Astrophys* **342**, L13.

Kurth, W.S., Gurnett, D.A., Scarf, F.L. and Poynter, R.L.: 1984, Detection of a radio emission at 3 kHz in the outer heliosphere, *Nature* **312**, 27.

Lallement, R.: 1998, Observations of the Local Interstellar Cloud, IAU colloquium N° 166, in: D. Breitschwerdt, M.J. Freyberg and J. Trumper (eds.), *The Local Bubble and Beyond*, Lecture Notes in Physics, **506**, 19–28.

Lallement, R.: 1999, Global structure of the Heliosphere, Solar Wind 9, in: Habbal, Esser, Hollweg and Isenberg (eds.), *AIP Conf. Proc.* **471**, pp. 205–210.

Lallement, R. and Bertin P.: 1992, Northern-hemisphere observations of nearby interstellar gas: possible detection of the local cloud, *Astron. Astrophys.* **266**, 479.

Lallement, R., Ferlet, R., Lagrange, A.M., Lemoine, M. and Vidal-Madjar, A.: 1995, Structure of the local interstellar clouds, *A&A* **304**, 461.

Lallement, R., Ferlet, R., Vidal-Madjar, A. and Gry, C.: 1990, Velocity structure of the local interstellar medium, in: S. Grzedzielski and E. Page (eds.), *Physics of the Outer Heliosphere*, COSPAR Colloquia Series, No. **1**, Pergamon Press, London, pp. 37–42.

Linsky, J.L. and Wood, B.E.: 1996, 'The alpha cen line of sight: D/H ratio, physical properties of local interstellar gas, and measurement of heated hydrogen (the hydrogen wall) near the heliopause, *Astrophys. J.* **463**, 254.

Moebius, E.: 1996, The local interstellar medium viewed through pickup ions, resent results and future perspectives, *Space Sci. Rev.* **78**, 375.

Parker, E.N.: 1963, *Interplanetary Dynamical Processes*, Interscience Publishers, New York.

Pesses, M.E., Jokipii, J.R. and Eichler, D.: 1981, Cosmic ray drift, shock wave acceleration, and the anomalous component of cosmic rays, *Astrophys. J.* **246**, L85.

Piskunov, N., Wood, B.E., Linsky, J.L., Dempsey, R.C. and Ayres, T.R.: 1997, Local interstellar medium properties and deuterium abundances for the lines of sight toward HR 1099, 31 Comae, beta Ceti, and beta Cassiopeiae, *ApJ* **474**, 315.

Quémerais, E., Bertaux, J.L., Lallement, R., Kyrola, E., Schmidt, W. and Berthé, M.: 1999, Interplanetary Lyman alpha Line Profiles derived from SWAN/SOHO H Cell measurements 1. The Full Sky Velocity Field, *J. of Geophys. Res.* **104**, 12585.

Sfeir, D., Lallement, R., Crifo, F. and Welsh, B.Y.: 1999, Mapping the contours of the Local Bubble: preliminary results, *Astron. Astrophys.* **346**, 785.

Von Steiger, R. *et al.*: 1996, *Space Sci. Rev.* **78**(1-2).

Witte, M., H. Rosenbauer, M. Banaszkiewicz, and Fahr, H.-J.: 1993, The Ulysses neutral gas experiment: Determination of the velocity and temperature of the interstellar neutral helium, *Adv. Space Res.* **13**, 121.

Witte, M., Banaszkiewicz, M. and Rosenbauer, H.: 1996, Recent results on the parameters of the interstellar helium from the Ulysses/GAS experiment, *Space Sci. Rev.* **78**, 289.

Wolff, B., Koester, D. and Lallement, R.: 1999, Evidence for an ionization gradient in the local interstellar medium: EUVE observations of white dwarf, *A&A* **346**, 969.

Wood, B.E., Alexander, W.R. and Linsky, J.L.: 1996, The properties of the local interstellar medium and the interaction of the stellar winds of epsilon INDI and lambda andromedae with the interstellar environment, *ApJ* **470**, 1157.

Zank, G.P. and Frisch, P.C.: 1999, Consequences of a change in the galactic environment of the sun, *ApJ* **518**(2), 965–973.

TURBULENCE AND WAVES IN THE SOLAR WIND FORMATION REGION AND THE HELIOSPHERE

IGOR VESELOVSKY
Institute of Nuclear Physics, Moscow State University, 119899, Moscow, Russia
E-mail: veselov@dec1.npi.msu.su, Fax: +7(095) 939 3553

Abstract. A short review is presented of the dissipative and radiative MHD, electromagnetic and plasma kinetic scaling approaches which are useful for sorting out different physical processes according to their relative importance. The 'velocity-emission' ratio Ve, the Faraday number F and 'Trieste numbers' T are described together with more familiar dimensionless parameters. The concepts of the turbosphere and turbopause around the Sun are outlined.

1. Introduction

The regular and irregular electromagnetic fields and plasma structures are ubiquitous in the solar atmosphere and the heliosphere. They are carrying a free magnetic and plasma energy in its different shapes being linearly and nonlinearly coupled and dynamically interacting in a complicated and rather unstable manner which is not sufficiently known, investigated, evaluated and understood in many instances. The mutual causal relations between self-organized laminar and turbulent structures are often not clear even in the best studied cases. The dimensionless scaling approach is very promising in this situation with nonlocal connections between different space-time scales. The aim of this paper is to present a short review of the dissipative and radiative MHD, electromagnetic and plasma kinetic scaling analysis for the solar corona and the heliosphere.

2. Dimensionless Scaling Approach

First of all, the Knudsen number $Kn = \lambda/L$ which is the ratio of the particle mean free path λ to the characteristic length scale of the problem L, delimits kinetic regimes with rare collisions $Kn \gg 1$ from macroscopic 'fluid' regimes (MHD etc.) with frequent collisions $Kn \ll 1$. The mean free path in the solar wind increases with the heliospheric distance r from values of an order of ~ 0.1 Mm in the corona up to ~ 0.1–1 AU at the Earth's orbit in the heliosphere. The internal state of the solar and heliospheric plasma appreciably deviates from the local thermodynamical equilibrium in many important instances because of not sufficient collisions. One of the actual unsolved problems of the coronal heating and solar wind acceleration

Astrophysics and Space Science **277:** 219–224, 2001.

is related to the question: what are kinetic structures and processes involved? How does all of this fit the local and global macroscopic or bulk plasma parameters?

Second, the macroscopic approximation and the dissipative MHD approach have their own applicability domain for sufficiently large scales, i.e. for $Kn \ll 1$. This approach was described earlier for the solar wind origin problem (Veselovsky, 1996). Here we recall only the main dimensionless parameters appearing in the mass, momentum and energy conservation equations. The Strouhal number $S = \frac{vt}{L}$, where v, t and L represents characteristic velocity, time and length scales of the phenomenon, delimits quasistationary ($S \gg 1$) and transient ($S \ll 1$) processes. Complicated multi-scale phenomena are typical for the most of solar and heliospheric morphological structures and processes. In this case, both quasistationary and transient features could be found and play an important role. Practically, we never met one-scale situations on the Sun. The Mach and Mach-Alfven numbers are well known quantities $M = v/C_s$, $M_A = v/V_A$, where C_s and V_A are the sonic and Alfven speed respectively. They are used most often in the solar literature and delimit subsonic ($M < 1$), supersonic ($M > 1$), subalfvenic ($M_A < 1$) and superalfvenic ($M_A > 1$) regimes. The Froude number $Fr = \frac{v^2}{Lg}$, delimits the flows with an essential ($Fr \gtrsim 1$) or negligible ($Fr \ll 1$) role of the gravity acceleration g.

Dissipative transport processes are represented by the kinematic viscosity ν, temperature conductivity χ and electrical conductivity σ. Correspondingly, the Reynolds number $Re \sim \frac{vL}{\nu}$ is often used to delimit laminar ($Re \ll 1$) and turbulent ($Re \gg 1$) regimes. In the simplest gas-dynamic approximation $\nu \sim \lambda v_T$, where v_T is the thermal speed which is of the same order of magnitude as C_s, one has $Re \sim M/Kn$. The macroscopic approximation is valid when $Kn \ll 1$, hence laminar regimes are strongly subsonic. To avoid possible confusions we remark that the definition of laminar and turbulent flows is not absolute, but relative and scale-dependent. The same flow often seems to be laminar for small-scale features and turbulent for sufficiently large scales. Opposite situations are also known. Moreover, the definition of the turbulence is not invariant, but reference frame dependent. There is no standard generally accepted description of the terms 'laminar' and 'turbulent'.

The dimensionless parameter Ve characterizes the relative importance of the bulk radiative gains (losses) due to atomic processes (excitation, recombination, ionization, bremsstrahlung etc.) in comparison with the kinetic energy losses (gains) from the given volume. It is the so called 'velocity-emission' ratio defined as

$$Ve = \frac{W}{W_{emission}}, \tag{1}$$

where W represents the plasma kinetic energy flux through the closed surface of the given volume, and $W_{emission}$ stands for the electromagnetic power emitted (absorbed) by the same volume. Optically thin cases are met in the solar corona and the heliosphere. The radiation is decoupled or plays a negligible role in the plasma energy and momentum balance for the coronal holes and the solar wind, where $Ve \gg 1$. The opposite situation, $Ve \ll 1$, is characteristic for the radiative zone and active regions. The magnetic Reynolds number $Re_m \sim \frac{\sigma L v}{c^2}$ and the Prandtl number $Pr \sim \nu/\chi$ together with the mentioned numbers S, M, M_A, Fr, Ve represent a set of physically independent dimensionless parameters used for the radiative MHD regimes classification scheme, which allow a better understanding of dominant physical processes in the solar wind formation region, where the ionization degree is high. Additional more detailed atomic parameters are needed to characterize the transition region and the chromosphere, where the excitation and ionization processes are taking place and the radiative losses dominate, $Ve \ll 1$. The energy balance equation includes several terms, which are scaled as follows. Let us put the kinetic energy flux density divergence as a reference quantity, which is arbitrary selected to be of an order of one, $\nabla \left(\rho \frac{v^2}{2}\vec{v}\right) \sim 1$. In this case, the divergence of the enthalpy flux density will be of an order of M^{-2}; the divergence of the viscous heat flux, $\sim Re^{-1}$; temperature conductivity term, $\sim Re^{-1}M^{-2}Pr^{-1}$; Joule dissipation, $\sim M_A^{-2}Re_m^{-1}$; electrodynamic pondermotive term, $\sim M_A^{-2}$; kinetic energy inertial term, $\sim S^{-1}$; internal energy inertial term, $M^{-2}S^{-1}$; radiation, $\sim Ve$. A similar scaling exists for the momentum balance equation (Veselovsky, 1996).

Another useful dimensionless parameter is given by the Faraday number F which is defined as follows. Let us consider the electric field

$$\vec{E} = -\vec{\nabla}\varphi - \frac{1}{c}\frac{\partial \vec{A}}{\partial t} \tag{2}$$

which is determined by the scalar potential φ and the vector potential $\vec{A}$. In this case, one defines

$$F = \frac{j}{\rho c}\frac{L}{ct}, \tag{3}$$

where j, ρ, L, t stand for the electric current and charge densities, space and time scales. In the situation when $F \gg 1$, induction fields dominate over potential fields. This situation is very common in the MHD and well investigated in the solar physics. An opposite situation, $F \ll 1$, is typical for the electrostatics. Double electric layers, particle acceleration etc. are taking place. The charging of the solar plasma is poorly known. It is often neglected without sufficient grounds. Nevertheless, the charges are very important for sufficiently small scales and large times. Asymptotically, at $t \to \infty$, the electrostatics is obviously dominating over the MHD. Because of this, a solution of the problem of the stationary plasma confinement should be searched taking into account the electric charging. A standard MHD approach is not sufficient for this purpose. The role of thin electrostatic layers and small cavities needs additional investigations.

The quantitative measure of the non-locality and the boundary condition influence on the physical processes and structures inside the given volume is represented by the so called 'Trieste numbers' $T_{\alpha\beta}$. They are introduced as follows (Veselovsky, 1999). Let us consider any vector field $\vec{v}(\vec{r}, t)$ (velocity, magnetic field, electric current density ...), which is composed of internal field lines $\vec{v}_i$, external field lines $\vec{v}_e$ and linking field lines $\vec{v}_l$, which are situated inside, outside or crossing the boundary of the considered volume $\vec{v} = \vec{v}_i + \vec{v}_e + \vec{v}_l$. 'Trieste numbers' $T_{\alpha\beta}$ represent the ratio of corresponding terms (i, e, l) to each other and characterize the opening degree of the system against the mass, momentum and energy transport. For example, the linking mass influx is given by $\dot{m}_l = \oint \vec{J} d\vec{s}$, where $\vec{J} = \rho\vec{v}$ is the mass flux density and the integration is performed on the boundary surface of the volume. This quantity is equal to $\dot{m}_l = \oint \vec{J}_l d\vec{s}$ by the definition. It is interesting to compare this value with the intensity of the internal mass circulation inside the volume, which can be represented by different characteristics. One of this internal characteristics is given by the mass flux $\dot{m}_i$ crossing an arbitrary surface S_i inside the volume, $\dot{m}_i = \int \vec{J}_i d\vec{s}$, where the integration is performed on the surface S_i. The ratio $T_{li} = \dot{m}_l/\dot{m}_i$ represents the simplest example of 'Trieste numbers' in this case. There is no mass exchange between the volume under consideration and its surroundings when $T_{li} = 0$. The exchange of mass is low when $T_{li} \ll 1$ and high when $T_{li} \gg 1$. The concept of an isolated system is valid only for sufficiently small T_{li} numbers defined for mass, momentum and energy fluxes.

3. Scales in the Turbosphere and Turbopause Around the Sun

The concepts of the turbosphere and turbopause were used for the description of the nonstationary and non-linear multiscale structures in the solar wind formation region (Veselovsky, 1998). The physical meaning of the turbosphere is very simple: it is a region around the Sun where turbulent velocities v_1 are higher than the regular mean flow velocity v. The turbopause delimits the turbosphere from the solar wind region where an opposite situation takes place, $v > v_1$. There is no downflows outside the turbopause. The alternative definition of the turbopause could be as the region where the condition $v_1 \approx v$ is fulfilled, or more formally: it is the manifold of the highest points, where the radial velocity component of the fluid element attains the last zero value along the streamline. This manifold includes isolated points, lines and surfaces. The heliocentric distance to the turbopause $r(\theta, \varphi, t)$ is very complicated and poorly known function of its arguments with many characteristic scales reflecting the solar and heliospheric activity. Globally, the turbopause is oblate during solar minimum years and more spherical, but more structured and dynamical during maximum ones. The dependence on the arguments is very strong and the turbopause is far from being a smooth spherically symmetric surface deviating occasionally from its average position at 2-3 solar radii up to the lowest solar atmosphere during the perturbed conditions. Both week (subsonic) and strong (supersonic) turbulent regimes are met in the turbosphere. Morphologically, they are represented by small inhomogeneities, oscillations and waves or by jets and shock waves respectively. Because of this, the turbopause does not coincide with the sonic surface and differs from it.

There are several important 'internal or physical' length scales in the turbosphere: the heliocentric distance, the mean free paths, Larmor radii, the Debye length. In addition to this, many 'external or geometry' length scales imposed by the inner and outer boundary conditions are manifested as different nonlocal morphological structures. Accordingly, 'intensive' and 'extensive' time scales and variables could be indicated. We consider the turbosphere as an open system with the energy, momentum and mass flows out of the local thermodynamical and mechanical equilibrium. The memory of the initial and boundary conditions and the proper self-organization and structuring processes due to the available free magnetic, gravity, thermal and mechanical energies are difficult to deconvolve. 'Trieste numbers' are useful from this point of view for sorting out the prominences, filaments, loops, arcades and jets according to their origins and the evolution. For example, if we take into account the observed continuos mass exchange between the 'quiescent' prominences and their surroundings, which happens many times during the life time of the prominence, we find $T \gg 1$. This means that theoretical models, which consider the quiescent prominence as an supported isolated system, are far from the adequate description of the situation. According to numerous observations, turbulent motions are enhanced during the lifting CME processes, but the outflow velocity increase is most striking. Because of this, the CME initiation process probably means a temporary local shift of the turbopause downwards. The dynamical role of the turbulence in the CME initiation process is not obvious and needs additional studies.

4. Conclusions

Independent dimensionless parameters allow to elucidate main physical processes in the turbulent solar and heliospheric plasma. Many of them are very common, but several important parameters, for example, (Ve, F, T) are not sufficiently investigated and evaluated. These parameters can be also used to delimit radiative dominated and plasma dominated energy transport regimes, inductive and electrostatic phenomena, local and non-local processes in many astrophysical and laboratory applications.

Acknowledgements

The work was partially supported by the RFBR grant 98-02-17660, the Federal Program 'Astronomy' project 1.5.6.2, the Program 'Universities of Russia' project 99 0600. The author is grateful to the Organizing Committee of the International Conference 'Physics of Space: Growth Points and Problems' for the travel support that made it possible to report this paper, to the referee and F. Pantellini for useful comments on the manuscript.

References

Veselovsky, I.S.: 1996, Nearly sonic and supersonic convective motions in the solar wind formation region, in: D. Winterhalter *et al.* (eds.), *Solar Wind Eight*, pp. 161–164. AIP, Woodbury, New York.

Veselovsky, I.S.: 1998, Turbosphere and turbopause in the solar wind formation region, in: S.R. Habbal *et al.* (eds.), *Solar Wind Nine*, pp. 381–384. AIP 382, Woodbury, New York.

Veselovsky, I.S.: 1999, *Scaling Approach in the Solar and Heliospheric Plasma*, The Abdus Salam International Centre for Theoretical Physics. SMR 1161/10, Trieste, Italy.

JETS FROM YOUNG STARS AND COMPACT OBJECTS ENVIRONMENTS

GUY PELLETIER
Laboratoire d'Astrophysique de l'Observatoire de Grenoble, FRANCE

Abstract. Several classes of cosmic objects, such as Young Stellar Objects, Active Galactic Nuclei, Micro-Quasars, Pulsars and probably Gamma Ray Bursts, display powerful winds and jets; for some of them the flow is even ultrarelativistic. For all these classes of objects, the magnetic field is supposed to play a major role in launching and collimating the flow, together with the angular momentum transfer. It probably plays an important role for the turbulent transport in accretion disks also. Regarding the high energy radiation of relativistic jets and the cosmic ray generation, the magnetic field is of course the acceleration agent and could produce the Ultra High Energy Cosmic Rays in some extragalactic objects. The main growth points of these topics are presented, mostly in the case of black hole environments; the case of Young Stellar Objects is more complicated because of the interaction of the stellar magnetosphere with the accretion disk, and the models for this interaction are not yet founded on a reliable theory.

1. Introduction

The development of interferometers has allowed us to observe the widespread phenomenon of winds and jets formation in the environment of several classes of objects, such as Young Stellar Objects (YSO), Active Galactic Nuclei (AGN), microquasars and also plausibly, although with no direct evidence, in Pulsars and in Gamma Ray Bursts (GRB). The analysis of jet formation in all these classes of objects tends to strongly favor the role of the magnetic field in the phenomenon; all other alternatives, based on radiative hydrodynamics, failed in describing powerful collimated jets. The main task of the magnetic field is to convert rotation energy into directed motions, either from a star (Mestel, 1965) or a black hole (Blandford and Znajek, 1977) in solid rotation, or from an accretion disk in Keplerian rotation (Blandford and Payne, 1982), through opened field lines anchored on the fast rotator and connected to the remote ambient medium. Many authors have elaborated with the same pattern, let me mention some other pioneering works: Michel (1975), Pudritz and Norman (1986), Lovelace *et al.* (1987), etc.

In 1962, Evry Schatzman proposed that the rotation of premain sequence stars could have been slowed down by the opened field lines unfolded in their wind and that can transfer the angular momentum. Only a tiny flux of matter from the corona is able to make the angular momentum transport. Thus all the progresses made in the understanding of the solar wind dynamics, especially in the corona, since Jean-

Astrophysics and Space Science **277:** 225–234, 2001.

Louis Steinberg and collaborators works, are very useful for the theory of ejection in remote astrophysics objects.

2. An Overlook of the Astrophysical Jets and Winds

In the case of YSOs, their matter content, their strong power correlated with the IR luminosity, and their velocity, are convincing arguments in favor of magnetized jets emanating from the internal region of an accretion disk. The kinetic power is comparable to the accretion power revealed by infra red luminosity. The motions of the clouds observed in the T-Tauri flows can be compared with the Keplerian motions in the innermost regions of the accretion disk. However the physical description is complicated by the interaction of the magnetosphere of the star with the accretion disk. For instance if the star has a dipolar field, the accretion flow is rerouted towards the polar caps along the magnetic lines. There is no clear answer to the question whether the interaction spins down the star or not, and whether a possible slowing down of the star rotation energizes the outflow. Moreover it seems that the magnetospheres are often oblique and therefore the accretion flow depends on favorable phase of the magnetosphere rotation.

Some AGN jets exhibit knots in central regions having superluminal velocities revealing relativistic motions with bulk Lorentz factors $\Gamma \sim 10$. At large scales, extragalactic jets do not seem to be highly relativistic; there is even evidence of subrelativistic motions in many cases. There are three possibilities to explain jets from AGNs. First, the conversion of part of the accretion power into Poynting flux, which, at MHD approximation is immediately converted into bulk motion. For a black hole of mass M_* and an accretion rate $\dot{M}_a$, the accretion power provided by matter flowing from infinity towards the innermost Keplerian orbit at r_i (last stable orbit at $r_i = 3r_G$ for a Schwarzschild black hole) is given by

$$P_a = \frac{1}{2}\dot{M}_a\frac{GM_*}{r_i} \simeq \frac{1}{12}\dot{M}_a c^2 \ . \tag{1}$$

This power can be measured in terms of Eddington luminosity, the black hole mass being thus bigger than this obtained from the Eddington limit:

$$P_a = \frac{\dot{M}_a}{\dot{M}_{Edd}}\frac{M_*}{10^8 M_\odot} \times 10^{46}\ erg/s \ . \tag{2}$$

The accretion rate at Eddington limit is proportional to the mass and is of the order of a solar mass per year for $10^8 M_\odot$. A ratio χ is converted into radiation (an UV-black body for an accretion disk of Shakura-Sunyaev type) and the other part into twin jet power, so that each jet has a power

$$P_j = \frac{1-\chi}{2}P_a \ . \tag{3}$$

Since only a tiny fraction of matter is rerouted from the accretion disk to the outflow, with a mass flux $\dot{M}_j \ll \dot{M}_a$, the jet velocity is larger than the Keplerian velocity of the foot point. When most of the power in the jet has become kinetic, the asymptotic bulk motion is such that

$$\Gamma \simeq 1 + \frac{1-\chi}{24}\frac{\dot{M}_a}{\dot{M}_j} . \tag{4}$$

Thus, since the power in the jets can be comparable to the luminosity of the disk, $\Gamma \sim 10$ is obtained when $\dot{M}_j \sim 10^{-2}\dot{M}_a$, whereas a more loaded jet with $\dot{M}_j \sim 10^{-1}\dot{M}_a$ is subrelativistic. As will be stated further on, a magnetic field of the order of its equipartition value in the vicinity of the black hole is necessary for the accretion disk to launch jets:

$$B \sim B_{eq} = (\frac{M_*}{10^8 M_\odot})^{-1/2} \times 10^3 G . \tag{5}$$

A similar general statement can be made for microquasars, the accretion-ejection laws can be obtained by simply changing the mass (of order $10 - 100 M_\odot$).

A spinning black hole can provide a strong Poynting flux as well (Blandford and Znajek, 1977); its rotation energy is tapped by a magnetic field maintained by the accretion disk (a black hole has no proper magnetic field). The maximum power liberated by the black hole slowing down is

$$P_{sd} = (\frac{a}{m})^2 (\frac{B_n}{10^3 G})^2 (\frac{M_*}{10^8 M_\odot})^2 \times 10^{41} erg/s , \tag{6}$$

where the reduced specific angular momentum a is such that $|a| \leq m$ (mass parameter) (see Camenzind 1990). Clearly this process of relativistic jets generation can be competitive with jet launching by the accretion disk only for black holes of mass larger than $10^9 M_\odot$ (the most powerful quasars), taking account of the scaling (5) of the magnetic field. The slowing down time scale is very long

$$t_{sd} \simeq (\frac{B_t}{10^3 G})^{-2} (\frac{M_*}{10^8 M_\odot})^{-1} \; 10^{12} years , \tag{7}$$

which can easily be longer than the Universe age. A spinning black hole, through the Kerr metric (Lens-Thiring effect) and dissipation, tends to enforce matter of the innermost region of the accretion disk to rotate in the equatorial plane orthogonal to the rotation axis of the black hole. The gyroscopic effect of the spinning black hole allows to maintain a stable axis for a long time.

This mechanism of relativistic jets production has similar features with the pulsar $e^+ - e^-$ wind (see V. Beskin talk in this proceedings). But because a pulsar has a strong proper field, $B \sim 10^{12} G$, the bulk Lorentz factor of the relativistic flow is very high: $\Gamma \sim 10^6$.

Another possibility to generate relativistic flows in a black hole environment is the Compton rocket effect (O'Dell, 1981). The anisotropic radiation field emanating from the accretion disk exerts a Compton force on electrons and positrons, that drags them if moving faster than some bulk velocity and, on the contrary, propels them if slowlier. A cold pair plasma is efficiently slowed down to a limit bulk Lorentz factor close to 2 (Phinney, 1982), whereas a pair plasma maintained at a high relativistic temperature by stochastic acceleration in the magnetic 'cauldron' can reach bulk Lorentz factors of order 10 (Henri and Pelletier, 1991). The magnetic "cauldron' is constituted by the inner magnetic surfaces of the MHD outflow from the accretion disk, and they are supposed to convey Alfvén turbulence necessary for particle acceleration.

The flaring behavior observed in the whole electromagnetic spectrum including gamma-rays up to $10 TeV$ in some class of AGNs, namely the 'blazars', is probably not compatible with the spinning black hole scenario which is very continuous. Those flares are likely due to instabilities in the accretion disk or in the high energy 'cauldron'.

The interesting new field of the Gamma Ray Bursts, presumably caused by either compact objects mergings or hypernovae, has aroused theoretical developments involving the formation of a black hole with an accretion disk in extreme regime, giving rise to a relativistic flow having a bulk Lorentz factor $\Gamma = 10^2 - 10^3$ (Rees and Meszaros, 1992). During the short time of the black hole formation, the magnetic field could exceed $10^{13} G$ and even reach $10^{15} G$.

3. The MHD Jet

All the previous examples of jet phenomena share common features: from the opened poloidal field lines anchored on a rotator, a toroidal field component develops because of differential rotation between field lines (Ω_*) and matter (Ω),

$$B_\phi = \frac{\rho r}{\kappa}(\Omega - \Omega_*) \ , \tag{8}$$

($\kappa \propto d\dot{M}_j(a)/2\pi da$ where $2\pi a$ is the magnetic flux across an horizontal disk of radius r centered on the axis at z) which has three major effects for stationary, axisymmetric flows. First, the magnetic torque transfers the angular momentum along the ideal MHD flow according to

$$\Omega r^2 - \frac{r B_\phi}{\mu_0 \kappa} = l_*(a) \equiv \Omega_* r_A^2 \ , \tag{9}$$

where r_A is an important theoretical ingredient, namely the Alfvén radius, defined on each magnetic surface by specifying the total specific angular momentum $l_*(a)$ and the field line rotation $\Omega_*(a)$. The Alfvén radius is related to the Alfvén surface where the local Alfvén number m ($m \equiv u_p/V_{Ap}$) reaches unity. This is a critical surface of the MHD flow; and the regularity condition at this surface implies that the flow must cross it at a distance r from axis exactly equal to the Alfvén radius. It is a measurement of the magnetic surface widening and the larger the Alfvén radius compared to the footpoint radius the more efficient the magnetic braking.

Second, a poloidal Poynting flux propels matter,

$$\vec{S}_p \equiv \frac{\vec{E} \times \vec{B}_\phi}{\mu_0} = -\Omega_* r \frac{B_\phi}{\mu_0} \vec{B}_p \, , \tag{10}$$

which modifies of the Bernoulli invariant (written in the nonrelativistic case):

$$\frac{1}{2}u^2 + h + \Phi \quad \Omega_* r \frac{B_\phi}{\mu_0 \kappa} - \mathcal{B}(u) \, . \tag{11}$$

When the lever arm is large enough ($r_A \gg r_0$, r_0 being the magnetic surface radius at the disk surface), the Bernoulli invariant is dominated by the magnetic braking contribution at the beginning of the outflow and $\mathcal{B}(a) \simeq \Omega_*^2 r_A^2$. Then, because of the flow widening, the poloidal kinetic energy dominates at infinity provided that the critical surfaces are crossed, and $\mathcal{B}(a) \simeq \frac{1}{2}u_{p\infty}^2$; which leads to $u_{p\infty} \simeq \sqrt{2}\Omega_* r_A$, which is larger than the velocity at the footpoint.

Third, the tension effect due to the toroidal field confines the flow, the confinement being described by a generalized Grad-Shafranov equation, modified by the flow. It is a second order partial differential equation of the flux function a, that degenerates on the Alfvén surface.

In the subAlfvénic region, magnetic surface widening allows to convert rotation energy into poloidal kinetic energy. Cold matter can escape by centrifugal effect from an accretion disk provided that the field lines bend with an angle with respect to axis larger than 30^0 (Blandford and Payne, 1982). Beyond the Alfvén surface (see 1), acceleration is no longer significant and the growth of the toroidal field, due to rotation slowing down, allows self-collimation of the flow; Heyvaerts and Norman (1989) showed that a current carrying jet undergoes cylindrical collimation.

The stationary flow is described by the system of the two couple equations (Pelletier and Pudritz, 1992) for the magnetic flux function $a(r, z)$ and the Alfvén number $m(r, z)$, namely Bernoulli and Grad-Shafranov equations, it displays critical surfaces, a priori unknown, and changes its type at other unknown surfaces (see Sauty *et al.*, 1998 for a good review on these issues). That kind of numerical analysis is still an unsolved problem from a mathematical viewpoint; there are nevertheless numerical simulations, but using a temporal scheme.

4. Accretion Disk Launching Jets: The Transport Issues

How does an accretion disk launch jets and how does it change its properties? It turns out that the magnetic field must not be too low compared to the equipartition

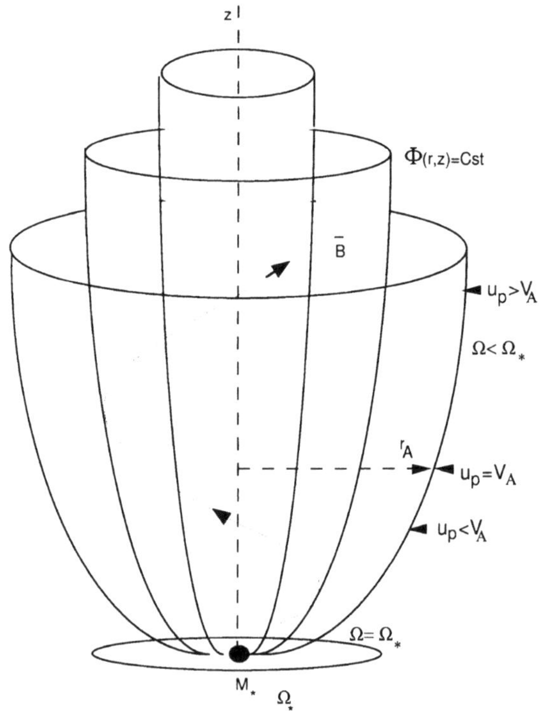

Figure 1. The plasma flows along widening magnetic surfaces, in quasi isorotation in the subAlfvénic region, is accelerated up to the Alfvén surface, passes through it and then self-collimates when losing rotation energy.

value, because the Keplerian flow would destroy the opened configuration. It must not be too strong, because then the appropriate bending of the opened field lines would not be set up. A necessary condition to launch jets is that the pressure ratio $\beta_p \equiv P/B^2/2\mu_0$ be between 0.5 and 2 (Ferreira and Pelletier, 1995). The MHD outflow is then able to extract the whole angular momentum allowing then accretion. However some magnetic diffusivity, and therefore some resistivity, is necessary for quasi steady accretion, otherwise matter could not flow across the magnetic surfaces toward the central object. Also Magnetic Reynolds numbers, defined by $\mathcal{R}_m \equiv ru_r/\nu_m$, in the range $20 - 200$ are required. Under the simplest description of the disk heating, namely through viscous and Joule heating, it turns out that the angular momentum is mostly transferred by the magnetic field despite the viscous torque contribution, and the dissipation, that maintains a black body radiation, leads to a luminosity very smaller than jet power. The jets take most of the accretion power and their mass flux is of order of ten percent of the accretion rate. Such regimes could be encountered in YSOs.

Any additional heating, due to microturbulence for instance, additional to the heating described by appropriate transport coefficients, turns out to change significantly the conditions to launch jets. Indeed Casse and Ferreira (2000) showed that a tiny fraction of additional heating, especially in the corona but not necessarily only in it, allows more flexible launching conditions. In particular the mass outflux can be as tenuous as a few percent of the accretion rate, luminosity and jet power can be comparable, and the magnetic Reynolds number can be as low as unity. It becomes also possible with this description to pass from a standard accretion disk to an advection dominated disk. However the condition of rough equipartition for the magnetic field is always required, and makes inner regions of the accretion disk more favorable for jet launching.

Standard accretion disks of Shakura-Sunyaev type need a turbulent viscosity in order to transfer the angular momentum, thus allowing accretion. The expected viscosity scales like $\nu_t = \alpha C_s h$ (C_s is the sound velocity of the accretion disk, and h its half width) and the turbulence parameter α would range between $10^{-3} - 10^{-1}$. As previously stated, opened field lines can transfer the angular momentum, and turbulent viscosity would not be useful. But magnetic diffusivity is necessary and a scaling similar to turbulent viscosity is required: $\nu_m = \alpha_m C_s h$. Indeed turbulence tends to generate an effective viscosity and an effective magnetic diffusivity of the same order of magnitude. A turbulent thermal conductivity is also generated and useful for a complete description of the accretion disk.

What is the source of the turbulence? A Keplerian disk is stable with respect to hydrodynamical perturbations according to Rayleigh criterion which states stability when $r\frac{\partial \Omega r^2}{\partial r} > 0$, except possibly with nonlinear perturbations (Zahn, 1991). In 1992, Balbus and Hawley investigated the capability of the magneto-rotational instability (Velikhov, 1959; Chandrasekhar, 1960) to supply turbulence in accretion disk and to solve the transport problem. As a consequence of the isorotation theorem of ideal MHD, magnetic surfaces threading a disk enforce matter to keep Keplerian rotation when they suffer Alfvén perturbations. Therefore moving away fluid elements, that rotate faster than necessary to balance gravity, pull magnetic surfaces outwards, whereas approaching fluid elements, that rotate slowlier than necessary to balance gravity, push magnetic surface inwards. This instability develops when the tension effect of the magnetic field is not too strong:

$$k_{\parallel}^2 V_A^2 + r\frac{\partial \Omega^2}{\partial r} < 0 \,. \tag{12}$$

For a given disk width h ($k_{\parallel min} = h/\pi$), the magnetic field must be weaker than a critical value B_c, such that $B_c^2/2\mu_0 = \rho|r\frac{\partial \Omega^2}{\partial r}|h^2/2\pi^2$, to get the instability and it turns out that this critical value is roughly the equipartition value. The maximum growth rate is thus $\gamma_i \sim \Omega(B_c^2 - B^2)^{1/2}/B_c$, and the instability can grow in few Keplerian periods for low enough magnetic field. It seems that the magnetic field can be arbitrary low, which would be contradictory with the Rayleigh criterion. In fact, the magneto-rotational instability stems from another branch (Alfvénic) of the dispersion relation and cannot be considered in ideal MHD approximation at vanishing magnetic field. Numerical simulations suggest that the transport problem could be solved with that instability (Balbus and Hawley) and Kolmogorov phenomenology of turbulence leads to a satisfactory estimate of the parameter α (Heyvaerts *et al.*, 1996).

However this is satisfactory as long as the issue of jet launching is not considered. Unfortunately, as previously seen, this requires a field intensity close to the equipartition value, which means close to the critical value B_c that quenches the instability. Thus to conciliate accretion and ejection, which seems possible with phenomenological transport coefficients, becomes a problem from the viewpoint of fundamental transport theory based on an instability generated turbulence.

Non steady accretion-ejection flows has been observed in microquasars (their proximity and smallness compared to AGNs allow to follow their variations) and non steady theories are being developed (Tagger and Pellat, 1999). Sandpile type avalanche behavior seems to happen in these flows. Interchange and ballooning instabilities are currently investigated for these accretion-ejection flows in this spirit (Kersale *et al.*, 2000).

Dynamo action in accretion disks is another important issue. These accretion-ejection machines work with a bipolar magnetic configuration. A quadripolar configuration does not extract the angular momentum efficiently. The bipolar configuration can simply be maintained by the accretion flow itself that carries and compresses the magnetic field. But the question raises: 'what happens if dynamo action works? Does it destroy the bipolar configuration?' The usual $\alpha\Omega$-dynamo, applied to the Sun for instance, is known to favor the quadripolar configuration. But it has been recently shown that the dynamo parameter α related to helicity could take negative values off the equatorial plane of the disk (Brandenburg, 1997) and then the bipolar configuration is favored. A first model of jet launching from a disk subject to that kind of dynamo action has been elaborated by Campbell (Campbell, 1999).

5. New Topics in Astrophysical Jets

Besides the important issues of instabilities and turbulent transport in these magnetized accretion-ejection flows, I talked about, the high energy radiation aspect deserves being mentioned. YSOs produce intense X-rays radiation, probably related to the interaction of their magnetosphere with the accretion disk and its outflow. This is an important observational topic with the satellites Chandra and XMM.

The relativistic flares in Blazars emit an intense gamma rays spectrum (up to 10TeV for some of them). Gamma ray astronomy revealed that extragalactic jets are efficient particle accelerators and they probably nourish the cosmic ray spectrum beyond $10^{15}eV$, the limit energy for the acceleration by supernovae remnants.

The origin of the Ultra High Energy Cosmic Rays is an important enigma of modern astrophysics for which new 'astroparticle' observatories are being built (Auger for the detection of UHE Cosmic Rays around $10^{20} eV$, Antares and Amanda for the detection of very high energy neutrinos produced by UHE Cosmic Rays colliding with photons).

For the 'bottom up' scenario to succeed in explaining the UHR Cosmic Rays spectrum (that even extends beyond $10^{20} eV$!..), opposed to the 'top down' scenario based on the disintegration of exotic particle physics objects, an extreme regime of Fermi acceleration must be considered. Particle acceleration in the relativistic plasma of relativistic jets is still an interesting and opened field of theoretical investigation.

GRBs are likely produced by ultrarelativistic winds, possibly collimated, in which the magnetic field plays an important role as well. They are considered also as good candidates to produce the UHE Cosmic Rays.

This brief talk has shown, I hope, that the topics of astrophysical jets and winds have considerably enriched nowadays. These investigations incite new theoretical developments of MHD, kinetics in relativistic plasmas, high energy astrophysics and new kind of instrumental developments, like for instance the 'astroparticles' observatories. They are related to fundamental issues of modern astrophysics like black holes, neutrons stars, UHE Cosmic Rays and high energy phenomena in the Universe. The knowledge accumulated in the investigation of the solar corona, the solar wind and its interaction with planetary magnetospheres is very precious for these new adventures in astronomy.

References

Balbus, S.A. and Hawley, J.F.: 1992, *ApJ.* **392**, 662.
Blandford, R.D. and Payne, D.G.: 1982, *MNRAS* **199**, 883.
Blandford, R.D. and Znajek, R.L.: 1977, *MNRAS* **179**, 433.
Beskin, V.: 2000, this proceedings.
Brandenburg, A. and Donner, K.J.: 1997, *MNRAS* **288**, L29.
Camenzind, M.: 1990, Reviews in Modern Astronomy **3**, in: G. Klare (ed.), *Accretion and Winds*, Springer-Verlag, Berlin.
Campbell, C.G.: 1999, *MNRAS* **310**, 1175C.
Casse, F. and Ferreira, J.: 2000, *Aand A* (in press).
Chandrasekhar, S.: 1960, *Proc. N.A.S.* **46**, 56.
Ferreira, J. and Pelletier, G.: 1995, *Aand A* **295**, 807.
Henri, G. and Pelletier, G.: 1991, *ApJ* **383**, L7.
Heyvaerts, J. and Norman, C.Z.: 1989, *ApJ* **347**, 1055.
Heyvaerts, J., Priest, E. and Bardou, A.: 1996, *ApJ* **473**, 403.
Kersalé, E., Longaretti, P.-Y. and Pelletier, G.: 2000, *Aand A* (submitted).
Lovelace, R.V.E., Wang, J.C.L. and Sulkanen, M.E.: 1987, *ApJ* **315**, 504.
O'Dell, S.L.: 1981, *ApJ* **243**, L147.
Pelletier, G. and Pudritz, R.: 1992, *ApJ* **394**, 117.
Phinney, E.S.: 1982, *MNRAS* **198**, 1109.

Pudritz, R. and Norman, C.: 1983, *ApJ* **274**, 677.
Rees, M.J. and Meszaros, P.: 1992, *MNRAS* **258**, 41.
Sauty, C., Tsinganos, K. and Trussoni, E.: 1998, *Apand SS.* **261.151S**.
Schatzman, E.: 1962, *Ann. d'Ap.* **25**, 18.
Tagger, M. and Pellat, R.: 1999, *Aand A* **349**, 1003.
Velikhov, E.: 1959, *Sov. Phys. JETP* **36**, 1398.
Zahn, J.P.: 1991, in: C. Bertout, S. Collin-Souffrin and J.P. Lasota (eds.), *Structure and Emission Properties of Accretion Disks*, Proc. IAU Colloq. **129**, Editions Frontieres, p. 87.

IV

PLASMAS AROUND PULSARS, PLANETS AND MOONS

PLASMAS AUTOUR DES PULSARS, PLANÈTES ET SATELLITES

RADIO PULSARS – WHAT IS TO BE DONE

VASILY S. BESKIN
P.N. Lebedev Physical Institute, Moscow, 117924, Russia

Abstract. In this review I discuss briefly the main theoretical problems concerning radio pulsars and formulate the observations aimed at a verification of the current models of pulsar magnetosphere and the radio emission mechanism.

1. Introduction

Radio pulsars, i.e., sources of the cosmic pulsed radio emission with a characteristic period $P \sim 1$ s, were discovered in 1967 (Hewish *et al.*, 1968). This discovery was one of the most important events in astrophysics of the twentieth century. Indeed, the cosmic sources associated with neutron stars predicted in the 1930s were first discovered. Such compact objects (with a mass of the order of the solar mass $M_{\odot} \approx 2 \times 10^{33}$ g, and a radius R of only 10 – 15 km) may arise owing to the collapse of normal massive stars at a late stage of their evolution or, for example, of white dwarfs whose mass exceeded the Chandrasekhar limit $M_{\rm Ch} \approx 1.4 M_{\odot}$ as a result of accretion. Many other cosmic sources (e.g., X-ray pulsars, X novae) discovered afterwards have shown that neutron stars are actually some of the most numerous objects in the Galaxy. Thus, it is not surprising that A. Hewish was awarded the 1974 Nobel prize in physics for the discovery of pulsars.

The physical parameters typical of neutron stars (superstrong gravitational and magnetic fields, high-energy particles) are inaccessible to ground-base laboratories. Therefore, pulsars allowed an investigation of the properties of matter in extreme conditions. Moreover, radio pulsars have been successfully employed as probes of the interstellar medium. One can mention, for example, the direct measurements of the electron density and the magnetic field in the Galaxy (Lyne and Graham-Smith, 1998) and of the stellar wind transillumination in binary systems (Johnston, 1995).

Finally, radio pulsars made it possible to verify the predictions of general relativity. Indeed, at present we know six close binary systems that include a radio pulsar and another neutron star. Since radio pulsars resemble an exact watch moving in the time-dependent gravitational field of its companion, they provide unique information on the spacetime curvature. In particular, a change in the orbital period was registered in two cases, and the origin of this change is naturally associated with the energy loss due to gravitational wave radiation. In other words, the predictions of the general theory of gravity were first confirmed (at least indirectly) to

Astrophysics and Space Science **277:** 237–246, 2001.

a higher order (c^{-5}) than the post-Newtonian corrections (c^{-2}). As is well known, R. Hulse and J. Taylor were awarded the 1993 Nobel prize in physics for their achievements in this field. Indeed, the discovery of a binary system with a lifetime of 200 million years, i.e., much shorter than that of the Universe, became one of the most important events of the past decades. This means that neutron stars must merge rather frequently. The model of the merging of two neutron stars is now considered to be one of the highly probable causes of cosmological gamma bursts; the same process seems to be extremely promising for the discovery of gravitational waves.

2. What is Known

The basic physical processes responsible for the observed activity of these unusual objects were clarified soon enough. It became immediately obvious that the exceedingly regular pulsations of the observed radio emission was caused by the neutron star's rotation (Gold, 1968). Some pulsars exhibit frequency stability on the scale of several years even exceeding the stability of some atomic standards, and so a new pulsar time scale is now being developed (Ilyasov *et al.*, 1998).

Next, the energy source of radio pulsars is their rotational energy, and the energy release mechanism is due to their superstrong magnetic field $B_0 \sim 10^{12}$ G (Pacini, 1967). Indeed, the energy loss estimated by the simple magnetodipole formula

$$W_{\rm md} = -J\Omega\dot{\Omega} \approx \frac{1}{6}\frac{B_0^2\Omega^4R^6}{c^3}\sin^2\chi, \qquad (1)$$

where $J \sim MR^2$ is the moment of neutron star inertia, χ is the inclination of a magnetic dipole to the rotational axis, and $\Omega = 2\pi/P$ is the angular velocity of rotation, makes up 10^{31}–10^{34} erg/s for the majority of pulsars. Such energy release leads to the observed deceleration rate $\mathrm{d}P/\mathrm{d}t \sim 10^{-15}$ which corresponds to a braking time $\tau_{\rm D} = P/2\dot{P} \sim 1$–10 million years.

Radio pulsars are thus the only cosmic objects whose evolution is determined by electromagnetic forces. Recall that the radio-frequency radiation itself makes up only 10^{-4}–10^{-6} of the total energy loss. As a result, pulsar radio luminosities range within 10^{26}–10^{28} erg/s, which is by five to seven orders of magnitude less than the solar luminosity $L_{\odot} \approx 3 \times 10^{33}$ erg/s. At the same time, the extremely high brightness temperature $T_{\rm b} \sim 10^{25}$–10^{28} K is a direct evidence in favor of the coherent mechanism of pulsar radio emission (Ginzburg and Zheleznyakov, 1975).

More than 1200 radio pulsars have been discovered by 2000. Most of them are single neutron stars and only 60 are members of binary systems. But the total number of neutron stars in our Galaxy may reach 10^9–10^{10}. This fact is mostly due to a very short (on a cosmic scale) lifetime $\tau_{\rm D}$ within which a neutron star can manifest itself as a radio pulsar. Hence, we are only able to register the youngest single neutron stars. Practically non of the radio pulsars radiate in other ranges of the electromagnetic spectrum. Only nine radio pulsars are seen in the optical, and seven – in the gamma range. Thirty five radio pulsars emit X-rays, but this radiation is not at all always of a clearly pronounced pulsed character.

A superstrong magnetic field of a neutron star has a number of important consequences. First of all, the duration of synchrotron radiation, $\tau_s \sim (m_e c^3)/(e^2 \omega_B^2) \sim 10^{-15}$ s ($\omega_B = eB/m_e c$), appears to be much smaller than the time within which the particle escapes from the magnetosphere. Consequently, the motion of a charged particle in the magnetosphere of a neutron star will be presented as a sum of their motion along magnetic field lines and the electric drift in the transverse direction.

The importance of the one-photon conversion $\gamma \rightarrow e^+ + e^-$ in a superstrong magnetic field, which takes place when a photon propagates at the sufficiently large angle to the external magnetic field, has been understood (Sturrock, 1971; Ruderman and Sutherland, 1975). Hence, in the dipole magnetic field of a neutron star, the necessary gamma-quanta can be emitted by primary particles moving along a curved magnetic field. This means that the magnetosphere of a neutron star must be efficiently filled with an electron-positron plasma which screens the longitudinal electric field, and the charge density

$$\rho_{\rm GJ} = -\frac{\Omega \mathbf{B}}{2\pi c} \tag{2}$$

in the magnetosphere must be nonzero (Goldreich and Julian, 1969). Such a redistribution of electric charges, as is known from the examples of the Earth and Jovian magnetospheres, leads to plasma's solid corotation with the star. Clearly, such a corotation is impossible outside the so-called light cylinder $R_{\rm L} = c/\Omega$.

Hence, in the magnetosphere of a radio pulsar, two essentially different regions must be formed, namely, the regions of open and closed field lines. Particles in the field lines not intersecting the light cylinder appear to be trapped, whereas the plasma in the lines crossing the light cylinder may go to infinity. In this case, the size of the region near the neutron star magnetic poles, which is crossed by open field lines

$$R_0 \approx R\left(\frac{\Omega R}{c}\right)^{1/2}, \tag{3}$$

makes up only several hundred meters for ordinary pulsars.

It is of importance that the outgoing plasma also carries away the electric charge (2). That is why the strong electric currents

$$I \sim I_{\rm GJ} = \pi R_0^2 c \rho_{\rm GJ}, \tag{4}$$

which are closed in the neutron star magnetosphere, must flow in the magnetosphere of a radio pulsar. A curious fact concerning this point became clear: the characteristic current losses, i.e., the intensity of the energy release due to the

Ampére action of the electric currents flowing in the magnetosphere and closing on the neutron star surface,

$$W_{\text{cur}} \approx IV \approx \frac{B_0^2 \Omega^4 R^6}{c^3} \frac{I}{I_{\text{GJ}}}, \tag{5}$$

coincide to an order of magnitude with the magnetodipole loses (1). Here $V \approx eEL \approx eB_0(\Omega R_0/c)R_0$ is the characteristic potential drop across the polar cap. For this reason the analysis of the statistical characteristics of radio pulsars (Michel, 1991; Beskin *et al.*, 1993) yields similar results with concerning the magnetodipole and current losses, and so the observations do not allow us to choose now these two deceleration mechanisms.

The general properties of the secondary electron-positron plasma created near the magnetic poles and flowing out along open magnetic field lines appeared to be generally insensitive to the structural details of the acceleration region. For the majority of models (Ruderman and Sutherland, 1975; Arons, 1983; Gurevich and Istomin, 1985), both the density and the energy spectra of the outflowing plasma prove to be fairly universal. So, why one can say with confidence that a plasma streaming along open magnetic field lines consists of a beam of primary particles of energy $\mathcal{E} \sim 10^7$ MeV and a density close to the Goldreich-Julian one $n_{\text{GJ}} = |\rho_{\text{GJ}}|/|e|$ and also of a secondary electron-positron component whose energy spectrum has to a good accuracy the power-like form

$$N(\mathcal{E}) \propto \mathcal{E}^{-2}, \tag{6}$$

and the energies themselves lie within the range form $\mathcal{E}_{\text{min}} \sim$ 10–100 MeV to $\mathcal{E}_{\text{max}} \sim 10^4$ MeV. The total secondary plasma density is 10^3–10^4 times higher than the Goldreich-Julian density. Such a model has been considered in the overwhelming majority of papers devoted to the theory of pulsar radio emission.

Finally, on the basis of the picture presented above, the hollow cone model was proposed (Radhakrishnan and Cocke, 1969; Ruderman and Sutherland, 1975) in which radiation is emitted tangentially to the local magnetic field direction on open field lines. This model explained rather well the main properties of radio emission. Indeed, the creation of secondary plasma is impossible in a rectilinear magnetic field where, first, the intensity of curvature radiation is low and, second, the photons emitted by relativistic particles propagate at small angles to the magnetic field. Accordingly, in the central regions of open field lines one should expect a decreased density of secondary plasma and, therefore, a lowered intensity of radio emission in the center of the directivity pattern. Disregarding the details (the main profiles of pulsars actually have a fairly complicated structure), one should expect a single mean profile of pulsars for which the line of sight traverses the directivity pattern far from its center and a double profile when the line of sight intersects the center of the diagram. This picture is generally known to be observed in reality. As a result, all basic properties of pulsar radio emission, such as the variation of the position angle of linear polarization along the mean profile, the distribution of pulsars with single and double mean profiles, the width of the directivity pattern and even its statistical dependence on the pulsar period, have been explained. The latter fact is based on the assumption that all pulsars generate radiation at approximately the same distance $r_{\rm rad}$ from the neutron star. Hence, for the width of the directivity pattern we have

$$w_{\rm d} \approx \left(\frac{\Omega r_{\rm rad}}{c}\right)^{1/2} \approx 10^{\circ} P^{-1/2} \left(\frac{r_{\rm rad}}{10R}\right)^{1/2}, \tag{7}$$

i.e., $w_{\rm d} \propto P^{-1/2}$, which agrees with the observations (Lyne and Graham-Smith, 1998).

3. What is Unknown

Thus, the general picture of radio pulsar activity seems to have been established many years ago. At the same time, some principal questions are still far from being solved. First of all, as in the 1970s, there is no generally accepted standpoint concerning the physical nature of coherent radio emission of pulsars. In particular, it has not yet been decided whether the coherent mechanism of radio emission is of maser or antenna type. Furthermore, there is no common point of view about the structure of the neutron star magnetosphere. That is why there is no generally acknowledged model of the structure of longitudinal currents circulating in the magnetosphere, which is necessary for the solution to the problem of neutron star braking, particle acceleration, and energy transfer outside the light cylinder. As we see, at present there is no general view about the main questions of the theory:

What is the physical nature of coherent radio emission?

and

What is the structure of pulsar magnetosphere?

Nevertheless, the number of papers devoted to these key problems has recently decreased sharply in spite of the fact that other, sometimes very refined questions, such as particle creation and propagation of gamma quanta in superstrong magnetic field (Mészáros, 1992; Usov and Melrose, 1996; Baring and Harding, 1997), the effects of general relativity on the pair creation region (Muslimov and Tsygan, 1990; Beskin, 1990; Muslimov and Harding, 1998; Mestel, 1999), the theories of the pulsar wind (Li *et al.*, 1992; Gallant and Arons, 1994; Bogovalov, 1997), and the high-frequency emission of radio pulsars (Chieng and Romani, 1994; Sturner *et al.*, 1995) have been under intensive investigation. Several recent papers on the theory of radio emission (Lyubarskii, 1992; Asséo, 1993; Luo *et al.*, 1994; Gedalin *et al.*, 1998) and the theory of the magnetosphere (Lyubarskii, 1990; Mestel and Shibata, 1994; Beskin and Malyshkin, 1998; Contopoulos *et al.*, 1999) have not essentially changed the general picture.

In particular, in spite of the great interest (especially in the 1970s-1980s) in the problem of generation of coherent radio emission, the theory has been developed up to concrete quantitative predictions accessible for its direct verification only in some exceptional cases. Therefore, great additional efforts are needed, including efforts in the framework of the models already constructed.

4. What is to be Done

Unfortunately, for radio pulsars there has been a manifest discrepancy between the theory and observations for many years. And if, for example, the verification of the general theory of relativity has played the key role in organizing the observations of binary pulsars, practically no research has been directed to the verification of modern theories of radio emission and the theory of pulsar magnetosphere. Now the increase of observational material is often not at all connected with theoretical works. That is why we recall the main observational tests which may clarify the fundamental basis of the physical processes responsible for the radio pulsar activity.

4.1. Particle creation

As has already been emphasized, if we understand correctly the physical nature of the activity of single neutron stars leading to the observed radio emission, it is due to the relativistic electron-positron plasma generated near the magnetic poles and streaming out along open magnetic field lines. The direct information on the processes proceeding in the particle creation region might first of all be obtained from 'relic' photons, i.e. hard gamma quanta with an energy insufficient for conversion into an electron-positron pair. The possibility of a direct recording of such photons has long been discussed (moreover, the spectra and intensities of the anticipated radiation have been determined for many models, see e.g. Beskin *et al.* (1993); Baring and Harding, (1997)), but the situation is not yet quite clear. The point is that in those rare cases where radio pulsars are at the same time sources of pulsed gamma radiation, their gamma-ray emission from the particle generation region cannot apparently compete with other mechanisms of gamma-ray emission, for instance, with radiation from an 'outer' gap.

Next, the relation

$$H(P, B) < R_0(P), \tag{8}$$

where $H(P, B)$ is the gap height depending on the pair creation mechanism, may be thought of as the 'ignition condition' separating active and passive parameter ranges, when the neutron star does not manifest itself as a radio pulsar. That is why, the physical processes in the pair creation region determine the maximum radio pulsar period $P_{\max}(B_0)$, see Zhang *et al.* (2000) for a review.

Finally, direct information on the potential drop must be contained in the subpulse drift velocity because it is directly related to the potential drop across the gap. For example, the recent analysis reported by Vivekanand and Joshi, (1999); Deshpande and Rankin, (1999) seems to testify in favor of Ruderman-Sutherland type models. However, further studies are needed here.

4.2. RADIO PULSAR MAGNETOSPHERE

The most convincing evidence in favor of the absence of magnetodipole loss was the discovery of time-dependent optical radiation from a companion in some close binary systems containing radio pulsars (Djorgovsky and Evans, 1988). Such an optical radiation with periodicity exactly coincident with the orbital period of the binary system is naturally associated with heating the part of the companion facing the radio pulsar. It has turned out that the energy re-emitted by the companion is practically coincident with the total energy emitted by the radio pulsar into the corresponding solid angle. Clearly, this fact cannot be understood from the model of the magnetodipole radiation. Only if the considerable portion of the energy is connected with a relativistic particle flux, will the heating of the star surface be sufficiently effective.

Another direct test offering some insight into the mechanism of radio pulsar braking centers around the determination of the so-called braking index $n_{\rm br} = \Omega\ddot{\Omega}/\dot{\Omega}^2$ which can only be found if the second derivative $\ddot{\Omega}$ of the angular velocity of rotation is known. However, since it takes much time to determine this quantity (not to mention the fact that in many cases the second derivative $\ddot{\Omega}$ cannot be distinguished against the background of low-frequency disturbances), it has been measured for only five radio pulsars.

It has turned out that in all the five cases $n_{\rm br} < 3$ (Lyne and Graham-Smith, 1998), whereas allowing for magnetodipole loss we must have $n_{\rm br} = 3 + 2\cot^2\chi$ (Michel, 1991). This circumstance can already be regarded as a direct contradiction between the model of magnetodipole loss and observations. As to the mechanism of current-induced braking, one can obtain $n_{\rm br} = 1.93 + 1.5\tan^2\chi$ (Beskin *et al.*, 1993), which is in good agreement with observations. In any case, the determination of the braking index for other radio pulsars would appreciably clarify the nature of the radio pulsar energy loss.

Finally, we recall that the determination of the evolution of the inclination angle χ may become a direct test. Since for current losses the braking torque **K** is directed oppositely to the magnetic moment of the neutron star, the Euler equation leads to a conservation of the value $\Omega \sin\chi$ during the evolution. Hence, the angle χ must increase upon a current loss (rather than decrease as in the case of magnetic dipole radiation), and the typical time of its evolution must coincide with the characteristic time $\tau_{\rm D}$ of the pulsar period variation. Unfortunately, no method has yet been found to determine the evolution of the inclination angle $\dot{\chi}$ for individual pulsars.

4.3. Radio emission

As stressed above, the theory of the radio emission has been developed up to concrete quantitative predictions only in some exceptional cases. For this reason it is rather difficult to propose the observations directed to the determination of physical processes in the generation region. The theory (Beskin *et al.*, 1993) we constructed ten years ago remains, in fact, the only example of passing the whole way from the consistent model of neutron star magnetosphere, the theory of particle creation and generation of coherent radio emission to a comparison of the quantitative predictions of the theory with observations, which showed a good agreement.

On the other hand, the theory of radio emission must include not only the generation mechanism itself, but the whole complex of questions involved in the formation of the directivity pattern. These are the problems of propagation and absorption of normal waves in the magnetosphere and those concerning the limiting polarization upon radiation ejection from the magnetosphere. In the last years several papers (von Hoensbroech *et al.*, 1998; Lyubarskii and Petrova, 1999) appeared devoted to radio wave propagation, so one may hope that a consistent theory (at least in this part of it) will be constructed in the nearest future. It will give the possibility, integrating backward, to determine the place of the radio waves generation.

5. Conclusion

Thus, the structural theory of pulsar magnetosphere and especially the theory of radio emission are still far from being completed. And although we now understand most of the key processes proceeding in a neutron star magnetosphere, a reliable quantitative description has only been given to some separate, although principal, elements. For this reason, the observations aimed at the establishment of the key physical properties of radio pulsars should be continued. They may include, for instance:

- determination of the braking index $n_{\rm br}$ for new pulsars,
- determination of the direction of the evolution of the inclination angle $\dot{\chi}$ for individual pulsars,
- refinement of the statistical dependence for the subpulse drift velocity,
- monitoring of the relic photons generated in the pair creation regions,
- two-parametric analysis of the main properties of the radio emission directivity pattern (which was impossible previously because of the lack of observation data).

Of course, those are the possible tests only. But what is definitely necessary to do is to diminish the gap between observers and theoreticians.

Acknowledgements

The author is greatly indebted to the Organizing Committee of the International Conference 'Physics of Space: Growth Points and Problems' for invitation and the Observatory Paris-Meudon for hospitality.

References

Arons, J.: 1983, *Astrophys. J.* **266**, 215.
Asséo, E.: 1993, *Mon. Not. R. Astron. Soc.* **264**, 940.
Baring, M.G. and Harding, A.K.: 1997, *Astrophys. J.* **482**, 372.
Begelman, M.C. and Li, Zh.-Yu.: 1994, *Astrophys. J.* **426**, 269.
Beskin, V.S.: 1990, *Sov. Astron. Lett.* **16**, 286.
Beskin, V.S., Gurevich, A.V. and Istomin, Ya.N.: 1993, *Physics of the Pulsar Magnetosphere*, Cambridge University Press, Cambridge.
Beskin, V.S. and Malyshkin, L.M.: 1998, *Mon. Not. R. Astron. Soc.* **298**, 847.
Bogovalov, S.V.: 1997, *Astron. Astrophys.* **327**, 662.
Chieng, J. and Romani, R.: 1994, *Astrophys. J.* **436**, 754.
Contopoulos, I., Kazanas, D. and Fendt, C.: 1999, *Astrophys. J.* **511**, 351.
Deshpande, A.A. and Rankin, J.M.: 1999, *Astrophys. J.* **524**, 1008.
Djorgovsky, S. and Evans, C.R.: 1988, *Astrophys. J. Lett.* **335**, L61.
Gallant, Y.A. and Arons, J.: 1994, *Astrophys. J.* **435**, 230.
Gedalin, M., Melrose, D.B. and Gruman, E.: 1998, *Phys. Rev.* **E57**, 3399.
Ginzburg, V.L. and Zheleznyakov, V.V.: 1975, *Ann. Rev. Astron. Astrophys.* **13**, 511.
Gold, T.: 1968, *Nature* **218**, 731.
Goldreich, P. and Julian, W.H.: 1969, *Astrophys. J.* **157**, 869.
Gurevich, A.V. and Istomin, Ya.N.: 1985, *Sov. Phys. JETP* **62**, 1.
Hewish, A., Bell, S.J., Pilkingston, J.D., *et al.*: 1968, *Nature* **217**, 709.
Ilyasov, Yu.P., Kopeikin, S.M. and Rodin, A.E.: 1998, *Astronomy Letters* **24**, 228.
Johnston, S.: 1995, *Current Science* **69**, 521.
Li, Zh.-Yu., Chiueh, T. and Begelman, M.C.: 1992, *Astrophys. J.* **394**, 459.
Luo, Q., Merrose, D.B. and Machabeli, G.Z.: 1994, *Mon. Not. R. Astron. Soc.* **268**, 159.
Lyne A., and Graham-Smith F.: 1998, *Pulsar Astronomy*, Cambridge University Press, Cambridge.
Lyubarskii, Yu.E.: 1990, *Sov. Astron. Lett.* **16**, 16.
Lyubarskii, Yu.E.: 1992, *Astron. Astrophys.* **261**, 544.
Lyubarskii, Yu.E. and Petrova, S.A.: 1999, *Astron. Astrophys.* **333**, 181.
Mestel, L.: 1999, *Stellar magnetism*, Clarendon Press, Oxford.
Mestel, L. and Shibata, S.: 1994, *Mon. Not. R. Astron. Soc.* **271**, 621.
Mészáros, P.: 1992, *High-Energy Radiation from Magnetized Neutron Stars*, Universty of Chicago Press, Chicago.
Michel, F.C.: 1991, *Theory of Neutron Star Magnetosphere*, Universty of Chicago Press, Chicago.
Muslimov, A.G. and Harding, A.K.: 1998, *Astrophys. J.* **508**, 328.
Muslimov, A.G. and Tsygan, A.I.: 1990, *Sov. Astron.* **34**, 133.
Pacini, F.: 1967, *Nature* **221**, 567.
Radhakrishnan, V. and Cocke, D.J.: 1969, *Astrophys. Lett.* **3**, 225.
Ruderman, M.A. and Sutherland, P.G.: 1975, *Astrophys. J.* **196**, 51.
Sturner, S.J., Dermer, C.D. and Michel, F.C.: 1995, *Astrophys. J.* **445**, 736.
Sturrock, P.A.: 1971, *Astrophys. J.* **164**, 529.
Usov, V.V. and Melrose, D.: 1996, *Astrophys. J.* **464**, 306.

Vivekanand, M. and Joshi, B.C.: 1999, *Astrophys. J.* **515**, 398.
von Hoensbroech, A., Lesch, H. and Kunzl, T.: 1998, *Astron. Astrophys.* **336**, 209.
Zhang, B., Harding, A.K. and Muslimov, A.G.: 2000, *Astrophys. J. Lett.* **531**, 45.

BASICS OF ROTATING MAGNETOSPHERES: EQUILIBRIUM AND STABILITY

KATIA M. FERRIÈRE
Observatoire Midi-Pyrénées, 14 avenue E. Belin, 31400 Toulouse, France

Abstract. We present the basic characteristics of a rotating magnetosphere. More specifically, we describe its overall equilibrium state, explain the 'subcorotation' phenomenon resulting from plasma production or from outward transport, discuss the conditions under which a magnetospheric system is driven unstable, and comment on the nature of the unstable motions, with emphasis on the interchange motions believed to be responsible for the outward plasma transport.

1. Introduction

The magnetosphere of an astrophysical body (planet, star, pulsar, radio galaxy ...) is the region of space controlled by this body's magnetic field. It usually contains several populations of plasma, the main sources of which are the central body itself, its satellites and their torus of neutral gas and dust, and the environment in which the magnetosphere is embedded (e.g., the stellar wind in the case of planets).

Plasma motions in the magnetosphere are usually driven either by the environment or by the central body's rotation. In particular, the solar wind constitutes the primary source of motions in most of the terrestrial magnetosphere (more specifically, outside the innermost region known as the plasmasphere), whereas the bulk of the Jovian and Saturnian magnetospheres (away from the magnetotail) are largely dominated by the rapid planetary rotation.

The purpose of the present paper is to review the main properties of rotation-dominated magnetospheres (or regions thereof). We will be working in the framework of ideal magnetohydrodynamics (MHD) and use SI units throughout. After some preliminary considerations, we will successively discuss the equilibrium state of a rotating magnetosphere, departures from strict corotation with the central body, stability properties, and motions of the interchange type.

2. Preliminary Definitions

Consider an astrophysical object, say a planet to fix ideas, which possesses an intrinsic magnetic field **B** and rotates at angular velocity Ω. For simplicity, we will assume that the magnetosphere is symmetric with respect to the equatorial plane and that the equilibrium magnetic field, $\mathbf{B}_0$, is poloidal and axisymmetric

Astrophysics and Space Science **277:** 247–257, 2001.

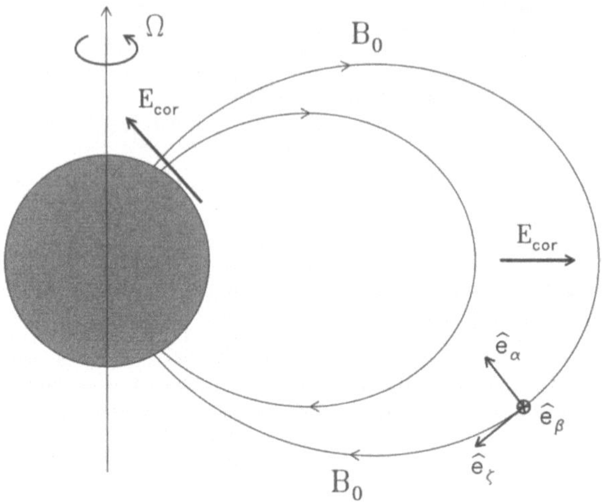

Figure 1. Cross-section of a planet rotating at angular velocity Ω, together with two lines of force of its equilibrium magnetic field, $\mathbf{B}_0$. $\widehat{e}_\alpha$, $\widehat{e}_\beta$, $\widehat{e}_\zeta$ are the unit vectors of an orthogonal curvilinear coordinate system based on the Euler potentials, α and β, and on a field-aligned coordinate, ζ. Here, β has been set equal to the azimuthal angle, φ. $\mathbf{E}_{\rm cor}$ is the corotational electric field.

about the spin axis (see Figure 1). Unless explicitly stated otherwise, the words 'perpendicular' and 'parallel', as well as the associated symbols $\perp$ and $\|$, will refer to the direction of $\mathbf{B}_0$.

When discussing rotating magnetospheres, it often proves convenient to assign a (perpendicular) velocity $\mathbf{v}_{\rm B}$ to the magnetic lines of force. According to Newcomb (1958), one may pick any function satisfying

$$\frac{\partial \mathbf{B}}{\partial t} = \nabla \times (\mathbf{v}_{\rm B} \times \mathbf{B}) \cdot \tag{1}$$

A comparison with Faraday's equation indicates that, in the absence of parallel electric fields, the simplest solution to (1) reads

$$\mathbf{v}_{\rm B} = \frac{\mathbf{E} \times \mathbf{B}}{B^2} , \tag{2}$$

where $\mathbf{E}$ is the electric field. On the other hand, in ideal MHD the plasma velocity $\mathbf{v}_{\rm i}$ satisfies

$$\mathbf{E} + \mathbf{v}_{\rm i} \times \mathbf{B} = 0 \tag{3}$$

(e.g., Cowling, 1957), so that its perpendicular component is given by

$$\mathbf{v}_{{\rm i},\perp} = \frac{\mathbf{E} \times \mathbf{B}}{B^2} \cdot \tag{4}$$

Equation (3) is customarily referred to as the frozen-in approximation, for its direct implication is that the field line velocity $\mathbf{v}_{\rm B}$ may be identified with the plasma perpendicular velocity $\mathbf{v}_{{\rm i},\perp}$, both being equal to the $\mathbf{E} \times \mathbf{B}$ drift velocity.

We will also find it helpful to adopt an orthogonal curvilinear coordinate system (α, β, ζ), where (α, β) is one of the many possible combinations of Euler potentials associated with $\mathbf{B}_0$ (by definition, $\mathbf{B}_0 = \nabla\alpha \times \nabla\beta$, so α and β are constant along the lines of force of $\mathbf{B}_0$) and ζ represents the field-aligned coordinate. Our hypothesis that $\mathbf{B}_0$ is poloidal and axisymmetric allows us to set one of the Euler potentials, say β, equal to the azimuthal angle, φ (see Figure 1).

3. Equilibrium State

Let us first examine the question of how the whole magnetosphere is driven into corotation (the following discussion is directly inspired from Hill, 1979). In a first step, the planetary rotation is transmitted to the neutral atmosphere through viscous stresses. In a second step, neutrals located in the ionospheric layer of the atmosphere transfer angular momentum to ions via ion-neutral collisions: these collisions exert a friction force (eastward in Figure 1) on the ions, and the associated (equatorward) drift current polarizes the plasma in such a way that the ensuing (poleward) electric field gives rise to an $\mathbf{E} \times \mathbf{B}$ drift in the direction of the neutrals' velocity; in the framework of the frozen-in approximation, this $\mathbf{E} \times \mathbf{B}$ drift corresponds to the velocity of both ions and magnetic field lines (see (4) and (2)). In the absence of processes generating a drag on the plasma (such as mass loading or outward flow in the magnetosphere), ions together with the frozen-in field lines are accelerated up to the corotation velocity, $\mathbf{v}_{\text{cor}} = \Omega \times \mathbf{r}$, which means that the electric field reaches its corotation value, $\mathbf{E}_{\text{cor}} = -\mathbf{v}_{\text{cor}} \times \mathbf{B}$. In a third step, the corotational electric field created in the ionosphere is transmitted to the magnetosphere along magnetic field lines, thereby forcing the magnetospheric plasma into corotation (see Figure 1).

It is readily demonstrated that field lines rotate rigidly once a steady equilibrium state is achieved. Indeed, by letting $\partial/\partial t = 0$ and $\mathbf{B} = \mathbf{B}_0$ in (1) and writing the right-hand side in curvilinear coordinates, we find that the β-component reduces to $\partial\dot{\beta}_{\text{B}}/\partial\zeta = 0$; with $\beta = \varphi$, this implies that the angular velocity of field lines is constant all along them. Consequently, to the extent that the ionospheric footpoints corotate with the planet, field lines as wholes together with the plasma tied to them also corotate with the planet.

We now turn to the impact of the overall rotation on the magnetospheric equilibrium configuration. In the purely corotating case, it suffices to consider the effects of the centrifugal force, $\mathbf{F}_{\text{cent}}$, both parallel and perpendicular to the magnetic field. The parallel component of $\mathbf{F}_{\text{cent}}$, directed toward the centrifugal equator, tends to confine the magnetospheric plasma to a disk centered on the equatorial plane. The disk thickness as well as the details of the plasma field-aligned distribution are determined by a balance between the centrifugal force ($\propto \Omega^2$) and the thermal pressure gradient force ($\propto T$). For instance, an isothermal plasma has a Gaussian distribution with a scale height squared $\propto T/\Omega^2$ (Hill and Michel, 1976).

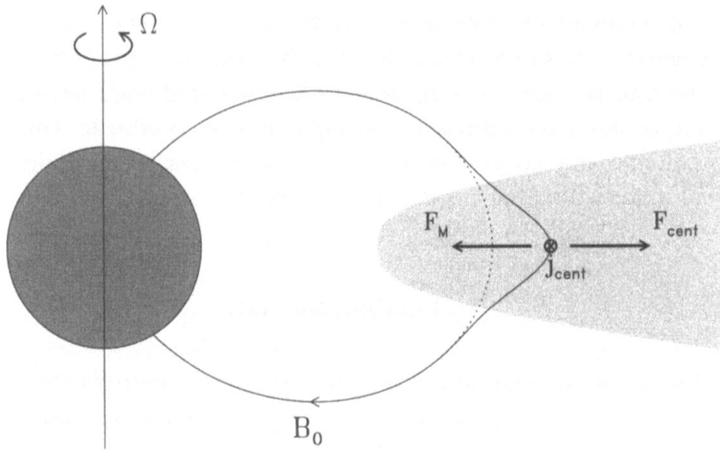

Figure 2. Cross-section through a magnetosphere rotating at angular velocity Ω. The light-shaded area represents the disk to which the magnetospheric plasma is confined. A typical line of force of the equilibrium magnetic field, $\mathbf{B}_0$, is deformed from the dotted line to the solid line by the centrifugal force, $\mathbf{F}_{\rm cent}$. $\mathbf{j}_{\rm cent}$ is the drift current associated with $\mathbf{F}_{\rm cent}$ and $\mathbf{F}_{\rm M}$ is the magnetic force resulting from field line deformation.

The perpendicular component of $\mathbf{F}_{\rm cent}$ gives rise to an azimuthal drift current, $\mathbf{j}_{\rm cent}$. If the magnetic field is initially dipolar (so that $\nabla \times \mathbf{B} = 0$), $\mathbf{j}_{\rm cent}$ induces a transient electric field in the opposite direction (as required by Ampère-Maxwell's equation), which causes field lines to move radially outward by an amount increasing toward the equator. The resulting field line displacement (illustrated in Figure 2) modifies the magnetic structure in such a manner that $\nabla \times \mathbf{B} = \mu_0 \, \mathbf{j}_{\rm cent}$ after the transient phase. From a MHD point of view, the magnetic force, $\mathbf{F}_{\rm M} \equiv \mathbf{j}_{\rm cent} \times \mathbf{B}$, which counterbalances $\mathbf{F}_{\rm cent}$, can be understood in terms of the superposition of a magnetic pressure gradient force arising from the radial compression of field lines and a more important magnetic tension force arising from their radial stretching.

4. Departures from Corotation

As first pointed out by Hill (1979), both ion production and radial transport in the magnetosphere entail departures from strict corotation. The basic physical idea, described in Hill's paper (see also Pontius and Hill, 1982), goes as follows. Neutrals present in the magnetosphere rotate on average at the local Keplerian speed, which is less than the local corotation speed. When a magnetospheric neutral gets ionized (for instance, by electron impact or by interaction with a photon), the new ion is instantly 'picked up' by the corotational electric field, which imparts to it an azimuthal $\mathbf{E} \times \mathbf{B}$ drift equal to the corotation velocity. In the process, the new ion is

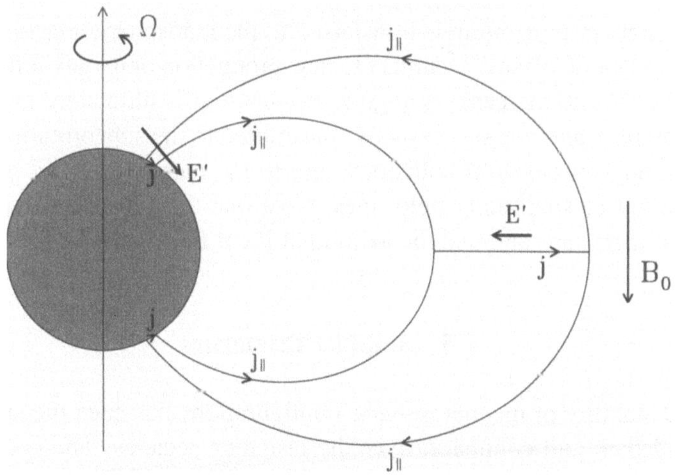

Figure 3. Schematics showing the electrodynamic coupling between the magnetosphere and the ionosphere of a rotating planet. $\mathbf{B}_0$ is the equilibrium magnetic field, $\mathbf{E}'$ the electric field in the corotating frame, $\mathbf{j}$ the transverse drift current, and $j_{\parallel}$ the field-aligned current.

also set into gyromotion, and its guiding center is displaced radially outward by one gyroradius. The global effect of successive ion pick-ups is to generate an electric current in the outward radial direction. This radial 'pick-up' current can also be regarded as the drift current associated with the azimuthal inertial drag force exerted by the newly-created ions. Likewise, if the magnetospheric plasma undergoes an outward radial motion, it is acted upon by the Coriolis force, which points in the azimuthal direction and has a drift current in the outward radial direction.

The outward magnetospheric drift current, be it from the inertial drag force or from the Coriolis force, induces a growing electric field $\mathbf{E}'$ in the corotating frame, which is opposite to the corotational electric field (in accordance with Ampère-Maxwell's equation; see Figure 3). The ensuing $\mathbf{E} \times \mathbf{B}$ drift, therefore, corresponds to a corotation lag. The electric field $\mathbf{E}'$ together with the corotation lag rapidly propagate along field lines inward into the ionosphere. There, the 'lagging' ions experience a friction force due to ion-neutral collisions, whose drift current is oriented in the same sense as $\mathbf{E}'$ (Pedersen current). The corotation lag increases until the ionospheric Pedersen current has grown strong enough to match the imposed magnetospheric drift current, the link between both currents being realized by field-aligned currents flowing between the magnetosphere and the ionosphere, in the manner depicted in Figure 3.

In a steady state, the drag ('pick-up' or Coriolis) force, which requires an increase in the angular momentum of the magnetospheric plasma, is in exact balance with the magnetospheric $\mathbf{j} \times \mathbf{B}$ force, the torque exerted by the magnetospheric $\mathbf{j} \times \mathbf{B}$ force is equal and opposite to the torque exerted by the ionospheric $\mathbf{j} \times \mathbf{B}$ force, and the latter is in exact balance with the ion-neutral collision friction force, by which planetary angular momentum is transferred to the ionospheric plasma. Thus, when the magnetospheric plasma is subject to new production or to outward transport, the torque necessary to increase its angular momentum is ultimately provided by the planet's rotation, and it is successively transmitted to the subcorotating ionospheric plasma, through ion-neutral collisions, and to the magnetospheric plasma, via an electric circuit closing along field lines. Note that this electric circuit bends field lines out of meridians and that the azimuthal $\mathbf{j} \times \mathbf{B}$ force acts via magnetic tension.

5. Stability Properties

The MHD stability of magnetospheric configurations has been the subject of considerable debate and confusion over the last four decades. The energy principle established by Bernstein et al. (1958) does not apply to rotating equilibria, so the logical way to proceed is to perform a normal-mode analysis of the linearized MHD equations for small-amplitude perturbations. In this approach, perturbations are assumed to vary in time like $\exp(-i\omega t)$, and by eliminating all perturbations in the momentum equation in favor of the Lagrangian displacement, $\delta\mathbf{r}$, one obtains a wave equation of the form

$$-\omega^2\, \delta\mathbf{r} = \mathbf{F}(\delta\mathbf{r}) \; , \tag{5}$$

where $\mathbf{F}$ is a linear and homogeneous function of $\delta\mathbf{r}$ and its spatial derivatives, whose coefficients are themselves functions of the equilibrium parameters and possibly of ω. For any given physical system, this wave equation, subject to the proper boundary conditions, is to be solved as an eigenvalue problem for ω. In particular, one may conclude that the system is stable against MHD perturbations if and only if all eigenvalues are found to have an imaginary part ≤ 0.

In order to gain some physical insight into the nature of the different MHD modes, we first consider the one-dimensional case in which the equilibrium parameters depend only on the cross-field coordinate α, and we restrict our attention to perturbations with a wavelength much smaller than the typical scale height of the equilibrium. The Lagrangian displacement may then be assigned a spatial dependence $\propto \exp(i\mathbf{k}\cdot\mathbf{r})$, which, upon substitution into the wave equation (5), leads to a dispersion relation of sixth degree in ω.

In the absence of rotation and stratification, one recovers the classic dispersion relation of MHD waves in a uniform medium (a third-degree equation in ω^2), the three solutions of which correspond to the standard fast, slow, and Alfvén modes. Let us simply recall that all three modes have $\omega^2 \geq 0$ and are, therefore, stable, although stability becomes marginal ($\omega^2 \to 0$) for the slow and Alfvén modes in the limit of perpendicular propagation ($k_\parallel/k \to 0$).

The presence of stratification modifies the frequency of the three standard modes by a relatively small amount on the order of the magnetic Rayleigh-Taylor (or magnetic Brunt-Väisälä) frequency, ω_0, and the only modes liable to become unstable

($\omega^2 < 0$) are the slow and Alfvén modes at $k_\parallel \ll k_\perp$. For reference, ω_0 is defined by

$$\omega_0^2 = \mathbf{g} \cdot \left(\frac{\nabla \rho_0}{\rho_0} - \frac{\nabla P_{T0}}{P_{\gamma 0}} - \frac{V_A^2}{c_s^2 + V_A^2} \, \mathbf{c} \right) + 2\, V_A^2 \, \mathbf{c} \cdot \left(\frac{\nabla B_0}{B_0} - \frac{\nabla P_{T0}}{P_{\gamma 0}} + \frac{c_s^2}{c_s^2 + V_A^2} \, \mathbf{c} \right) , \tag{6}$$

where all quantities on the right-hand side pertain to the equilibrium state: **g** is the gravitational acceleration, **c** the magnetic curvature vector (both in the α-direction orthogonal to $\mathbf{B}_0$), ρ_0 the plasma mass density, B_0 the magnetic field strength, P_0 the plasma thermal pressure, $P_{M0} \equiv B_0^2/2\mu_0$ the magnetic pressure, $P_{T0} \equiv P_0 + P_{M0}$ the total pressure, $P_{\gamma 0} \equiv \gamma P_0 + 2 P_{M0}$, $c_s \equiv \sqrt{\gamma P_0/\rho_0}$ is the sound speed, and $V_A \equiv \sqrt{2 P_{M0}/\rho_0}$ the Alfvén speed.

The modified slow and Alfvén modes in the regime $k_\parallel \ll k_\perp$ correspond to the so-called 'quasi-interchange' modes discussed by Newcomb (1961), who divided them into two types according to their behavior in the limit $k_\parallel \to 0$: the type 1 mode goes into pure interchange (with strictly perpendicular motion) and has $\omega^2 \to (k_\beta^2/k_\perp^2)\,\omega_0^2$, while the type 2 mode goes into pure translation (with strictly parallel motion) and has $\omega^2 \to 0$. Besides, it can be shown that the Alfvén mode connects with the type 1 mode and the slow mode with the type 2 mode when $\omega_0^2 > 0$, and vice-versa when $\omega_0^2 < 0$.

The type 1 mode is a straightforward generalization to a magnetized medium of the classic gravity mode in which fluid parcels move up and down in a gravitational field under the action of the buoyancy force. In the type 1 mode at $k_\parallel = 0$, magnetic flux tubes as wholes are set into transverse motion, and the driving buoyancy force has a gravitational component due to a density perturbation in a gravitational field (represented by the first term on the right-hand side of (6)) plus a magnetic component due to a field strength perturbation in a curved magnetic field (second term on the right-hand side of (6)). When $k_\parallel \neq 0$, field lines ripple, and their bending produces a magnetic tension restoring force which adds to the driving buoyancy force. Not surprisingly, the stability criterion of the type 1 mode is set by its behavior at $k_\parallel = 0$, and simply given by

$$\omega_0^2 \geq 0 \,. \tag{7}$$

In the type 2 mode, magnetic flux tubes are not displaced as wholes; instead, they ripple and their enclosed plasma undergoes predominantly parallel motions under the combined action of a thermal pressure gradient and a gravitational force component along the deformed field lines. The type 2 mode is particularly well-known in the interstellar medium community, where it is traditionally referred to as the Parker instability mode (Parker, 1966); this instability is believed to play an important role in the formation of interstellar clouds (as a result of the interstellar gas sliding down along rippled field lines), in the escape of cosmic rays from the

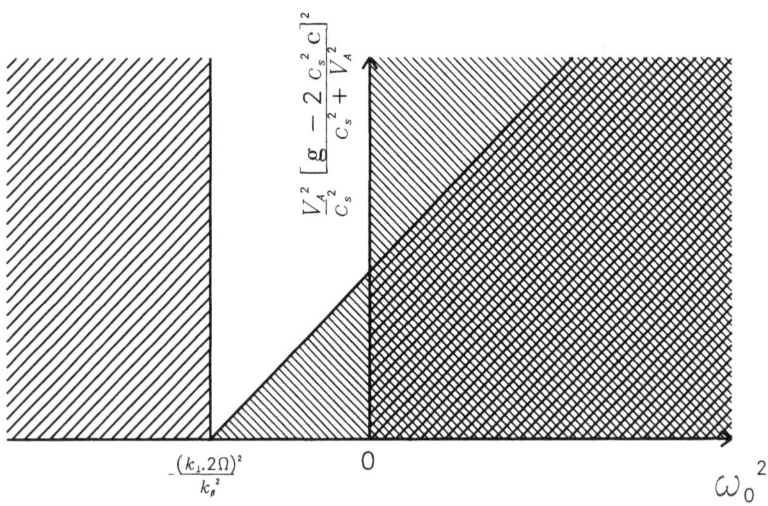

Figure 4. Stability domains of the type 1 (\\\) and type 2 (///) quasi-interchange modes in the plane $\left(\omega_0^2,\ (V_A^2/c_s^2)\,|\mathbf{g} - 2\,c_s^2\,\mathbf{c}|^2/(c_s^2 + V_A^2)\right)$, for a rotating one-dimensionally stratified system. The different symbols are defined in the text.

galactic disk (through the unburdened rising field line crests), and in the destruction of the Galactic magnetic field. Another manifestation of the type 2 mode is the ballooning instability frequently discussed in the context of planetary magnetospheres: triggered by a thermal pressure gradient in an unfavorable magnetic field curvature, the ballooning instability has been invoked to explain the substorm phenomenon in the Earth magnetotail (Roux, 1985) and plasma depletions in the Jovian magnetosphere (McNutt et al., 1987). The stability criterion of the type 2 mode, again set by the limit $k_\parallel \to 0$, can be written as

$$\omega_0^2 \geq \frac{V_A^2}{c_s^2}\,\frac{|\mathbf{g} - 2\,c_s^2\,\mathbf{c}|^2}{c_s^2 + V_A^2} \qquad \text{or} \qquad \omega_0^2 \leq 0\,\cdot \tag{8}$$

Combining (7) and (8) leads to the conclusion that a non-rotating one-dimensionally stratified system is stable against small-scale MHD perturbations if and only if

$$\omega_0^2 \geq \frac{V_A^2}{c_s^2}\,\frac{|\mathbf{g} - 2\,c_s^2\,\mathbf{c}|^2}{c_s^2 + V_A^2}\,. \tag{9}$$

If in addition to being stratified, the system rotates at angular velocity Ω, two new forces come into play: the centrifugal force, which may simply be included into the gravitational force, and the Coriolis force, which gives rise to a truly new term in the momentum equation and causes the three MHD modes to split into two. Modifications in the mode frequencies are on the order of Ω, which is typically $\lesssim \omega_0$, so that the 'dangerous' region for stability is again the region $k_\parallel \ll k_\perp$ on the slow and Alfvén branches. The connection between the slow and Alfvén modes on the one hand and the type 1 and type 2 quasi-interchanges on the other hand is inherently more intricate than in the non-rotating case, with the possibility that one branch of the slow and Alfvén modes goes over to the type 1 mode and the other branch to the type 2 mode (see Figure 3 in Ferrière et al., 1999). The stability domains of both quasi-interchanges are displayed in Figure 4, for a given value of the parameter $(\mathbf{k}_\perp \cdot 2\Omega)^2/k_\beta^2$, which represents the effect of the Coriolis force relative to the buoyancy force. It clearly emerges that the Coriolis force is always stabilizing (or neutral) for the type 1 mode, may either stabilize or destabilize the type 2 mode, and altogether enlarges the domain of stability against both modes. On the other hand, there appears a region wherein neither mode is stable, owing to a destabilizing coupling between the buoyancy force and the Coriolis force at finite parallel wavelengths.

The notion of quasi-interchanges may be extended to the general three-dimensional case, the two different types being then distinguishable by their lowest state in the parallel direction (the three-dimensional equivalent of the limit $k_\parallel \to 0$). In this lowest state, the type 1 mode consists of an interchange between neighboring flux tubes, and the type 2 mode consists of a translation of the plasma along flux tubes. In general, however, neither mode is pure (i.e., strictly perpendicular or strictly parallel), insofar as field-aligned gradients in the equilibrium parameters entail an automatic coupling between perpendicular and parallel motions. For instance, a parallel motion along a density gradient produces density perturbations which, acted upon by the transverse gravitational field, give rise to perpendicular motions. Conversely, a perpendicular motion perturbs the plasma density and pressure at a variable degree along field lines, and the resulting thermal pressure gradients lead to parallel motions.

6. Interchange Motions

There exists ample observational evidence that some kind of large-scale radial transport takes place in the rapidly rotating magnetospheres of Jupiter and Saturn: the observed homogeneity of the magnetospheric plasma distribution between its broadly separated source and sink regions, the detection of plasma originating from identified sources (e.g., Io in the Jovian magnetosphere) at long distances from these sources, velocity measurements revealing a corotation lag increasing away from the planet, etc.

Although the details of the radial transport process are still poorly understood, it is now well accepted that transport is driven by the centrifugal instability and that it proceeds through motions of the interchange type. It is, therefore, important to have a good grasp of the fundamental properties of interchange motions. Their stability was discussed in the previous section, while a detailed description of their intrinsic physical characteristics can be found in Ferrière et al. (2001).

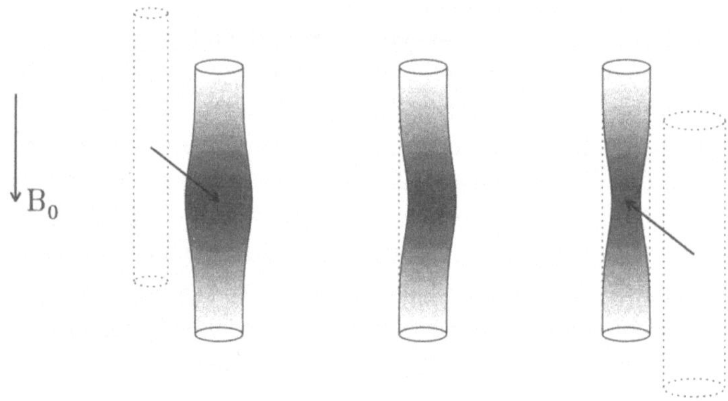

Figure 5. Schematics illustrating the local contraction or expansion of magnetic flux tubes during an interchange motion in a rotating magnetosphere, as they adjust to their new surrounding total pressure. Only the tubes' segment across the magnetospheric disk is drawn and their initial curvature is neglected to simplify the plot.

Here, we will be content to disprove the widely held view that interchange motions leave the magnetic field unaltered (Gold, 1959; see also Southwood and Kivelson, 1987). A first reason why this may not be true stems from the requirement that a displaced flux tube must come into total pressure balance with its new surroundings, which generally implies perturbations in thermal and magnetic pressures separately; in consequence, the displaced flux tube experiences a local contraction ($\delta B > 0$) or expansion ($\delta B < 0$), which is typically most pronounced near the equatorial plane where the background thermal pressure is highest (see Figure 5). A second reason is that the transverse displacement itself varies along the flux tube, not only because the buoyancy force responsible for the transverse motion as well as the mass density dictating the plasma response to the applied force depend on position along the tube, but also because of the friction force exerted by ion-neutral collisions at the ionospheric footpoints (see Figure 6).

Understanding the behavior of magnetic flux tubes undergoing interchange motions and being able to quantify their deformations in terms of both their displacement and the ambient parameters will undoubtedly prove crucial in the interpretation of the numerous observational data from the different space missions launched in the last decade. Amongst these missions, let us mention the Galileo spacecraft, which has been in orbit about Jupiter since December of 1995, and the Cassini-Huygens spacecraft, which is scheduled to fly by Jupiter at the end of 2000 and to reach Saturn in July of 2004.

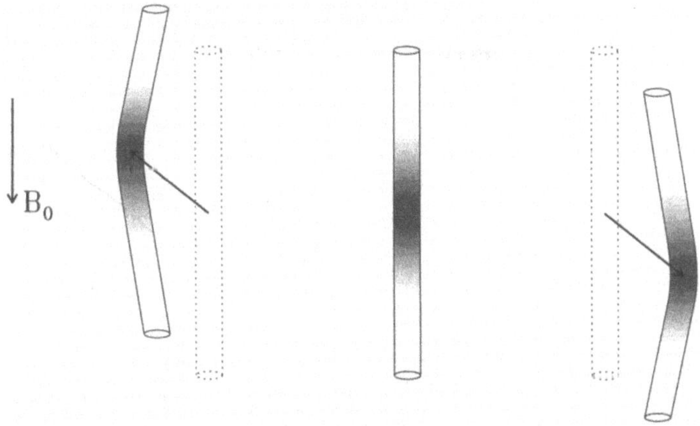

Figure 6. Schematics illustrating the bending of magnetic flux tubes during an interchange motion in a rotating magnetosphere, under the effect of the friction force exerted by ion-neutral collisions at the ionospheric footpoints. The initial curvature of magnetic field lines is neglected to simplify the plot.

References

Bernstein, I.B., Frieman, E.A., Kruskal, M.D. and Kulsrud, R.M.: 1958, An energy principle for hydromagnetic stability problems, *Proc. R. Soc. London* **244**, 17.

Cowling, T.G.: 1957, *Magnetohydrodynamics*. Interscience, New York, USA.

Ferrière, K.M., Zimmer, C. and Blanc, M.: 1999, Mhd waves and gravitational/centrifugal instability in rotating systems, *J. Geophys. Res.* **104**, 17335.

Ferrière, K.M., Zimmer, C. and Blanc, M.: 2001, Quasi-interchange modes and interchange instability in rotating magnetospheres, *J. Geophys. Res.* **106**, 327.

Gold, T.: 1959, Motions in the magnetosphere of the Earth, *J. Geophys. Res.* **64**, 1219.

Hill, T.W.: 1979, Inertial limit on corotation, *J. Geophys. Res.* **84**, 6554.

Hill, T.W. and Michel, F.C.: 1976, Heavy ions from the Galilean satellites and the centrifugal distortion of the Jovian magnetosphere, *J. Geophys. Res.* **81**, 4561.

McNutt, R.L., Jr., Coppi, P.S., Selesnick, R.S. and Coppi, B.: 1987, Plasma depletions in the Jovian magnetosphere: Evidence of transport and solar wind interaction, *J. Geophys. Res.* **92**, 4377.

Newcomb, W.A.: 1958, Motion of magnetic lines of force, *Ann. Phys. N.Y.* **3**, 347.

Newcomb, W.A.: 1961, Convective instability induced by gravity in a plasma with frozen-in magnetic field, *Phys. Fluids* **4**, 391.

Parker, E.N.: 1966, The dynamical state of the interstellar gas and field, *Astrophys. J.* **145**, 811.

Pontius, D.H. and Hill, T.W.: 1982, Departure from corotation of the Io plasma torus: local plasma production, *Geophys. Res. Lett.* **9**, 1321.

Roux, A.: 1985, Generation of field-aligned current structures at substorm onsets, *Proceedings of ESA Workshop on Future Missions in Solar, Heliospheric and Space Plasma Physics*, ESA Spec. Publ. **235**, 151.

Southwood, D.J. and Kivelson, M.G.: 1987, Magnetospheric interchange instability, *J. Geophys. Res.* **92**, 109.

THE PHYSICS AND CHEMISTRY OF SPUTTERING BY ENERGETIC PLASMA IONS

R.E. JOHNSON and F. LEBLANC
Engineering Physics, University of Virginia, USA; E-mail: rej@virginia.edu

Abstract. Energetic ions from the solar wind, local pick-up ions or magnetospheric plasma ions impact the atmospheres and surfaces of a number of solar system bodies. These energetic incident ions deposit energy in the gas or solid. This can lead to the ejection of atoms and molecules, a process referred to as sputtering. In this paper we first describe the physics and chemistry of atmospheric and surface sputtering. We then apply this to the production of a thin atmosphere on Europa by magnetospheric ion bombardment of Europa's surface and show that Europa loses more Na atoms than it receives from the Jupiter magnetosphere. The loss of atmosphere from Mars in earlier epochs by pick-up ion sputtering of that atmosphere is also calculated.

1. Introduction

The flow of the solar wind plasma, a plasma trapped in a planetary magnetic field or a local pick-up ion plasma onto a surface or onto the exobase of an atmosphere can cause chemistry, heating and the ejection of atoms or molecules. These processes, which can affect the evolution of a solar system object are often referred to as sputtering. In the laboratory, sputtering is a procedure in which a flux of heavy ions produces a vapor above a material that would otherwise have a low vapor pressure. This is often done for controlled vapor deposition of a thin coating on a glass or on an electronic device. Here we consider the ejection of atoms and molecules from a natural surface or atmosphere by an energetic incident plasma ion or electron (Johnson, 1990). The effect of solar wind and solar flare ion sputtering of the lunar surface has been of long term interest and may contribute to the observed lunar atmosphere (Sprague *et al.*, 1992). Similarly the observed sodium cloud at Io was suggested as being produced by the energetic ion bombardment of Io's surface (Matson *et al.*, 1974). Based on their initial laboratory experiments, Lanzerotti, Brown and co-workers first suggested that large amounts of surface are sputtered and an atmosphere is produced over the surface of Europa by ion bombardment (Lanzerotti *et al.*, 1978). Later this atmosphere was shown to be dominated by sputter-produced O_2 (Johnson *et al.*, 1982) which was recently observed (Hall *et al.*, 1995). By analogy with the ejection of material from surfaces, Haff and Watson (1979) thought that atmospheres might also be sputtered by energetic ions. This has since been shown to be an important evolutionary process at Mars (Jakosky *et al.*, 1994), Io (Smyth and Combi, 1988) and Europa (Saur *et al.*, 1998).

Astrophysics and Space Science **277:** 259–269, 2001.

In this article honoring the work of J.L. Steinberg we review the physics and chemistry of sputtering since, not only does it affect the evolution of solar system objects, but it is often the source of the local plasma. For instance, atmospheric sputtering plays an important role in forming the Io plasma torus and the sputtering of the icy Saturnian satellites and the E-ring is the likely source of the trapped plasma in Saturn's inner magnetosphere. We first outline the physics and chemistry of sputtering, relying heavily on an earlier monograph (Johnson, 1990) and previous reviews (Johnson, 1994, 1996, 1998). We then consider examples for which we have recently made progress in relating space observations to the plasma sputtering processes.

2. Physics and Chemistry of Sputtering

Sputtering is a process by which an energetic ion deposits its energy in a material initiating a cascade of events which eventually leads to the ejection of atoms or molecules from the surface. Ions lose their energy by direct knock-on (momentum transfer) collisions with an atom in the material and by electronically exciting the atoms and molecules in the material. The effects produced by the incident ion are typically described using the stopping power of the material for a given ion, dE/dx, which is the energy lost per unit path length. Therefore, dE/dx can be roughly written in terms of knock-on (n) and electronic excitation (e) contributions: $(dE/dx)_n + (dE/dx)_e$.

In the sputtering of refractory solids for vapor deposition, it is knock-on collisions that cause the ejection of atoms from the surface. Each atom set in motion (a recoil) by the incident ion in turn collides and produces additional recoils, so there is a cascade of collisions. It is straight forward to show that, if the cascade of collisions evolves fully, the spectra of recoil energies, E, produced has the form $\beta E_i/E^2$, where E_i is the incident ion energy. Sputtering then occurs when an atom or molecule in the material reaches the surface (exobase for an atmosphere) with sufficient energy to overcome the energy barrier (gravity for an atmosphere). If the cascade of collisions evolves sufficiently to become nearly isotropic, then the energy spectrum above can be used to estimate the number of recoils which are sputtered per ion incident, the yield, Y (Johnson, 1990; 1994): $Y \approx \alpha\beta\,[S_n \,/\, U\,\sigma_d]$. Here S_n is the stopping cross section $[(dE/dx)_n = n\,S_n$, where n is the material number density] giving the knock-on contribution to the energy loss. The quantity U is the barrier to escape from the material [a chemical barrier for a solid and the gravitational escape energy for an atmosphere], and σ_d is the collision cross section between the surface (exobase) species and the exiting atoms or molecules. Finally, α contains the angular factors and corrects for the fact that the cascades are not isotropic. The remarkable aspect of this expression, noted by Haff and Watson (1979), is that it is independent of density! This allows it to be applied to both atmospheres and solids. Monte Carlo simulations of atmospheric sputtering confirm the above form for Y (Johnson *et al.*, 2000). However, the parameters α and σ_d have been estimated incorrectly in many applications. The form for Y has been confirmed empirically for surface sputtering (e.g., Andersen and Bay, 1981). The recoil energy distribution, if applicable at the surface leads to a spectrum of ejected atom energies $f(E) \approx 2EU/(E+U)^3$. This expression has also been established experimentally for surface sputtering. Because of the importance of 'edge' effects it decreases more slowly with increasing E in atmospheric sputtering (Smyth and Combi, 1988; Johnson *et al.*, 2000).

Since the Jovian plasma ions do not bombard a refractory, lunar-like surface, but rather the Gallilean satellite surfaces which are composed of low-temperature condensed gas solids and salts, a new experimental effort was launched (Nash and Fanale, 1977; Brown *et al.*, 1978). These materials typically have much smaller surface binding energies. As they are also electric insulators, they can retain the electronic excitations produced by the incident ions long enough to produce luminescence and non-radiative energy release events. The energy release events can cause bond breaking and, often, the direct desorption of a species like Na (Yashinsky and Madey, 1999). In addition, when the excitation density is high, sufficient energy may be released to cause large amounts of material loss. This is the case for water ice as shown in Figure 1 for bombardment by energetic H^+ and O^+. The hump at low velocities is due to knock-on collisions but the much larger peak in the yield at high velocities is due to the electronic excitations and ionizations. Brown, Lanzerotti and co-workers (e.g., Brown *et al.*, 1978) referred to this as electronic sputtering. Because the energy flux for the Jovian plasma ions peaks in the 10 keV-1 MeV region (Cooper *et al.*, 2000), the sputtering of the icy Galilean satellites is predominantly electronic.

For the ions and ices of interest in the Jovian plasma the dependence on $(dE/dx)_e$ was shown to be $Y \approx c[(l/U)\ (dE/dx)_e]^2$ where $l = n^{-1/3}$ (Johnson 1990; 1998). This expression for Y is clearly dependent on the density and, therefore, is not applicable to atmospheres. The equivalent atmospheric process is ionization followed by dissociative recombination near the exobase leading to loss. This is similar to an individual surface desorption event (Johnson, 1990; 1994). The quadratic dependence in Y indicates it is the combined effect of many ionizations and excitations that produce the large yields seen in Figure 1. In both knock-on and electronic sputtering the yield depends inversely on the binding energy U. Because bond breaking occurs, new molecules can be formed and ejected. Therefore, for an atmosphere, fragments (low U) are preferentially ejected, but for a solid the most volatile species formed (low U) will be sputtered preferentially. This is seen to be the case in Figure 2 for deuterated ice. Whereas bond breaking initially forms OD + D or D_2 + O, the O, D and OD have strong binding energies to the solid. Therefore the principal ejecta are D_2, D_2O and O_2. That is, even though decomposition of ice is inefficient, D_2 when formed is readily lost leaving an oxidizing surface so that O_2 gradually forms (Reimann *et al.*,1984). For ice the principal ejecta are H_2O, H_2, and O_2. At Europa, H_2 which is light, directly escapes whereas the H_2O and O_2 do

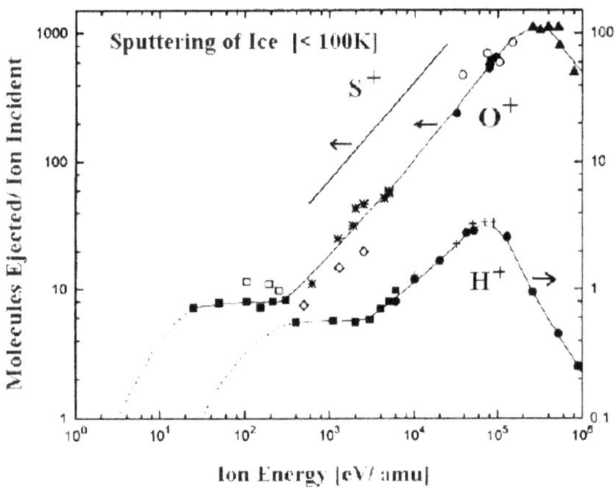

Figure 1. The sputtering yield (equivalent H_2O molecules ejected per ion incident) for water ice at ~77K for incident O^+, S^+ (left hand axis) and H^+ ions (right hand axis). Lines are model fits of data indicated by points.

not. Because the H_2O sticks when it returns to the surface but O_2 does not stick at the temperatures at Europa, a very thin O_2 atmosphere was predicted (Johnson *et al.*, 1982) and recently detected (Hall *et al.*, 1995). Of course, bond breaking, decomposition and chemistry occur in all of the materials suggested for the icy Galilean and Saturnian satellites (Johnson, 2001). Therefore, other decomposition products are also present in Europa's atmosphere (Johnson *et al.*, 1998).

3. Surface Sputtering

3.1. Europa

Although sputtering is the likely source of the Na atoms observed in the atmosphere of Europa, the ultimate origin of the sodium is not fully understood. Based on their first observations, Brown and Hill (1996) suggested that source of the Na was ions from Io implanted into Europa's surface. Johnson (2001) used a model for sputtering and implantation to show that decomposition of Europa's subsurface materials by the incident plasma more likely to be the principal source of the sodium. This conclusion was also drawn by comparing the Na to K ratios at Europa and Io (Brown, 2001) and it bears on whether material from the putative underground ocean has reached Europa's surface in geologically recent times.

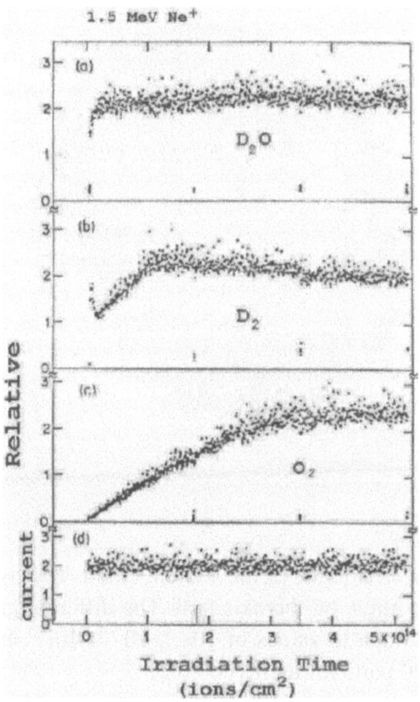

Figure 2. The sputtering rate vs. irradiation time for sputtering of low temperature D_2O ice by 1.5 MeV Ne^+ ions (Reimann *et al.*, 1984). These ions have a $(dE/dx)_e$ equivalent to energetic ions in the Jovian magnetosphere (Cooper *et al.*, 2001).

Sodium atoms are sputtered from Europa's surface by incident ions according to an energy distribution like $f(E)$ given earlier (Weins *et al.*, 1997). The ejected Na cross a thin atmosphere consisting mainly of O_2 and move along bound or escape ballistic trajectories determined by the gravity of Europa and Jupiter. Although a simple spherically-symmetric, analytic model of a sputtered atmosphere was shown to roughly reproduce the averaged observations (Johnson, 2000), N-S and E-W asymmetries observed more recently by Brown require a more accurate three dimensional simulation of the motion of the Na atoms. Such a calculation also allows one to extract more detail on the spatial and energy distribution of the sputter source. Each particle ejected from the surface is followed along its trajectory until either it reaches a maximum distance from Europa, is ionized by energetic electrons trapped in Jupiter's magnetosphere, or reimpacts Europa's surface. We typically follow several hundred thousand particles and compute the density of the generated sodium cloud. The flux ejected from the Europa surface is chosen to be distributed from trailing to leading hemispheres following a cosine law with a maximum on the trailing hemisphere and a null flux at the leading hemisphere (Popieszalska and Johnson, 1989). The probability of ionization is integrated over the trajectory following Smyth and Combi (1988). Figure 3 reproduces the shape of the Na cloud around Europa obtained for this set of parameters. The emission of the Na atoms along a line of sight as seen from the Earth is then calculated and

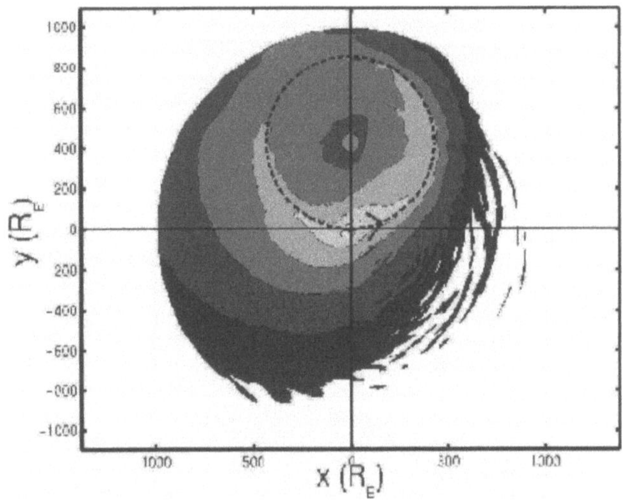

Figure 3. Cut along the Europa's orbit plane of the sodium cloud. The figure is centered on Europa and Jupiter is represented by the circle on the axis x=0. The different grays correspond to density from 10^{-2} Na/cm^{-3} close to Europa to values of 10^{-3}, 10^{-4}, 10^{-5} and 10^{-6} Na/cm^{-3} for the darker ones. We followed 500,000 particles until 1000 R_E.

compared with recent observations made by Brown. The best agreement found has been obtained for a flux of ~6x10^7 Na/cm^2/s with a binding energy U equal to 0.055 eV. This implies a loss of ~6 10^6 Na/cm^2/s for Europa which is 7 to 30 times higher than the estimate flux of Na impacting Europa from the magnetosphere. Therefore, the source of ejected Na atoms from the Europa surface is likely to be endogenic. In addition, we show that direct ejection of atoms from the surface, although dominant, is not the only source of the observed Na, and that the effective binding to the surface is lower than that measured by Weins *et al.* (1997) because the surface is porous and the Na is ejected from icy as well as refractory regions.

3.2. Saturn's icy satellites and E-ring

A model for the extensive laboratory data for the sputtering of low temperature water ice was recently constructed and used to calculate the sputtering of the icy satellites and E-ring grains by Saturn's magnetospheric plasma ions (Jurac *et al.*, 2001). At the temperatures of interest, the principal ejecta is H_2O. H_2O is dissociated in Saturn's plasma more rapidly than it is ionized producing an OH cloud, originally observed by Shemansky *et al.* (1993). Since, unlike Europa, the icy satellites are small the escape fraction is large, so the sputter contribution from the icy satellites and E-ring grains can be modeled using the initial sputtered-molecule energy distribution [$f(E)$ with $U \approx 0.055eV$] and the lifetime for dissociation and ionization. Although sputtering is likely to be the ultimate source of the observed OH as well as the magnetospheric plasma, Jurac *et al.* (2001) found that a detailed calculation, including all enhancements, could not account for the observations.

Either a large amount of additional surface is required, particularly near Enceladus, or the surfaces are not simply pure water ice but some more volatile mix. This issue will be resolved by the plasma ion measurements in the Saturnian magnetosphere by CASSINI.

4. Atmospheric Sputtering: Mars

The mechanism of sputtering of the Martian atmosphere by pick-up ions is described in Figure 4. On the dayside (left side on Figure 4), a part of the neutral atmosphere is ionized by the solar EUV and the photo-electrons from the sun. These newly ionized particles are 'picked-up' by the magnetic field lines initially frozen in the solar wind but distorted near Mars, and they are accelerated along gyroradial trajectories in the Martian tail direction. The gyroradius around the magnetic field lines of these pick-up ions has been estimated to be of the order of Mars radius (Luhmann and Kozyra, 1991). Some of these particles reimpact the neutral atmosphere with sufficient energy to generate new ejecta and ballistic particles. The net loss of atmosphere has been estimated in several papers (Jakosky *et al.*, 1994; Kass and Yung, 1995; Johnson *et al.*, 2000) and has been identified as one of the mechanism for loss of Martian water and CO_2. It should be particularly efficient (Johnson and Luhmann, 1998) during the period following the disappearance of the Mars magnetic field which has been estimated to occur before a period of solar intensity 3 times higher than the present intensity (3EUV) defined by Zhang *et al.* (1993). The method for calculating this loss is complex. First, the trajectories of impacting particles need to be calculated between the time of their ionization and the time they are neutralized by charge exchange collisions, which typically occurs above the atmospheric exobase (Luhmann and Kozyra, 1991). This involves describing the draping of the magnetic field around Mars' ionopause, and the motions of electrons and ions in the vicinity of Mars. The hybrid code developed by Brecht (1997) describes this phase self-consistently.

Second, the trajectories of newly neutralized pick-up ions close to the exobase are tracked in the neutral atmosphere accounting for collisions and gravity. Although the Mars atmosphere near the exobase is primarily O and CO_2 (Bougher *et al.*, 1999), we describe it as purely atomic composed only of O at the correct total atomic density, which is realistic above the exobase. We developed a 3-D Monte Carlo model. Each incident particle ($\sim$keV) is followed in the atmosphere ($\ll$1eV), describing all the collisions until either it reaches a region where the density is so high that it has no effect (around 120 km for Mars) or it again crosses the exobase.

We calculated the loss of atmosphere and the increase in the coronal density due to the bombardment by the flux of pick-up ions suggested by Zhang *et al.* (1993) for the 3EUV epoch. The initial neutral population is described using the results of Bougher *et al.* (1999). We first calculated the heating produced by the O^+ pick-up ions using a 2D Direct Simulation Monte Carlo. This new atmosphere was then

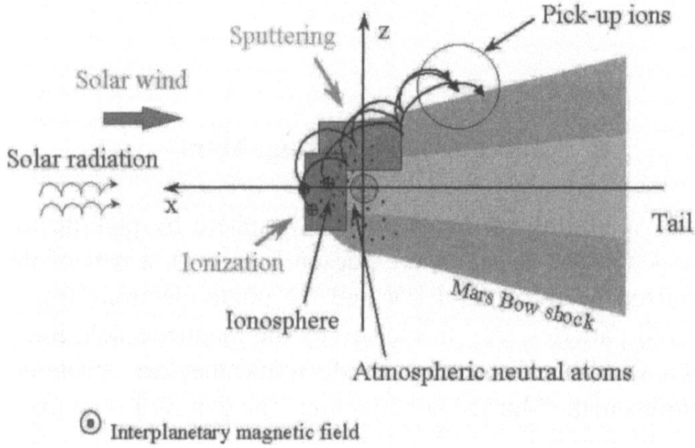

Figure 4. Sputtering of the Mars neutral atmosphere by pick-up ions dragged by the solar wind.

used in the 3D code to calculate the increase of the corona density, considering those recoil particles which have sufficient energy to escape or to populate the upper corona where pick-up ions are formed. In Figure 5, density is seen to be equivalent to that estimated for dissociative recombination of the O_2 molecules for this epoch (Zhang *et al.*, 1992) except in the polar regions. Therefore, feedback will be important because the additional O atoms reaching high altitudes can be ionized by the EUV flux and become pick-up ions reimpacting the neutral atmosphere (Johnson and Luhmann, 1998). Ignoring feedback, the sputtering yield is equal to 3.6 atoms/ion, slightly lower than the 4.2 obtained in a 1D model (Johnson *et al.*, 2000) because of our more accurate description of the collisions above the exobase. We carried out a similar study for the present epoch and then extrapolated to obtain the total loss due to sputtered O atoms over the history of Mars using data in Luhmann *et al.* (1992). We find a loss of the equivalent of 4 m of water, to which should be added the larger loss by ionization in the corona estimated earlier (Luhmann *et al.*, 1992). These calculations show that sputtering is effective primarily because it changes the density in the corona where atoms are ionized and swept away.

5. Conclusions

In many solar system environments space plasmas have been shown to effect the evolution of objects. The bombarding plasma sputters either the surface material or the gravitationally bound gases. This bombardment not only leads to the ejection of

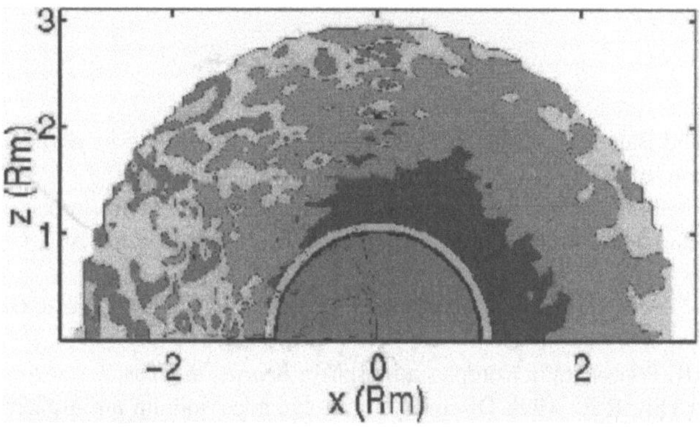

Figure 5. Corona density due to a flux of pick-up ions as defined by Zhang *et al.* (1993) for the 3EUV epoch. The different grays from darker to lighter correspond to 10^4, 10^3, 10^2 and 10^1 particles / cm^3. The sun is at the right of the figure. Mars is the gray half-circle in the center of the figure, the dark half-ring is the atmosphere below 120 km, the clear half-ring the atmosphere where the incident pick-up ions collide with the neutral atmosphere and above this circle the domain is supposed to be collisionless.

atoms and molecules but also to chemistry. Solar wind and solar flare ions sputter grains ejected from comets and the surfaces of the moon and Mercury. Pick-up ions erode the Martian atmosphere and magnetospheric plasma ions erode the atmospheres of Io, Titan and Triton, the surfaces of the other moons and the grains in planetary rings. In this article we briefly described sputtering and how Monte Carlo models can be used to describe observations for a few caes: the surface of Europa is a net source of Na; there is likely missing material in Saturn's magnetosphere; and atmospheric sputtering effects the morphology of Mars' corona and, hence, atmospheric loss by pick-up ion formation. With the recent improvements in space borne ion mass spectrometers, there will be a wealth of new data from CASSINI and from the planned missions to Mars, comets and Pluto. The critical need for interpreting this data will be the availability of good models for the flow of plasma onto an object. In addition, a new laboratory and calculational effort is needed to produce the sputtering and decomposition yields for plasma bombardment of realistic surface materials and for determining collision cross sections in the critical energy regime ($\sim$100eV - 10keV) for ions and neutrals impacting molecules.

Acknowledgement

We would like to acknowledge the support of NASA's Planetary Atmosphere Program.

References

Andersen, H.H. and Bay, H.L.: 1981, in: D. Behrisch (ed.), *Sputtering by particle bombardment I*, Springer-Verlag, Berlin.

Bougher, S.W., Engel, S., Roble, R.G. and Foster, B.: 1999, Comparative terrestrial planet thermospheres. 2. Solar cycle variation of global structure and winds at equinox, *J. Geophys. Res.* **104**, 16591–16611.

Brecht, S.H.: 1997, Solar wind proton deposition into the Martian atmosphere, *J. Geophys. Res.* **102**, 11287-11294.

Brown, M.E.: 2001, Potassium in Europa's atmosphere, *Icarus*, in press.

Brown, M.E. and Hill, R.E.: 1996 Discovery of an extended sodium atmosphere around Europa, *Nature* **380**, 229–231.

Brown, W.L., Lanzerotti, L.J., Poate, J.M. and Augustyniak, W.M.: 1978, Sputtering of ice by MeV ions, *Phys. Rev. Lett.* **49**, 1027–1030.

Cooper, J.H., Johnson, R.E., Mauk, B.H. and Gehrels, N.: 2001, Energetic ions and electron irradiation of the icy Galilean satellites, *Icarus* **149**, 133–159.

Haff, P.X. and Watson, C.C.: 1979, The erosion of planetary and satellite atmospheres. *J. Geophys. Res.* **84**, 8436–8442.

Hall, D.T., Strobel, D.F., Feldman, P.D., McGrath, M.A. and Weaver, H.A.: 1995, Detection of an oxygen atmosphere on Jupiter's moon Europa, *Nature* **373**, 677–679.

Jakosky, B.M., Pepin, R.O., Johnson, R.E. and Fox, J.L.: 1994, Mars atmospheric loss and iostropic fractionation by solar-wind induced sputtering and photochemical escape, *Icarus* **111**, 271–288.

Johnson, R.E.: 1990, *Energetic charged-particle interactions with atmospheres and surfaces*, Springer Verlag, Berlin.

Johnson, R.E.: 1994, Plasma-ion sputtering of an atmosphere, *Space Sci. Rev.* **69**, 215–253.

Johnson, R.E.: 1996, Sputtering of ices in the outer solar system. *Rev. Modern Phys.* **68**, 305–312.

Johnson, R.E.: 1998, Sputtering and desorption from icy surfaces, in: B. Schmitt, C. Debergh and M. Festou (eds.), *Solar Sytem Ices,* pp. 303–334.

Johnson, R.E.: 2000, Sodium at Europa, *Icarus* **143**, 429–433.

Johnson, R.E.: 2001, Surface chemistry in the Jovian radiation environment, in: R. Dessler (ed.), *Chemical Dynamics in Extreme Environments*, Chap. 8, pp. 390–419.

Johnson, R.E., Killen, R.M., Waite, J.H. and Lewis, W.S.: 1998, Europa's surface composition and sputter-produced atmosphere. *Geophys. Res. Lett.* **25**, 3257–3260.

Johnson, R.E. and Luhmann, J.G.: 1998, Sputter contribution to the atmospheric corona on Mars, *J. Geophys. Res.* **103**, 3649–3653.

Johnson, R.E., Lanzerotti, L.J. and Brown, W.L.: 1982, Planetary applications of ion sputtering of ices, *Nucl. Instrum. Methods* **198**, 147–157.

Johnson, R.E., Schnellenberger, D. and Wong, M.C.: 2000, The sputtering of an oxygen thermosphere by energetic O^+, *J. Geophys. Res.* **105**, 1659–1670.

Jurac, S.: 2001, Saturn's E-ring and the production of a neutral torus, *Icarus*, **149**, 384–396.

Kass, D. M. and Yung, Y.L.: 1995, Loss of atmosphere from Mars due to solar-wind induced sputtering, *Science* **268**, 697–699.

Lanzerotti, L.J., Brown, W.L., Poate, J.N. and Angustyniak, W.M.: 1978, On the contribution of water products from Galilean satellites to the Jovian magnetosphere, *Geophys. Res. Lett.* **5**, 155–158.

Luhmann, J.G. and Kozyra, J.U.: 1991, Dayside pick-up oxygen in precipitation at Venus and Mars: Spatial distribution, energy deposition and consequences, *J. Geophys. Res.* **96**, 5457–5467.

Matson, D.L., Johnson, T.V. and Fanale, F.P.: 1974, Sodium D-line emission from Io: sputtering and resonant scattering hypothesis, *Astrophys. J.* **192**, L43–L46.

Nash, D. and Fanale, F.P.: 1977, Io's surface composition based on reflectance spectra of sulfur/salt mixtures and proton-irradiation experiments, *Icarus* **31**, 40–80.

Popieszalska, M.K. and Johnson, R.E.: 1989, Magnetospheric ion bombardment profiles of satellites: Europa and Dione, *Icarus* **78**, 1–13.

Reimann, C.T., Boring, J.W., Johnson, R.E., Garett, J.W., Farmer, K.R. and Brown, W.L.: 1984, Ion-induced molecular ejection from D_2O ice, *Surf. Sci.* **147**, 227–240.

Saur, J., Strobel, D.F. and Neubauer, F.M.: 1998, Interaction of the Jovian magnetosphere with Europa: Constraints on the neutral atmosphere, *J. Geophys. Res.* **103**, 19947–19962.

Shemansky, D.E., Matherson, P., Hall, D.T., Hu, H.-Y. and Tripp, T.M.: 1993, Detection of the hydroxil radiacal in the Saturn magnetosphere, *Nature* **363**, 329–332.

Smyth W.H. and Combi, M.R.: 1988, A general model for Io's neutral gas clouds. II Application to the sodium cloud, *Astrophys. J.* **328**, 888–918.

Sprague, A.L., R.W.H. Kozlowski, D.M. Hunten, W.K. Wells, and F.A. Grosse: 1992, The sodium and potassium atmosphere of the Moon and its interaction with the surface, *Icarus* **96**, 27–42.

Weins, R.C., Burnett, D.S., Calaway, W.F., Hansen, C.S., Lykkem, K.R. and Pellin, M.L.: 1997, Sputtering products of sodium sulfate: Implications for Io's surface and for sodium bearing molecules in the Io torus, *Icarus* **128**, 386–397.

Yakshinskiy, B.V. and Madey, T.E.: 1999, Photon-stimulated desorption as a substantial source of sodium in the lunar atmosphere, *Nature* **400**, 642–644.

Zhang, M.H.G., Luhmann, J.G., Bougher, S.W. and Nagy, A.F.: 1993, The ancient oxygen exosphere of Mars: Implication for atmospheric evaluations, *J. Geophys. Res.* **98**, 10915–10923.

THE ATMOSPHERE OF IO: ABUNDANCES AND SOURCES OF SULFUR DIOXIDE AND ATOMIC HYDROGEN

DARRELL F. STROBEL[1,2] and BRIAN C. WOLVEN[2]
[1]*Department of Earth and Planetary Sciences*
[2]*Department of Physics and Astronomy*
The Johns Hopkins University, Baltimore, MD 21218 USA
E-mail: strobel@jhu.edu

Abstract. An analysis and interpretation of reflected solar Lyman α intensity data acquired with the Hubble Space Telescope (HST) implies an equatorially confined atmosphere with SO_2 column densities $\sim 1-2 \times 10^{16}$ cm^{-2}. Poleward of 30° the SO_2 density must decrease sharply reaching an asymptotic polar value of $< 10^{15}$ cm^{-2} at 45° to achieve the observed 2 kR intensity peaks. The corresponding surface reflectivities must be either a constant 0.047 for higher equatorial SO_2 or a variable reflectivity of 0.027 with lower SO_2 densities at the equator increasing to a polar value of $\sim$ 0.05. The average residence time for an atmospheric SO_2 molecule is $\sim$ 2–3 days for the canonical mass loading rate of the Io plasma torus = 10^{30} amu s^{-1}. With atomic hydrogen in the atmosphere and corona constrained by the HST observations, it is estimated that a pickup proton density ratio of 0.25–0.4% can be sustained by a supply of Io plasma torus protons neutralized in Io's atmosphere/exosphere, if protons constitute 7% of the total torus ion density, which is close to the Chust *et al.* (1999) pickup proton density ratio and under the widely quoted 10% proton content of the torus.

1. Introduction

In this article honoring the work of J.L. Steinberg, Io represents a meeting place for the overlap of his research interests with ours and where interdisciplinary science is mandatory. Bigg (1964) noted that Io controlled Jupiter's decametric radiation and hinted that it implied the existence of an atmosphere. Pioneer 10 detected an ionosphere confirming the prospect of an atmosphere (Kliore *et al.*, 1975). The discovery of active volcanism on Io coupled with the Voyager detection of SO_2 IR band absorption in a localized volcanic plume atmosphere (Pearl *et al.*, 1979), and measurement of a dense plasma torus in the inner Jovian magnetosphere dominated by heavy ions of oxygen and sulfur (Broadfoot *et al.*, 1979; Bridge *et al.*, 1979) provided initial evidence for an SO_2 atmosphere. Millimeter observations by Lellouch *et al.* (1990) and Lellouch *et al.* (1992) of SO_2 rotational line emission followed by HST observations of UV sunlight reflected from Io's surface with numerous SO_2 absorption bands (Ballester *et al.*, 1994) supplied persuasive evidence for an SO_2 atmosphere. These initial observations had limited success in defining the spatial distribution of Io's SO_2 atmosphere. At millimeter wavelengths spatial resolution requires interferometric techniques. From the shape and contrast of the

Astrophysics and Space Science **277:** 271–287, 2001.

SO_2 rotation line profiles one can infer fractional areal coverage of the atmosphere over Io's disk, but without any precision as to the actual distribution/location of the atmosphere (Lellouch *et al.*, 1992). Similarly Ballester *et al.* (1994) were able to obtain a better fit to the atmospheric SO_2 absorption band signatures in their HST data with localized, patchy atmospheres, but with no ability to locate these regions.

Recently McGrath *et al.* (2000) used the same HST FOS instrument as Ballester *et al.* (1994), but at higher spatial resolution to resolve regions on Io such as the Pele volcano. With both spatial resolution to resolve Io's disk and spectral resolution to distinguish atmospheric SO_2 from surface SO_2, McGrath *et al.* inferred a range of SO_2 column densities at three locations of $(7 - 32) \times 10^{15}$ cm^{-2}, about an order of magnitude smaller than previously inferred by Sartoretti *et al.* (1996) and Spencer *et al.* (1997) from HST data at longer wavelengths where atmospheric SO_2 could not be spectrally differentiated from surface albedo variations in reflected UV sunlight. The McGrath *et al.* (2000) column densities were considerably below values inferred by Hendrix *et al.* (1999) with the Galileo UV spectrometer in orbit around Jupiter. The latter data possess neither the required spectral and spatial resolutions necessary to separate atmospheric SO_2 variations from surface albedo variations and are clearly inferior to Earth orbiting HST data. Spencer *et al.* (2000) note in their paper on the first atmospheric detection of S_2 in the Pele plume that anomalously high SO_2 densities reported in the past could be due to S_2 absorption features interpreted as SO_2 absorption at long wavelengths. The lower McGrath *et al.* (2000) column densities are consistent with millimeter observations when the latter are interpreted in terms of dynamic plume atmospheres, but inconsistent when interpreted as a hydrostatic atmosphere on Io (Lellouch, 1996).

With the addition of Space Telescope Imaging Spectrograph (STIS) on board HST, Roesler *et al.* (1999) demonstrated and Feldman *et al.* (2000) confirmed in a convincing manner that Lyman-α imaging is a powerful method to map globally the atmospheric SO_2 distribution on Io. At Lyman-α, SO_2 is a continuum absorber of radiation without any of the attendant problems in the analysis of SO_2 absorption in electronic band systems longward of 1960 Å with spectrally unresolved vibrational-rotational lines. However, there is the residual problem common to all observations of reflected sunlight, namely, the ability to distinguish uniquely the atmospheric signature of SO_2 from the surface reflectivity signature. As will be demonstrated below Lyman-α intensity variation over the disk of Io must be mainly due to atmospheric SO_2.

Another important issue concerning Io is the presence of hydrogen bearing constituents on the surface. There now is credible evidence for pickup protons in the vicinity of Io from the Galileo plasma spectrometer (PLS, Frank and Paterson, 1999) and from the Galileo Plasma Wave System (PWS, Chust *et al.*, 1999). To date there is no unambiguous evidence for the two most probable hydrogen bearing molecules, namely water and hydrogen sulfide, as discussed in Roesler *et al.* (1999).

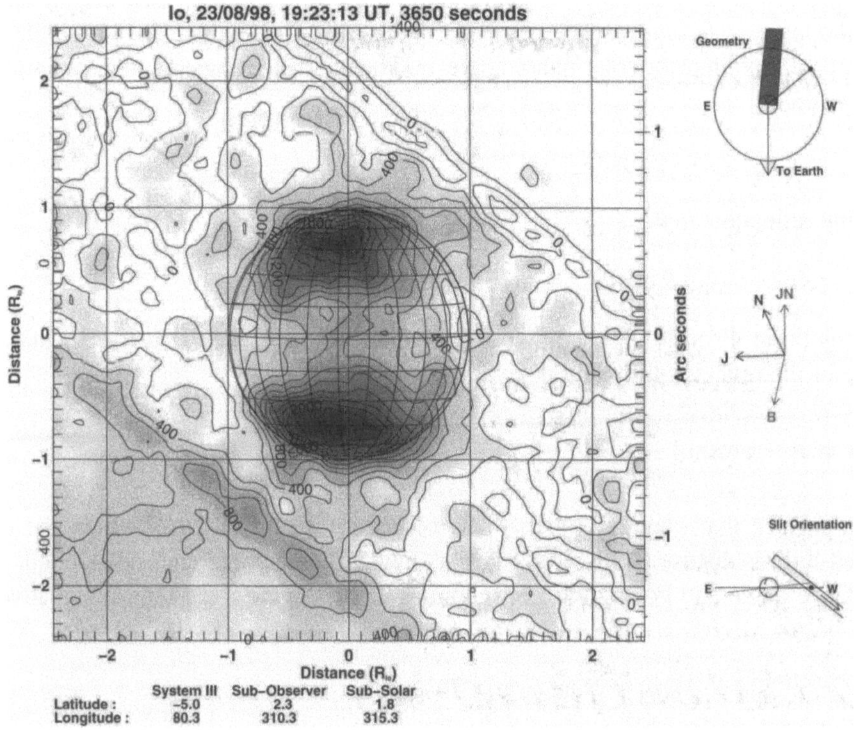

Figure 1. The Lyman α intensity in kiloRayleighs based on HST observations with the STIS instrument on 23 August 1998 and the analysis of Feldman *et al.* (2000). Note latitudes are spaced proportional to sin(latitude) about the equator. Relevant observation information such as slit orientation, directions of Jupiter from Io, magnetic field, Jupiter north and magnetic dipole north, and latitudes and longitudes are also given.

2. The Atmospheric Distribution of SO_2

Feldman *et al.* (2000) analyzed three HST Lyman-α images that showed both similarities and substantial differences. The brightest regions were located at middle to high latitudes with peak brightness occurring at $\sim 45^{\circ}$ latitude, whereas the low latitude equatorial regions were always dark. For the 1998 West exposure a contour plot of the Lyman-α intensity in kilorayleighs (kR) was generated and is displayed in Figure 1. Note that the peak brightness is $\sim$ 2 kR at $\sim 45^{\circ}$ latitude, with an intensity of $\sim$ 0.7 kR at the center of Io's disk.

We now show that the major features of this intensity pattern can be understood by a latitudinal variation of the SO_2 column densities and the geometry associated with incident and reflected solar Lyman-α radiation off Io's spherical surface. Below we outline the standard description of the two way transmission of light through the atmosphere with surface reflection based on standard references (e.g. Harris, 1961; Chamberlain and Hunten, 1989). On a great circle passing through the sub-Earth point E and the subsolar point S, the longitude, ϕ, is measured from E and the latitude, θ, is measured perpendicular to this great circle. The longitude of the subsolar point is the solar phase angle, α. The incidence angle of solar radiation, i, is given by

$$\mu_0 = \cos i = \cos\theta \cos(\phi - \alpha) \tag{1}$$

and the reflection angle, ε, is

$$\mu = \cos\varepsilon = \cos\theta \cos\phi \tag{2}$$

For any point, P, on Io, let the angle E P S equal $\pi - \psi$, where ψ is the azimuthal angle of the reflected light and is given by

$$\psi = \pi - \arcsin\left(\frac{\sin\theta \sin\alpha}{\sin\varepsilon \sin i}\right) \tag{3}$$

For a surface that obeys Lambert's law of diffuse reflection from which the reflected light is diffuse and isotropic in the upward hemisphere and independent of incidence angle and polarization of incoming light, the reflected intensity is given by

$$I(\mu, \psi, \mu_0) = \lambda_0\, \mu_0\, F\, Tr_i(\psi, \mu_0)\, Tr_r(\mu) \tag{4}$$

where λ_0 is the reflection albedo of the surface, πF is the solar Lyman-α flux, and $Tr(\mu, \psi, \mu_0)$ is the atmospheric transmission function. In terms of the SO_2 absorption cross section at Lyman-α, $\sigma_{SO_2} = 3.9 \times 10^{-17}$ cm^2 (Manatt and Lane, 1993), and the SO_2 vertical column density, $N(\theta, \phi)$, the transmission functions for incident and reflected light are

$$Tr_i(\theta, \psi(\alpha), \phi) = \exp\left[-\left(\frac{N}{\mu_0}\right)\sigma_{SO_2}\right], Tr_r(\theta, \phi) = \exp\left[-\left(\frac{N}{\mu}\right)\sigma_{SO_2}\right] \tag{5}$$

For completeness we define the following additional quantities. The integrated light diffusely reflected toward the Earth by Io at a given phase angle is

$$j(\alpha) = \int_{\alpha-\frac{\pi}{2}}^{\frac{\pi}{2}} \int_{-\frac{\pi}{2}}^{\frac{\pi}{2}} I(\mu, \psi, \mu_0) \cos^2\theta \cos\phi \, d\theta d\phi \tag{6}$$

The Bond albedo, A, is

$$A = \frac{2}{\pi F}\int_0^{\pi} j(\alpha) \sin\alpha \, d\alpha = pq \tag{7}$$

where p is the geometric albedo and q is the phase integral

$$p = \frac{j(0)}{\pi F}, \quad q = 2\int_0^{\pi} \Phi(\alpha) \sin\alpha \, d\alpha, \quad \text{where } \Phi(\alpha) = \frac{j(\alpha)}{j(0)} \tag{8}$$

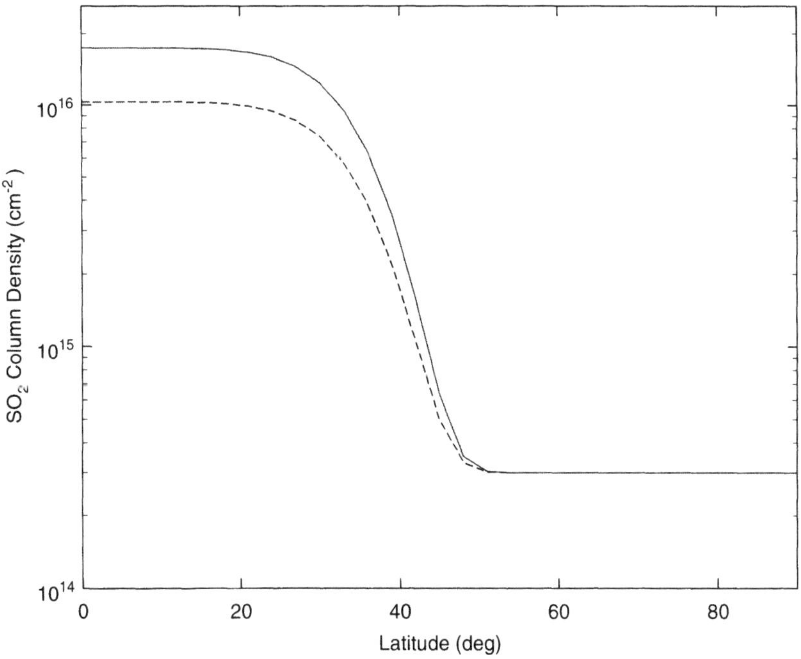

Figure 2. The SO_2 column density variation with latitude given by Eq. (9) with equatorial values $N_{eq} = 1.7 \times 10^{16}$ cm^{-2} (solid line) and $N_{eq} = 1.0 \times 10^{16}$ cm^{-2} (dashed line).

Given that the Lyman-α brightness is highly variable on Io, our principal interest is to understand the essential pattern displayed in Figure 1. The phase angle is small for this observation, only 5°, and not strikingly evident in the data. It is included in the calculations, but could be set to zero with negligible error. As a working hypothesis longitude variations arise from geometry and latitude variations are generated mostly by latitudinal variations in the SO_2 column density from the equator to midlatitudes and by geometry at high latitudes. Consistent with this hypothesis is the following latitudinal variation of SO_2 column density

$$N(\theta) = N_0 + N_{eq} \exp\left[-\left(\frac{\theta}{\delta\theta}\right)^6\right] \tag{9}$$

where $N_0 = 3 \times 10^{14}$ cm^{-2} and $N_{eq} = 1.7 \times 10^{16}$ cm^{-2} are the polar and equatorial column densities of SO_2, $\delta\theta = 0.625$ radian, and the overall variation is illustrated in Figure 2. At the time of the Figure 1 observation the solar Lyman-α flux at 1 AU was 4×10^{11} photon cm^{-2} s^{-1} and Jupiter was at 4.97 AU. The surface reflection albedo is unknown and chosen to be $\lambda_0 = 0.047$ to achieve the peak Lyman-α brightness of 2 kR at midlatitudes. In the absence of an SO_2 atmosphere, Io's geometric albedo would be $p = 2/3\lambda_0 = 0.031$. This latter albedo may be compared with the UV value inferred by Feldman *et al.* (2000) at longer wavelengths (1520–1700 Å) of 0.019. Whether the unknown surface composition could have

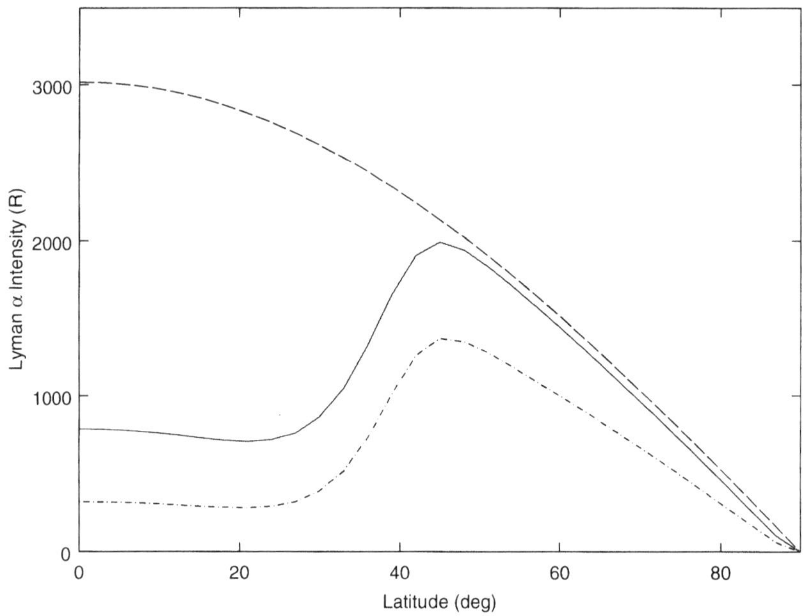

Figure 3. The Lyman α intensity in Rayleighs as a function of latitude calculated at the central meridian without an atmosphere (dashed line) and with an atmosphere (solid line). The dot-dashed line is the intensity variation with an atmosphere at longitude $\pm 45^{\circ}$. All calculations with constant surface reflectivity $\lambda_0 = 0.047$. The calculations with an atmosphere had $N_{eq} = 1.7 \times 10^{16}$ cm^{-2}.

this wavelength variation is uncertain, but will be discussed below. In Figure 3 the calculated latitudinal variation of the reflected solar Lyman-α intensity is displayed along the central meridian with and without an SO_2 atmosphere and at a longitude 45° from the center of the disk. Without an SO_2 atmosphere the Lyman-α intensity at the center of the disk would be 3000 R rather than the observed and computed values of $\sim$ 700 R with model SO_2 given by (9). To achieve the observed intensity of 2000 R at 45° requires a rapid drop in the SO_2 column density between 35° and 45° latitude. This decrease is consistent with the latitudinal boundaries of the most active equatorial areas of volcanism. Also the horizontal density length scale at these boundaries ($\sim$ 100 km) associated with (9) is consistent with Ingersoll's (1989) horizontal averaging length $\sqrt{(2\pi)}H/\alpha$, where H is the atmospheric scale height and α is the SO_2 sticking coefficient. For H = 20 km and α = 0.5 for midlatitude surface temperatures, this length scale is $\sim$ 100 km. In Figure 4 a model contour plot is shown for comparison with the observations in Figure 1. At the center of the disk the intensities are slightly larger than 700 R and increase with latitude to peak brightnesses of 2000 R at 45° and then decline, as expected by geometry, to zero at the poles and everywhere along the edge of the disk. This azimuthally symmetric model of latitudinal variation of SO_2 column density clearly captures the essence of the HST data and suggests that Io 'wears' its atmosphere as an equatorial collar with spatial dimensions that coincide with regions of active

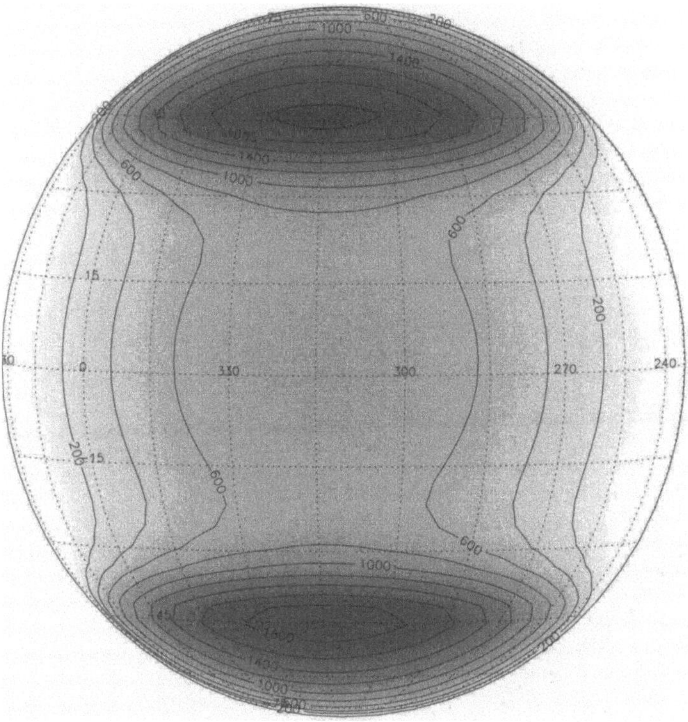

Figure 4. Contour plot of the Lyman-α intensity in Rayleighs over Io disk for assumed atmosphere with $N_{eq} = 1.7 \times 10^{16}$ cm^{-2} and surface reflectivity $\lambda_0 = 0.047 (p = 0.031)$.

volcanism. When this atmosphere is averaged over the entire surface, an average column density of $\sim 1 \times 10^{16}$ cm^{-2} is obtained. This value is very close to the hemispheric average value of 6×10^{15} cm^{-2} obtained by Ballester *et al.* (1994), in the middle of the range of values inferred by McGrath *et al.* (2000), and in line with the volcanic plume atmosphere interpretation of millimeter observations (Lellouch, 1996). It should be noted that polar SO_2 column densities could be lower than $N_0 = 3 \times 10^{14}$ cm^{-2}, without any discernable effect in Figure 4 at high latitudes. For this value of N_0, the intensity at 75° is 91% of the intensity in the absence of an SO_2 atmosphere. If the polar SO_2 column densities were $N_0 = 6 \times 10^{14}$, 9×10^{14}, and 12×10^{14} cm^{-2}, then the relative Lyman-α intensity would be 84, 76, and 70%, respectively. Given the large Lyman-α intensities observed in the polar regions as shown in Figure 1, a conservative upper limit on the polar SO_2 column density is $N_0 \sim 10^{15}$ cm^{-2}.

We now return to a discussion of the required surface reflection albedo. Whereas a wavelength variation from 0.019 observationally inferred at ~ 1600 Å to 0.031 model inferred at 1216 Å is possible, other scenarios may be more plausible. For example, if certain volatiles such as SO_2 are preferentially deposited on colder middle and high latitude surfaces and absent at low latitudes, a case could be made

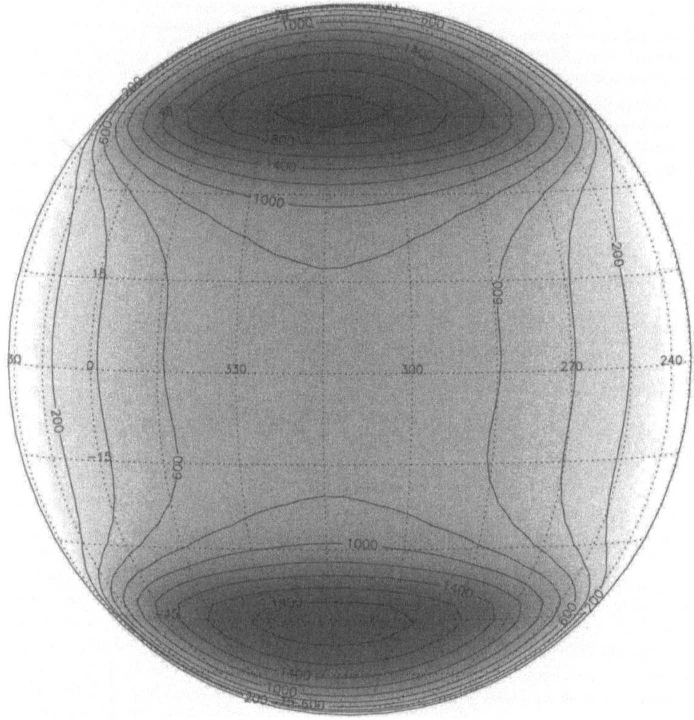

Figure 5. Contour plot of the Lyman-α intensity in Rayleighs over Io disk for assumed atmosphere with $N_{eq} = 1.0 \times 10^{16}$ cm^{-2} and variable surface reflectivity $\lambda_0 = 0.027$ (equator) to 0.054 (poles) ($p = 0.022$).

for a latitudinally varying surface reflection albedo. If λ_0 varies from 0.027 at the equator with a rapid increase at middle latitudes, where the SO_2 column density declines sharply, to a value of 0.054 at the poles, then the equatorial SO_2 column density can be decreased to $N_{eq} = 1.0 \times 10^{16}$ cm^{-2} and the Lyman-α geometric albedo, in the absence of an atmosphere, would decrease to 0.022, in comparison to 0.019 at longer wavelengths, and a more conceivable wavelength dependence for surface albedo. The globally averaged column density is now 6×10^{15} cm^{-2} and an equally acceptable fit, comparable to Figure 4, to the HST data in Figure 1 is realized, as illustrated in Figure 5. Over Io's equatorial belt with latitudinal width of $\pm 30^\circ$ an average pressure $\sim$ 0.2-0.35 nbar is implied by the HST data. At middle and high latitudes where Lyman-α is bright a surface reflectivity $\sim$ 0.05 at 1216 Å is required, whereas at low latitudes there is a trade off between higher/lower SO_2 column densities and higher/lower surface reflectivity.

3. Volcanic Plume Atmospheres

Given a reasonably good estimate of atmospheric abundance of SO_2, we now address the question of whether volcanic plume atmospheres in sufficient numbers could account for the inferred SO_2 abundance. Although alternative explanations are available such as a sublimation atmosphere driven by plasma heating (Wong and Johnson, 1995), our intend is to demonstrate that volcanic plume atmospheres are an equally viable candidate. For purposes of answering this question the discussion in the excellent chapter by Kieffer (1982) on the dynamics and thermodynamics of Io's volcanic eruption and plumes is adopted. In Figure 6 the diagram shows a volcanic system that consists of a reservoir, conduit, crater, and plume. The reservoir is characterized by two quantities pressure, p_0, and temperature, T_0. The conduit delivers high pressure, p^*, high temperature, T^*, SO_2 supersonically with Mach number, M^*, to the crater and is characterized by a critical radius, which is defined to be its actual radius and presumed to be much shorter than its length. Based on Kieffer (1982), this radius, r^*, is set equal to 5 m. The violent nature of volcanic eruptions produces a funnel-shaped crater at the surface, with radius, r_c, and depth/height, h. Kieffer (1982) argues that a reasonable estimate for the inclination or slope angle is 60° and for a nominal height of 1 km. The SO_2 flow in the crater is assumed to be quasi-one-dimensional with variable cross sectional area and described by pressure, p_{sur}, temperature, T_{sur}, Mach number, M_{sur}, and velocity, v_{sur}.

The complexities in the calculation of plume atmospheres are discussed in Kieffer (1982). For our purposes a crude description will suffice. Only a single-phase fluid is considered - the vapor phase. Strom and Schneider (1982) showed that Voyager data on Io volcanic plumes could be adequately interpreted by gas and entrained particles following ballistic trajectories. Thus we assume that volcanic SO_2 gas follows ballistic trajectories. Then the maximum distance traveled by a parcel will be one that leaves the surface at an 45° angle. This maximum distance for a calculated v_{sur} determines the radius, r_p, of the plume atmosphere and its areal coverage over Io's surface. From the crater area, the surface velocity, v_{sur}, the SO_2 mass density at the surface from the pressure, ρ_{sur}, and temperature, T_{sur}, and the maximum flight time, t_{flight}, on a ballistic orbit, an average pressure, $< p >$, can be calculated with the assumption that the mass leaving the crater in time, t_{flight}, is uniformly distributed over the plume area, πr_p^2. Mathematically, the average pressure is given by

$$< p > \; = g \left(\frac{\pi r_c^2 \rho_{sur} \mathrm{v}_{sur} t_{flight}}{\pi r_p^2} \right) \tag{10}$$

where g is the gravitational acceleration. Implicit in this calculation of the average pressure is the assumption that the SO_2 sticking coefficient on the surface is one, i.e., it condenses.

VOLCANIC ERUPTIONS AND
PLUME ATMOSPHERES ON IO

PLUME PROPERTIES						
$N(10^{17}\text{cm}^2)$	1.8	2	15	0.18	0.18	0.18
$< p >$(nbar)	3.5	3.8	28	0.35	0.35	0.35
r_p(km)	920	850	320	320	920	190
$^1F(10^{29}\text{s}^{-1}\text{volc}^{-1})$	46	46	77	1	4.6	0.45
^{2}# of plumes				66	8	185
CRATER PROPERTIES						
r_c(km)	1.7	0.17	1.7	1.7	1.7	1.7
h(km)	1	0.1	1	1	1	1
T_{sur}(K)	30	116	11	11	30	6
$p_{sur}(\mu\text{bar})$	1.6	640	1.6	0.02	0.16	0.01
M_{sur}	18	8.7	18	18	18	18
$\text{v}_{sur}(\text{km s}^{-1})$	1.3	1.2	0.75	0.75	1.3	0.58
CONDUIT PROPERTIES						
r^*(m)	5	5	5	5	5	5
M^*	2.8	2.7	2.8	2.8	2.8	2.7
T^*(K)	1220	1220	440	440	1220	260
p^*(bar)	22	22	22	0.3	2.2	0.1
RESERVOIR PROPERTIES						
T_0(K)	1400	1400	500	500	1400	300
p_0(bar)	40	40	40	0.5	4	0.18

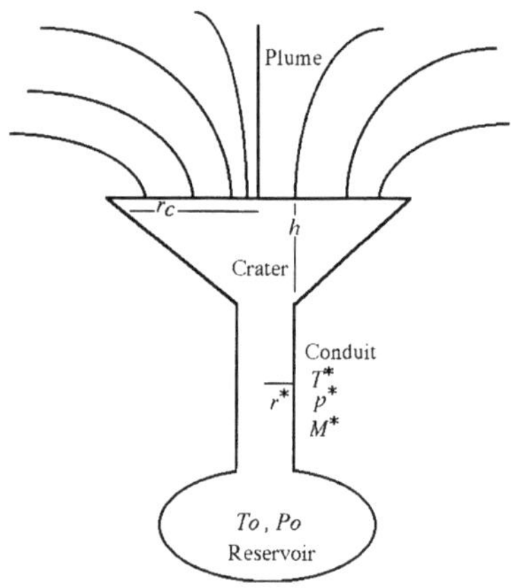

[1] The total number of SO_2 molecules ejected per volcano

[2] Number of volcanoes needed to cover Io's surface area between ±30 degrees

The total number of SO_2 molecules ejected which would sustain the HST inferred column density $\sim 5 \times 10^{30}\text{s}^{-1}$

Figure 6. Calculations of the properties of volcanic plume atmospheres as a function of the properties of the reservoir, conduit, and crater with the equations of Kieffer (1982). The number (#) of plumes is computed only for cases 4, 5, and 6, where the average pressures $< p >$ of the plume atmospheres are in agreement with the HST inferred value. Cartoon at the right indicates nomenclature and is not to scale.

Representative solutions for given reservoir conditions (T_0, p_0) are presented in Figure 6. The random kinetic energy in the reservoir measured by T_0 determines the surface velocity, v_{sur}, and the radius of the plume atmosphere, r_p, as may be seen by comparing columns (1, 2, 5) with (3, 4). The height/depth of the crater (and its radius) have a minimal impact on the plume properties by cross-comparing columns 1 and 2. Strom and Schneider (1982) inferred an ejection velocity range of 0.5-1.0 km s^{-1} with their ballistic model. Of course, it must be remembered that solutions based on Kieffer (1982) and given in Figure 6 are for a dissipationless, frictionless conduit and crater system. For a specified reservoir temperature, the calculated surface or ejection velocity is an ideal upper limit value. To account for the HST data an average pressure $\sim$ 0.2–0.35 nbar must be maintained over Io's equatorial belt with latitudinal width of ±30°. The approximate number of volcanic plumes needed is Io's surface area between ±30° divided by the plume area. For a high entropy case T_0 = 1400 K (Kieffer's Reservoir V), only 8 volcanoes (≥Pele size) would be needed. This class of volcanoes proceeds only in the vapor phase and was the basis of the 'stealth plume' scenario advocated by Johnson *et al.* (1995) as the most common on Io. Without entrained particulates, Johnson *et al.* argued that detection would be difficult. Whereas in the low to moderate entropy reservoirs with $T_0 = 500$ K, approximately 66 volcanoes would suffice. If the reservoir temperature were lowered to 300 K, then $v_{sur} \sim 0.6$ km s^{-1}, and 185 volcanoes would be needed. Clearly some admixture of volcano sizes and types consistent with Galileo spacecraft reports on some 100 active volcanic areas would account for our HST inferred equatorial collar atmosphere. The large number of inferred volcanic plumes is in agreement with Lellouch's (1996) gas plume model interpretation of millimeter wavelength observations.

Regardless of reservoir temperature, the total number of SO_2 molecules that must be ejected from all these sites (the product of the flux per volcano and the number of plumes – F and # of plumes under plume properties in Figure 6) is $\sim 5 \times 10^{30}$ s^{-1}, and insensitive to reservoir conditions. If the SO_2 sticking coefficient is considerably less than one, then far fewer active volcanoes would be needed to sustain the atmospheric SO_2 abundance. The principal conclusion from this crude calculation is that active volcanoes with their plume atmospheres could account directly for the bulk of the observed SO_2 in contrast to hot lava vaporizing SO_2 frost off the surface and into the atmosphere or an SO_2 frost buffered atmosphere in agreement with Lellouch's (1996) conclusions.

Although plume atmospheres are low pressure and supersonic, their molecules still experience collisions with the possibility of converting their ballistic kinetic energy into random kinetic energy as gas rebounds off the surface and generate shock waves that interfere with the motion of fresh emerging volcanic gases. Because it is improbable that all SO_2 molecules stick upon impact on the surface, it is likely that the atmosphere contains a quasi-hydrostatic component away from volcanic vents and at high altitudes if the gas eruptions are steady. A relevant time constant for the establishment of a hydrostatic atmosphere is the hydrostatic adjustment time constant, which is equal to the atmospheric scale height, H, divided by the speed of sound, c. Based on the 1D models of Strobel *et al.* (1994) and the MHD model of Saur *et al.* (1999) that demonstrates Joule heating approaches its theoretical maximum value on Io, the vertical range of temperature on Io is between 100-2000 K with a corresponding range for $c \sim 130$-550 m s^{-1}. The hydrostatic adjustment time varies as the $\sqrt{T}$, inversely with the gravitational acceleration, and has a range of ~ 100 s near the surface to 600 s at high altitudes and is shorter than the flight time for ballistic trajectories. In Figure 7, a plausible equatorial hydrostatic model atmosphere based on Strobel *et al.* (1994) is illustrated with an SO_2 column density of 1.35×10^{16} cm^{-2}, midway between the two models shown in Figure 2. The exobase for this model is located at 1400 km above the surface due to the very large scale height there associated with the 1800 K thermosphere and the large SO_2 collision diameter of 5.3 Å. Hence the vertical exospheric column density is only 10^{14} cm^{-2}.

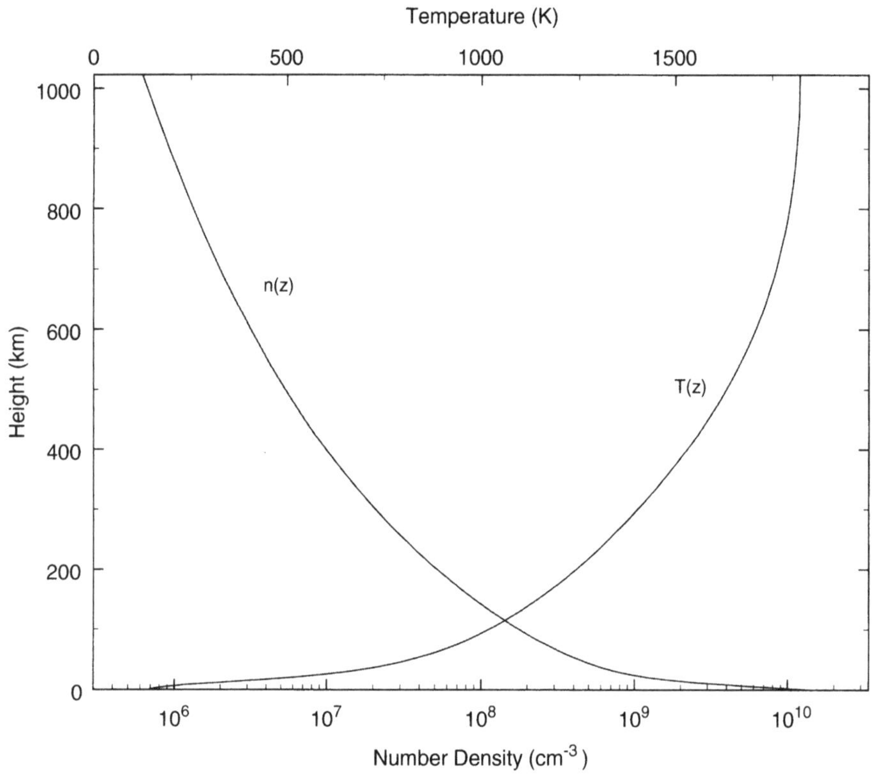

Figure 7. Vertical structure of the number density (lower scale) and temperature (upper scale) in the equatorial region if the atmosphere becomes quasi-hydrostatic away from volcanic vents. Exobase is at 1400 km.

To sustain the Io plasma torus, which is composed of mainly heavy ions of oxygen and sulfur most plausibly of Iogenic origin, a mass loading rate of $\sim 10^{30}$ amu s^{-1} has been estimated by a number of authors. This value, which has achieved canonical status, was inferred from torus energetics by Broadfoot *et al.* (1979), from corotation lag by Hill (1979), power dissipation in the Jovian magnetosphere by Dessler (1980), and from an ion physics model of the torus by Smith and Strobel (1985), to list a few authors. Whether this mass loading rate is truly invariant as its canonical stature would suggest is uncertain. The author is not aware of any analysis of Galileo spacecraft data to derive torus mass loading rates. If 10^{30} amu s^{-1} $= 1.6 \times 10^{28}$ SO_2 s^{-1} is taken as the average mass loading rate and set equal to the SO_2 atmospheric loss rate, then the residence time of SO_2 in Io's atmosphere can be estimated to be 2–3 day from HST inferred total atmospheric abundance of SO_2. Consequently Io's atmosphere is as transient as Europa's O_2 atmosphere, since the supply and loss rates calculated by Saur *et al.* (1998) yield a residence time of ~ 2 day. The SO_2 loss rate is only a small fraction (~ 0.003) of volcanic supply rate estimated above.

4. Atomic Hydrogen and Pickup Protons

Feldman *et al.* (2000) noted the absence of any limb brightening of Lyman-α in their HST observations and placed an upper limit of 300 R on the limb intensity. Is this HST observational constraint compatible with the evidence for pickup protons in the vicinity of Io from the Galileo plasma spectrometer (PLS, Frank and Paterson, 1999) and from the Galileo Plasma Wave System (PWS, Chust *et al.*, 1999)? Atomic hydrogen is assumed to be a minor constituent in the atmosphere and to form a corona around Io with a number density varying as r^{-2} out to $6R_{Io}$, approximately the location of the Hill sphere. With the surface H atom number density equal to 1.1×10^4 cm^{-3}, the calculated limb intensity of Lyman-α is 300 R, the observational upper limit. Optical depth effects are only very modest and not a critical element in the calculation. To compute the proton pickup flux, which is the quantity reported by Frank and Paterson (1999) from their PLS measurements, the following assumptions are made. The electrodynamic interaction of torus plasma results in strong shielding and the ionospheric electric field is reduced to ~ 0.1 of the corotational value based on the calculations of Saur *et al.* (1999, their calculated reduction factor was 0.07, but a slightly larger value is preferable based on unpublished work). As Goertz (1980) pointed out the effective geometric cross section of Io is reduced by this factor. As plasma flows around Io on the flanks it speeds up to perhaps $v_{plasma} \sim 15$ km s^{-1} and most of the relevant pickup protons may be generated by charge exchange on the flanks. Assume all the torus ions charge exchange with H atoms with the same cross section as $O^+ + H$ ($\sigma_{chex} \sim 1.6 \times 10^{-15}$ cm^2) and the torus ion density, n_{ion}, is 4000 cm^{-3}. The predicted pickup proton flux given by

$$\phi_{H^+}(r_{obs}) = \int_{streamline}^{r_{obs}} v_{plasma}\, \sigma_{chex}\, n_{ion}\, n_H\, ds \tag{11}$$

where the integration is performed along a plasma streamline starting far upstream and ending at the observation point, r_{obs}. When averaged over the altitude range 1000-2000 km or a cylindrical shell 1000 km thick, this flux is $\sim 2-3 \times 10^7$ cm^{-2} s^{-1}, approximately a factor of 40 lower than Frank and Paterson (1999) reported. More than an order of magnitude enhancement in the H atom density would be needed to bring our calculations in agreement with PLS measurements and accompanied by an order of magnitude violation of the HST upper limit constraint on limb brightness as stated by Frank and Paterson (1999) and Feldman *et al.* (2000). Alternatively, it is noted that Frank and Paterson presented PLS data at the 1999 Fall American Geophysical Union meeting from later Galileo spacecraft passes that suggest lower densities and fluxes of pickup protons than reported in Frank and Paterson (1999).

While it is always possible to invoke time dependent phenomena, especially on a satellite with active volcanism, to resolve conflicting data, we first address the plausibility of the PWS detection of pickup protons from harmonics of the local proton gyrofrequency with densities ≥ 20 cm^{-3} at $\sim 3R_{Io}(2R_{Io}$ above the surface). With the same assumptions given above, the pickup proton density at the PWS observation point can be calculated with expression (11) for the streamline passing through this point and is given by

$$n(H^+_{pickup}) = \frac{\phi_{H+}(r_{obs})}{\mathrm{v}_{plasma}} \tag{12}$$

of $\sim$ 9-17 cm^{-3} (depending on the proximity of the plasma streamline to Io at closest approach) and confirmation of the PWS detection. Note that in this calculation the plasma flow speed divides out and the torus ion density is the most critical input quantity. Because the PWS observational data are reported fundamentally as a density ratio ($\geq 0.5\%$), the calculated pickup proton mixing ratio is ~ 0.22–0.4%. It may be possible to achieve even better agreement by inclusion of actual torus ion composition and individual charge exchange cross sections rather than using the O^+ + H value. For example, the H^+ + H charge exchange cross section is a factor of 3 larger than the O^+ + H value. As a corollary to these calculations the total removal rate of atomic hydrogen from Io by charge exchange is estimated at $L_H \sim 4 \times 10^{25}$ s^{-1}, about a factor of 10 less than Chust *et al.* (1999) and mainly due to assumptions of flow speeds near Io.

This hydrogen loss rate is significant in light of our inability to detect any hydrogen-bearing molecules in the atmosphere. Could this hydrogen loss rate be supplied by torus protons, which comprise approximately 10% of the torus density and originate from Jupiter's upper atmosphere and the water ice on the other Galilean satellites? Based on our inferred SO_2 abundance, probable exobase height of $\sim$ 1000 km, and the calculations of Saur *et al.* (1999), Io's effective cross section to torus plasma is $A_{eff} \sim \pi(1.6R_{Io})^2$. The strong electrodynamic interaction reduces the plasma speed to 0.1 of the normal relative velocity (v_{rel} = 57 km s^{-1}) or $\sim$ 6 km s^{-1} and Io's effective cross section also by this factor of 0.1 (Goertz, 1980). Thus the required number density of torus protons needed to balance the H atom loss rate by charge exchange to fresh pickup protons is given by

$$n(H^+, torus) \sim \frac{L_H}{0.1\mathrm{v}_{rel}A_{eff}} \tag{13}$$

and approximately 280 protons cm^{-3}, or 7% of the assumed 4000 cm^{-3} torus ion density and within range of the widely quoted 10% proton content of the Io plasma torus (Bagenal, 1994; Crary *et al.*, 1996), especially when the sensitivity to the reduced plasma speed is noted. Thus it appears entirely plausible that the Io plasma torus supplies the hydrogen to Io and it is not necessary to invoke an Iogenic source.

5. Concluding Remarks

To quote Lellouch (1996) in his Urey Prize Lecture: 'If Io's observable atmosphere is primarily volcanic, then the task to model it completely will be formidable. All radiative-conductive, photochemical, and ionospheric models will have to be coupled with dynamical models in plume-like situations. ... Computing the velocity, pressure, and temperature fields in volcanic eruptions is indeed another challenge.' Unfortunately this may decidedly be the case as certainly the distribution of SO_2 density inferred from the HST Lyman-α intensities implies SO_2 coincides with the volcanically active equatorial regions of Io. Inferred SO_2 column densities from millimeter wave observations agree better with the SO_2 column densities derived in this paper, Feldman *et al.* (2000), and McGrath *et al.* (2000), when interpreted as rotational line emission from a plume atmosphere rather than a hydrostatic atmosphere. The bulk of the atmospheric SO_2 may be contained in plume atmospheres. Only regions laterally far away from volcanic vents and above the peak altitudes attained by escaping gas will have SO_2 molecules making the transition to a hydrostatically adjusted atmosphere. Remote sensing at millimeter and ultraviolet wavelengths probes preferentially the higher density plume atmospheres. High spatial resolution coupled with the ability to detect Doppler shifts associated with typical plume eruption velocities $\sim$ 0.5–1 km s^{-1} will be needed to unravel and map the location of these two types of atmospheres.

It appears entirely plausible that the Io plasma torus supplies sufficient hydrogen to Io in the form of protons which are neutralized by a collisionally thick equatorial atmosphere $\sim 1 - 2 \times 10^{16}$ cm^{-2} to maintain atomic hydrogen as a minor component of the atmosphere with surface mixing ratio $\sim$ 1 ppmv and an important constituent in an extended corona around Io. Charge exchange by ions in the Io plasma torus with this hydrogen corona generates pickup proton density ratios compatible with the Chust *et al.* (1999) PWS observations. The flux of elemental hydrogen through the local Io environment (torus protons into and pickup protons out of) is estimated at $\sim 4 \times 10^{25}$ s^{-1}. It is not necessary to invoke an Iogenic source.

Acknowledgements

This research was supported by NASA Grant NAG5-4168. The authors thank Paul Feldman for helpful discussions and referee Emmanuel Lellouch for constructive comments on the manuscript.

References

Bagenal, F., 1994, Empirical model of the Io plasma torus: Voyager measurements, *J. Geophys. Res.* **99**, 11,043–11,062.

Ballester, G.E., McGrath, M.A., Strobel, D.F., Zhu, X., Feldman, P.D. and Moos, H.W.: 1994, Detection of the SO_2 Atmosphere on Io with the Hubble Space Telescope, *Icarus* **111**, 2–17.

Bigg, E.K., 1964, Influence of the satellite Io on Jupiter's decametric emission, *Nature* **203**, 1008–1010.

Bridge, H.S. *et al.*: 1979, Plasma observations near Jupiter: Initial results from Voyager 1, *Science* **204**, 987–991.

Broadfoot, A.L. *et al.*: 1979, Extreme Ultraviolet Observations from Voyager 1 Encounter with Jupiter, *Science* **204**, 979–982.

Chamberlain, J.W. and Hunten, D.M.: 1987, *Theory of Planetary Atmospheres*, 2nd edition, pp. 160–225, Academic Press, Orlando.

Chust, T., Roux, A., Perraut, S., Louarn, P., Kurth, W.S. and Gurnett, D.A.: 1999, Galileo plasma wave observations of iogenic hydrogen, *Planetary Space Sci.* **47**, 1377–1387.

Crary, F.J., Bagenal, F., Ansher, J.A., Gurnett, D.A. and Kurth, W.S.: Anisotropy and proton density in the Io plasma torus derived from whistler wave dispersion, *J. Geophys. Res.* **101**, 2699–2706.

Dessler, A.J.: 1980, Mass-injection rate from Io into the Io plasma torus, *Icarus* **44**, 291–295.

Feldman, P.D., Strobel, D.F., Moos, H.W., Retherford, K.D., Wolven, B.C., McGrath, M.A., Roesler, F.L., Woodward, R.C., Oliverson, R.J. and Ballester, G.L.: 2000, Lyman-α Imaging of the SO_2 Distribution on Io, *Geophys. Res. Lett.* **27**, 1787–1790.

Frank, L.A. and Paterson, W.R.: 1999, Production of hydrogen ions at Io, *J. Geophys. Res.* **104**, 10,345–10,354.

Goertz, C.K.: 1980, Io's interaction with the plasma torus, *J. Geophys. Res.* **85**, 2949–2956.

Harris, D.L.: 1961, Photometry and colorimetry of planets and satellites, in: G.P. Kuiper and B.M. Middlehurst (eds.), *Planets and Satellites*, pp. 272–340, U. of Chicago Press, Chicago.

Hendrix, A.R., Barth, C.A. and Hord, C.W.: 1999, Io's patchy SO_2 atmosphere as measured by the Galileo ultraviolet spectrometer, *J. Geophys. Res.* **104**, 11,817–11,826.

Hill, T.W.: 1979, Inertial limit on corotation, *J. Geophys. Res.* **84**, 6554–6558.

Ingersoll, A.P.: 1989, Io meteorology: How atmospheric pressure is controlled locally by volcanoes and surface frost, *Icarus* **81**, 298–313.

Johnson, T.V., Matson, D.L., Blaney, D.L., Veeder, G.J. and Davies, A.G.: 1995, Stealth plumes on Io, *Geophys. Res. Lett.* **22**, 23–26.

Keiffer, S.W.: 1982, Dynamics and thermodynamics of volcanic eruptions: Implications for the plumes of Io, in: D. Morrison (ed.), *Satellites of Jupiter*, pp. 647–723, U. Arizona Press, Tucson.

Kliore, A.J., Fjeldbo, G., Seidel, B.L., Sweetman, D.N., Sesplaukis, T.T., Woiceshyn, P.M. and Rasool, S.I.: 1975, The atmosphere of Io from Pioneer 10 radio occultation measurement, *Icarus* **24**, 407–410.

Lellouch, E.: 1996, Io's atmosphere: Not yet understood, *Icarus* **124**, 1–21.

Lellouch, E., Belton, M., de Pater, I., Gulkis, S. and Encrenaz, T.: 1990, Io's atmosphere from microwave detection of SO_2, *Nature* **346**, 639–641.

Lellouch, E., Belton, M.J.S., de Pater, I., Paubert, G., Gulkis, S. and Encrenaz, T.: 1992, The structure, stability, and global distribution of Io's atmosphere, *Icarus* **98**, 271–295.

Manatt, S.L. and Lane, A.L.: 1993, A compilation of SO_2 absorption cross-sections from 106–403 nm, *J. Quant. Spectrosc. Radiat. Trans.* **50**, 267–276.

McGrath, M.A., Belton, M.J.S., Spencer, J.R. and Sartoretti, P.: 2000, Spatially resolved spectroscopy of Io' Pele plume and SO_2 atmosphere, *Icarus* **146**, 476–493.

Pearl, J., Hanel, R., Kunde, V., Maguire, W., Fox, K., Gupta, S., Ponnamperuma, C. and Raulin, F.: 1979, Identification of gaseous SO_2 and new upper limits for other gases on Io, *Nature* **280**, 755–758.

Roesler, F.L., Moos, H.W., Oliverson, R.J., Woodward, R.C. Jr., Retherford, K.D., Scherb, F., McGrath, M.A., Smyth, W.H., Feldman, P.D. and Strobel, D.F.: 1999, Far UV Imaging Spectroscopy of Io's Atmosphere with HST/STIS, *Science* **283**, 353–357.

Sartoretti, P., Belton, M.J.S. and McGrath, M.A.: 1996, SO_2 distributions on Io, *Icarus* **122**, 273–287.

Saur, J., Neubauer, F.M., Strobel, D.F. and Summers, M.E.: 1999, 3D Plasma Simulations of Io's Interaction with the Io Plasma Torus: Asymmetric Plasma Flow, *J. Geophys. Res.* **104**, 25,105–25,126.

Saur, J., Strobel, D.F. and Neubauer, F.M.: 1998, Interaction of the Jovian Magnetosphere with Europa: Constraints on the Neutral Atmosphere, *J. Geophys. Res.* **103**, 19,947–19,962.

Smith, R.A. and Strobel, D.F.: 1985, Energy Partitioning in the Io Plasma Torus, *J. Geophys. Res.* **90**, 9469.

Spencer, J.R., Jessup, K.L., McGrath, M.A., Ballester, G.E. and Yelle, R.V.: 2000, Discovery of Gaseous S2 in Io's Pele Plume, *Science* **288**, 1208 1210.

Spencer, J.R., Sartoretti, P., Ballester, G.E., McEwen, A.S., Clarke, J.T. and McGrath, M.A.: 1997, The Pele plume (Io): Observations with the Hubble Space Telescope, *Geophys. Res. Lett.* **24**, 2471–2474.

Strobel, D.F., Zhu, X. and Summers, M.E.: 1994, On the Vertical Thermal Structure of Io's Atmosphere, *Icarus* **111**, 18–30.

Strom, R.G. and Schneider, N.M.: 1982, Volcanic eruption plumes of Io. in: D. Morrison (ed.), *Satellites of Jupiter*, pp. 598–633, U. Arizona Press, Tucson.

Wong, M.C. and Johnson, R.J.: 1995, The Effect of Plasma Heating on Sublimation-Driven Flow in Io's Atmosphere, *Icarus* **115**, 109–118.

TWO-DIMENSIONAL NON-LINEAR ALFVÉN WINGS GENERATED BY THE ELECTRODYNAMIC INTERACTION BETWEEN CALLISTO AND THE JOVIAN MAGNETOSPHERE

ELENA S. BELENKAYA
Institute of Nuclear Physics, Moscow State University, Russia

Abstract. When a highly conducting magnetized plasma passes an object with lower conductivity, or a body with inhomogeneous conductivity, 2-D structures are formed, the so-called 'Alfvén wings'. These structures may arise, for example, at a Jovian moon without an intrinsic magnetic field (Callisto). In this case, Alfvén wings could be generated in the magnetized Jovian magnetospheric plasma flow owing to the inhomogeneity of the moon's ionosphere/atmosphere conductivity. Such Alfvén wings may be considered as a satellite magnetosphere; the satellite magnetospheric magnetic field is a disturbed field of the Jovian magnetospheric plasma flow. An analytical solution is obtained in a simple proposed model.

1. Introduction

Hydrodynamic motion of a conducting fluid (plasma) in a magnetic field induces electric fields which generate electric currents. In a magnetic field these currents are undergoing an action of forces; this leads to a change in the plasma motion. The magnetic field induced by electric currents disturbs in turn the background magnetic field $\vec{B}_0$. So, a complex interaction between magnetic and hydrodynamic phenomena arises. Theoretical studies and observations showed that in a magnetized highly conducting plasma, different objects can generate the 2-D structures, called Alfvén wings (Drell *et al.*, 1965; Neubauer, 1980; Lyatsky and Maltsev, 1983; Wright and Southwood, 1987; Kropotkin, 1989; Alexeev and Belenkaya, 1989). During recent years, a set of numerical 3-D MHD simulations of the interaction between Io and the Io plasma torus was developed (e.g. Linker *et al.*, 1998; Combi *et al.*, 1998; Saur *et al.*, 1999). While Linker *et al.* (1998) considered the plasma interaction both with and without an internal magnetic field of Io, Combi *et al.* (1998) only studied the case without internal field. Saur *et al.* (1999) concluded that, in the case of Io, the perturbations are mostly due to the electrodynamic interaction, but not to an intrinsic magnetic field. The basic physics governing the electrodynamic interaction of Galilean satellites with the Jovian magnetosphere have been discussed in details by Neubauer (1998). Here we focus on the case of satellites which have no internal magnetic field (Callisto). Using the technique of Wright and Southwood (1987), Alexeev and Belenkaya (1989, 1991) found two analytical solutions for the problem of Alfvén wings generation by an unmagnetized obstacle in a highly conducting magnetized plasma (one solution is essentially

Astrophysics and Space Science **277:** 289–292, 2001.

non-linear and corresponds to large perturbations, the other one describes small perturbations). This paper considers an application of these solutions to the case of Callisto.

2. 2-D Alfvén Wings Due to the Inhomogeneous Conductivity of Callisto

We consider the set of MHD-equations for a noncompressible magnetized plasma with infinite conductivity flowing past an obstacle without an internal magnetic field and with inhomogeneous surface conductivity:

$$(\vec{V} \cdot \vec{\text{grad}})\vec{V} = -(1/\rho)\vec{\text{grad}}(p + B^2/2\mu_0) + (1/\rho\mu_0)(\vec{B} \cdot \vec{\text{grad}})\vec{B} \quad (1)$$

$$(\vec{V} \cdot \vec{\text{grad}})\vec{B} = (\vec{B} \cdot \vec{\text{grad}})\vec{V}; \ \text{div}\vec{V} = 0; \ \text{div}\vec{B} = 0 \quad (2)$$

where $\vec{V}$ is the plasma flow velocity, $\vec{B}$ is the magnetic field, p and ρ are the plasma pressure and density, respectively. Two non-linear analytical solutions of equations (1, 2) were found by Alexeev and Belenkaya (1989, 1991) for the case of total pressure conservation: $p + B^2/2\mu_0$ =const. The transverse scale of the source of Alfvén wings should be larger than the ion Larmor radius (in the case of Callisto, the ion gyroradius ≈775 km and the satellite radius ≈2403 km (Neubauer, 1998)).

Alexeev and Belenkaya (1989, 1991) have shown that, under the mentioned conditions, the generation of Alfvén wings requires a decrease of the source conductivity Σ with respect to the outer region conductivity Σ_0:

$$\Delta\Sigma_P + \Delta\Sigma_H \Sigma_H/(\Sigma_P + \Sigma_W) < 0 \quad (3)$$

where $\Delta\Sigma = \Sigma - \Sigma_0$; Σ_P and Σ_H are the Pedersen and the Hall height-integrated surface conductivities of the source of Alfvén wings, respectively. $\Sigma_W = 1/(V_{A0}\mu_0)$ is the wave conductivity of the magnetospheric plasma for the Alfvén waves. Here $\vec{V}_{A0} = \vec{B}_0/\sqrt{\rho\mu_0}$, $\vec{B}_0$ is the magnetic field in the undisturbed flow; $V_0/V_{A0} = \tan\alpha = M_A$ is the Alfvén Mach number.

It was found that Callisto has no internal magnetic field (Khurana *et al.*, 1997; Neubauer, 1998, Kivelson *et al.*, 1999). The velocity $\vec{V}_0$ of the Jovian magnetospheric plasma flow was less than the Alfvén velocity, $\vec{V}_{A0}$, in the undisturbed plasma flow. Due to the mass concentration near the Jovian geographic equatorial plane, M_A increases when approaching this plane, and approximates to 0.94 at Callisto, which is located in the plasma sheet (Neubauer, 1998). Galileo measured the Jovian southward magnetic field $B_0 \sim$ 10nT in the vicinity of Callisto (e.g. Kivelson *et al.*, 1999). In the proposed model, $\vec{B}_0$ plays the role of an external magnetic field inducing the satellite's magnetosphere. The relative velocity of the rotating Jovian plasma at the Callisto orbit is V_0 ~175 km/s (Neubauer, 1998). The y-component of the modeled disturbed magnetic field b_y is oriented from Jupiter for all locations of Callisto on its orbit. It corresponds to the equivalent induced magnetic moment pointing towards Jupiter (in accordance with observations by Khurana *et al.* (1997)).

The surface conductivity is a significant parameter determining the character of the electrodynamic interaction between a satellite and the Jovian magnetospheric plasma flow. Every Galilean satellite possesses an atmosphere (Neubauer, 1998). Gurnett *et al.* (2000) showed that the electron density near Callisto is quite high (up to 400 cm^{-3}), much greater than the Jovian magnetospheric plasma density at the orbit of Callisto ($\approx$ 1 cm^{-3} (Neubauer, 1998)). This fact indicates that Callisto is probably surrounded by a dense ionospheric-like plasma. Gurnett *et al.* (2000) also concluded that the solar illumination plays a role in controlling the plasma density. Consequently, the inhomogeneous ionospheric conductivity should exist at Callisto, and Alfvén wings as discussed here can be generated. However, at this time the actual spatial and temporal distribution of plasma around Callisto is unknown, and we investigate only an hypothetical ionosphere/atmosphere.

Neubauer (1998) proposed a model of the typical variation of the Pedersen, Σ_P, and Hall, Σ_H, conductivity integrals along the magnetic field lines as a function of the transverse distance from the center of a Jovian satellite. The conductivities decrease from the edges of a satellite to its center due to the decrease of the magnetic field line lengths inside the atmosphere/ionosphere. The conductances peak near a circle with satellite's radius where the path length with significant conductivity contribution to the integration is the largest. Inside that circle, the conductances decrease, down to non-negligible values. We apply the Neubauer's, (1998) scheme of the surface conductivity distribution to our model for the estimation of magnetic perturbations in the Alfvén wings.

The value of the Callisto conductivity is unknown, and for a rough estimate we compare it with the conductivity of Io which is better known. Linker *et al.* (1998) noted that $\Sigma_I \sim 100 - 300$ S, Saur *et al.* (1999) estimated it to 100 - 200 S. Kivelson *et al.* (1999) mentioned that a peak ionospheric conductivity can be estimated from:

$$\sigma = Ne^2/m\omega_{ce}, \qquad (4)$$

where N is the electron density, m is the mass of the ionospheric ions, ω_{ce} is the electron gyrofrequency. If we assume that the ionospheric depths of Callisto, h_C, and Io, h_I, are of the same order, then their height-integrated conductivities ($\Sigma_C = \sigma_C h_C$ and $\Sigma_I = \sigma_I h_I$, respectively) will relate to each other as σ_C/σ_I. From (4) it follows that $\sigma_C/\sigma_I \sim N_C B_I/N_I B_C$(if the difference between the ionospheric ion masses for the two satellites is ignored). Magnetic field near Io, $B_I \sim 1835$ nT, and near Callisto, $B_C \sim 35$ nT (Neubauer, 1998). The peak electron density measured by Galileo during the C3 pass was about 100 cm^{-3} (Gurnett *et al.*, 2000); the electron density observed in the downstream wake of Io was about 4 10^4 cm^{-3} (Gurnett *et al.*, 1996, 2000). For the average value of $\Sigma_I \sim 200$ S, we obtain $\Sigma_C \sim 26$ S. Using the Neubauer's (1998) model for representation of Σ_{CP} and Σ_{CH} distributions, we derive the following estimations: $\Sigma_{CP0} \sim \Sigma_{CH0} \sim 26$S,

$\Delta\Sigma_{CH} \sim -0.5\Sigma_{CH0} \sim -13S$, $\Delta\Sigma_{CP} \sim -0.5\Sigma_{CP0} \sim -13S$, $\Sigma_W \sim 4.4S$. From solutions obtained in (Alexeev and Belenkaya, 1989, 1991) it follows that $b_y \sim -B_0\Delta\Sigma_{CH}\tan\alpha/(\Sigma_{CP0} + \Delta\Sigma_{CP} + \Sigma_W) \sim 7$ nT for these parameters.

Kivelson *et al.* (1999) calculated magnetic perturbations b_x, b_y, and b_z along several Galileo passes, including C3 (x along the direction of corotation, y radially inward toward Jupiter, and z parallel to Jupiter's rotation axis). The strength and orientation of the calculated induced magnetic field governed by the instantaneous local value of the time-varying component of Jupiter's magnetic field coincided well with observed data. However, the authors noted that corrections for the effect of Alfvén wing currents are needed. Calculated by Kivelson *et al.* (1999) and observed value of b_y is of the order of $\sim$ 7.5nT, that is in a good accordance with our results.

In conclusion, analytical solutions obtained by Alexeev and Belenkaya (1989, 1991) for the Alfvén wings generated in the flow of the magnetized Jovian magnetospheric plasma by a Jovian satellite without own magnetic field and possessing inhomogeneous ionosphere/atmosphere conductivity, allow to estimate the value of the induced magnetic field. For example, the Callisto magnetospheric magnetic field y-component, which is the disturbed Jovian dipole magnetic field, is found to be of the order of 7 nT.

References

Alexeev, I.I. and Belenkaya, E.S.: 1989, *Geomagn. Aeron.* **29**, 902–909.

Alexeev, I.I. and Belenkaya, E.S.: 1991, *J. Atmos. Terr. Phys.* **53** (11/12), 1099–1101.

Combi, M.R., Kabin, K., Gombosi, T.T., Zeeuw, D.L. and Powell, K.G.: 1998, *J. Geophys. Res.* **103**, 9071–9081.

Drell, S.D., Folley, H.M. and Ruderman, M.A.: 1965, *J. Geophys. Res.* **70**, 3131–3145.

Gurnett, D.A., Kurth, W.S., Roux, A., Bolton, S.J. and Kennel, C.: 1996, *Science* **274**, 391–392.

Gurnett, D.A., Persoon, A.M., Kurth, W.S., Roux, A. and Bolton, S.J.: 2000, *Geophys. Res. Letts.* **27**, 1867–1872.

Khurana, K.K., Kivelson, M.G., Russell, C.T., Walker, R.J. and Southwood, D.J.: 1997, *Nature* **387**, 262–263.

Kivelson, M.G., Khurana, K.K., Stevenson, D.J., Bennet, L., Joy, S., Russell, C.T., Walker, R.J., Zimmer, C. and Polanskey, C.: 1999, *J. Geophys. Res.* **104**, 4609–4625.

Kropotkin, A.P.: 1989, *Geomagn. Aeron.* **29**, 452–458.

Linker, J.A., Khurana, K.K., Kivelson, M.G. and Walker, R.J.: 1998, *J. Geophys. Res.* **103**, 19,867–19,877.

Lyatsky, V.B. and Maltsev, Yu.P.: 1983, *Magnetosphere-Ionosphere Coupling*, pp. 90, Nauka, Moscow, USSR.

Neubauer, F.M., 1980: *J. Geophys. Res.* **85**, 1171–1178.

Neubauer, F.M., 1998: *J. Geophys. Res.* **103** (E9), 19,843–19,866.

Saur, J., Neubauer, F.M., Strobel, D.F. and Summers, M.: 1999, *J. Geophys. Res.* **104**, 25,105–25,126.

Wright, A.N. and Southwood, D.J.: 1987, *J. Geophys. Res.* **92**, 1167–1175.

MAGNETICALLY-DRIVEN PLANETARY RADIO EMISSIONS AND APPLICATION TO EXTRASOLAR PLANETS

PHILIPPE ZARKA
DESPA, CNRS/Observatoire de Paris, 92195 Meudon, France

RUDOLF A. TREUMANN
Max-Planck Institute for Extraterrestrial Physics, D-85740 Garching, Germany
International Space Science Institute, CH-3012 Bern, Switzerland

BORIS P. RYABOV and VLADIMIR B. RYABOV
Institute of Radio Astronomy, Kharkov 310002, Ukraine

Abstract. At least six intense nonthermal planetary radio emissions are known in our solar system: the auroral radio emissions from the Earth, Jupiter, Saturn, Uranus and Neptune, and the radio bursts from the Io-Jupiter flux tube. The former are thought to be driven by the solar wind flow pressure or energy flux on the magnetospheric cross-section, while the latter is a consequence of the Io-Jupiter electrodynamic interaction. Although in the solar wind, the flow ram pressure largely dominates the magnetic one, we suggest that the incident magnetic energy flux is the driving factor for all these six radio emissions, and that it can be estimated in the same way in all cases. Consequences for the possible radio emission from extrasolar planets are examined. 'Hot Jupiters', if they are magnetized, might possess a radio emission several orders of magnitude stronger than the Jovian one, detectable with large ground-based low-frequency arrays. On the other hand, 'giants' analogous to the Io-Jupiter interaction in the form of a pair star/hot-Jupiter are unlikely to produce intense radio emissions, unless the star is very strongly magnetized.

1. Introduction

The Earth and the four giant planets Jupiter produce nonthermal cyclotron radio emissions in the kilometer to decameter wavelength range, depending on the planetary magnetic field intensity (see e.g. Zarka, 1998, 2000, and references therein). The most intense components originate from strongly magnetized regions of their magnetospheres, where the local electron plasma frequency (f_{pe}) is much smaller than the gyrofrequency (f_{ce}). These components are attributed to a cyclotron-maser process fed by unstable electrons distributions with characteristic energy of a few keV. Both prequisites ($f_{pe} \ll f_{ce}$ and presence of unstable keV electron distributions) are fulfilled along auroral, high magnetic latitude ($\geq 70^{\circ}$) field lines in both hemispheres, from slightly above the planetary surface up to a few radii. The precipitating electrons are thought to be accelerated in the magnetotail and/or at the magnetopause. But energetic electrons are also produced through the interaction of the Galilean satellite Io with the Jovian magnetosphere, and precipitate along Jovian field lines down to their footprints at northern and southern magnetic

Astrophysics and Space Science **277:** 293–300, 2001.

latitudes about $60^{\circ} - 65^{\circ}$, producing the intense 'Io-induced' decameter radio emission, also attributed to a cyclotron-maser process. A similar but less energetic phenomenon may occur in the interaction of Ganymede's magnetosphere with Jupiter's one (see below). We suggest here that the output power of these planetary 'radio generators' is ultimately related to the sweeping of an obstacle (planet or satellite) by a magnetized flow (solar wind or Jovian magnetospheric plasma), derive the corresponding 'efficiency', and extrapolate this result to the case of hot Jupiters.

2. Physical Properties and Emitted Power of Magnetospheric Radio Emissions

Decades of measurements and theoretical studies of these radio emissions allowed to derive their general characteristics. They are (1) generated near the local gyrofrequency f_{ce}, on the extraordinary (X) magneto-ionic mode; (2) very intense, with a brightness temperature up to 10^{15} K; (3) beamed at relatively large angle with respect to the local magnetic field ($\geq 30^{\circ}$, up to 90°), along the walls of a conical sheet a few degrees thick; and (4) covering a broad frequency range, from nearly zero to the maximum gyrofrequency close to the planetary surface (i.e. $\Delta f \sim < f >$). They are also 100% polarized, circularly or elliptically, but this property will not intervene in the following analysis. From their average spectrum width, flux density and beaming, it is possible to deduce their average radiated radio power Pr, found to cover 4 orders of magnitude, from 4.10^{6} W in the case of Neptune's kilometric radio emissions (North + South), to $\sim 10^{10}$ W for the Io-induced and 4.10^{10} W for the Jovian auroral components (Zarka *et al.*, 2000).

3. Solar Wind Influence and Radio Bode's Law

The fact that the solar wind 'controls' in some way the auroral radio emissions of the five 'radio planets' has been demonstrated through numerous correlations of the variations of the solar wind conditions (density, velocity, interplanetary magnetic field – IMF –, and especially ram pressure) at these planets with their output radio power (Zarka *et al.*, 2000; and references therein). Studies of the long-term variations of the Jovian radio and plasma waves observed by Ulysses (Reiner *et al.*, 2000) and Galileo (Louarn *et al.*, 1998) even suggest that the Jovian magnetospheric dynamics at large may be controlled by the solar wind. At Saturn, the auroral radio radiation is virtually turned off when Saturn's magnetosphere is engulfed in Jupiter's huge magnetotail (Desch, 1983). This led Desch and Kaiser (1984), followed by Zarka (1992), to establish an empirical scaling law based on the correlation of the average auroral radio power (P_r) of the five radio planets with

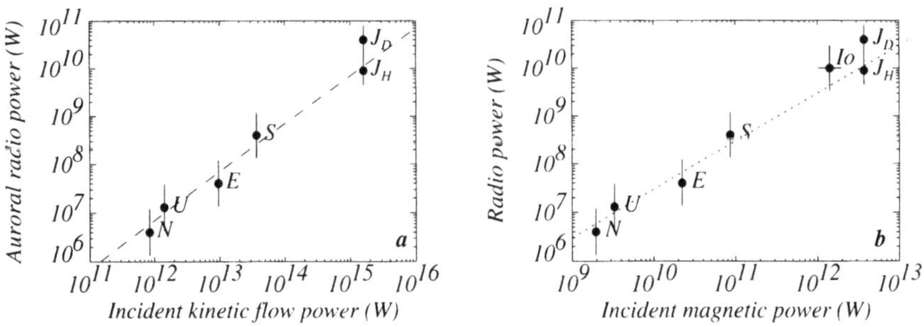

Figure 1. (a) Initial radio Bode's law for the auroral radio emissions of the five radio planets (Earth, Jupiter, Saturn, Uranus and Neptune) (Desch and Kaiser, 1984; Zarka, 1992). J_D and J_H correspond to the decameter and hectometer Jovian components, respectively. The dashed line has a slope of 1 with a proportionality constant of 7.10^{-6}. Error bars correspond to the typical uncertainties in the determination of average auroral radio powers. (b) Magnetic radio Bode's law with auroral and Io-induced emissions (see text). The dotted line has a slope of 1 with a constant of 3.10^{-3}.

the incident kinetic power (P_{ram}) due to the solar wind ram pressure (ε_{ram}) on their magnetosphere cross-sectional area. This law writes:

$$P_r = \alpha P_{ram} = \alpha \varepsilon_{ram} V \pi R_{mp}^2 \quad \text{with} \quad \varepsilon_{ram} = (N_o/d^2) m_p V^2 \tag{1}$$

with R_{mp} the dayside magnetopause distance, N_o the average solar wind density at 1 AU ($\sim$7 cm^{-3}), V the solar wind speed ($\sim$400 km/s), d the planet's orbital radius in AU, m_p the proton mass, and α the efficiency ratio (P_r/P_{ram}). Using for R_{mp} the values measured by the Voyager spacecraft (see Zarka, 1992), one obtains the excellent correlation illustrated in Figure 1a with $\alpha \approx 7.10^{-6}$ (dashed line), slightly higher than the value derived by Desch and Kaiser (1984). Note that although the planetary rotation is thought to play a role (e.g. via the centrifugal force) in particle acceleration and auroral processes, especially for rapidly rotating planets as Jupiter, no such correlation can be found between the auroral radio power and the planetary rotation (kinetic moment or typical corotation electric field).

The solar wind being magnetized, the term ε_{ram} should also include a contribution of the IMF pressure $\varepsilon_{imf} = B^2/2\mu_o$ with B the IMF amplitude perpendicular to the solar wind speed in the obstacle's frame (we neglect the thermal pressure term $Nk(T_e + T_i)$). Beyond 1 AU from the Sun, the IMF becomes nearly azimuthal, along the Parker spiral, so that $B \approx B_{az} \approx (B_s R_s^2 \Omega_s)/(Vr)$ with B_s the equatorial solar surface magnetic field ($\sim$1.5 Gauss), R_s the solar radius, Ω_s the angular frequency of the solar rotation ($2\pi \times 27$ days^{-1}) and r the radial distance in meters. It can be checked that beyond 1 AU, $\varepsilon_{ram}/\varepsilon_{imf} \approx 400$, which explains that the IMF pressure is generally neglected. However, as both pressure terms have the same dependance with the radial distance (i.e. in d^{-2} or r^{-2}), the correlation of Figure 1a still holds for the auroral radio power versus the incident magnetic energy flux, i.e.

$$P_r = \beta P_{imf} = \beta \varepsilon_{imf} V \pi R_{mp}^2 \tag{2}$$

but with an efficiency ratio $\beta \approx 3.10^{-3} \approx 400\alpha$. We conclude that if the conversion of incident magnetic energy into electron acceleration -for example through magnetic reconnection- is much more efficient ($> 400\times$) than the conversion of the incident solar wind kinetic energy, then the former may actually drive the processes leading to auroral radio emission generation.

4. Io-Induced Radio Emission

Because it is generated near Io's magnetic flux tube – IFT –, deeply embedded in Jupiter's internal magnetosphere, the Io-induced radio emission is – as expected – independent of solar wind conditions. The interaction of Io with the Jovian magnetic field is thought to generate Alfvén waves (Neubauer, 1980), whose associated parallel electric field may accelerate electrons to keV energies or more (Crary, 1997). The power dissipated in the Io-Jupiter electrodynamic circuit has been estimated in various ways. Observations of the IFT footprints in the far-UV (Prangé *et al.*, 1996) and IR (Connerney *et al.*, 1993) ranges allowed to estimate the energy budget of electron precipitations along the IFT, and to derive a precipitated power of a few $\times 10^{11}$ W per hemisphere. Crary (1997) obtained a similar value for the precipitated power of electrons accelerated by Alfvén waves. This value is consistent with the upper limit, about 10^{12} W per hemisphere, deduced from the product of the current flowing between Io and Jupiter (a few $\times 10^6$ A, based on Voyager magnetic field measurements (Acuña *et al.*, 1981)) by the voltage drop across Io ($\Delta V = 2R_{Io}E = 2R_{Io}|v \times B_J| \approx 5.10^5$ V, $v(\sim 57$ km/s) being the relative velocity between Io and the corotating Jovian magnetosphere and $B_J(\sim 0.02$ G) being the Jovian field amplitude at Io's orbit). Neubauer (1980), via a nonlinear MHD analysis of the Alfvén wing current, also derived a maximum dissipated power in the circuit: assuming a simple distribution of current around Io, he derived a maximum Joule dissipation $P_d = \pi R_{Io}^2 E^2 \Sigma_a$, with $E = -v \times B_J$ the background electric field and Σ_a the conductance of the Alfvén wave current tubes acting like an 'external load' for the circuit. For an incident flow perpendicular to B_J, $\Sigma_a = 1/\mu_o v_a (1 + M_a^2)^{1/2} = M_a/\mu_o v (1 + M_a^2)^{1/2}$. After some rearrangement, we get:

$$P_d = (2M_a/(1 + M_a^2)^{1/2})(B_J^2/2\mu_o)v\pi R_{Io}^2 \approx 10^{12}\mathrm{W} \tag{3}$$

with $M_a(\sim 0.3)$ the Alvénic Mach number of the Jovian magnetospheric flow past Io.

With a total dissipated power (North + South) of $1-2\,10^{12}$ W and an average Io-induced radio power $\sim 10^{10}$ W, we obtain an overall radio efficiency $\approx 5-10\,10^{-3}$, close to the above 'radio-to-magnetic' efficiency β of auroral radio emissions, as illustrated on Figure 1b. Comparing equations (2) and (3), it is interesting to note that the latter actually represents the magnetic energy flux incident on Io's cross sectional area (or rather that of its ionosphere). The only difference comes from the factor $2M_a/(1+M_a^2)^{1/2}$, which may take values between $\sim 2M_a$ (for $M_a < 1$) and 2 (for$M_a \gg 1$), and is thus always close to unity.

5. Magnetic Radio Bode's Law and Applications to Solar System Objects

It appears thus that (i) we can estimate the power dissipated in the interaction of a magnetized flow with an obstacle simply by computing the intercepted flux of magnetic energy

$$P_d \approx B^2/2\mu_o v\pi R_{obs}^2 \tag{4}$$

where B is the magnetic field amplitude perpendicular to the flow speed in the obstacle's frame ($B^2 v = B \cdot (v \times B)$ in (4)), and that (ii) accelerated electrons and the subsequent cyclotron-maser radio emission that they generate are produced with a quasi-constant efficiency, i.e.

$$Pr \approx \beta \times P_d \tag{5}$$

The fact that we found similar radio-to-magnetic efficiencies ($\beta \approx 3 - 10\,10^{-3}$) for apparently very different flow-obstacle interactions – solar wind with a magnetized planet, or rotating magnetosphere with a satellite's ionosphere – may seem coincidental, but it is probably due to more fundamental reasons as qualitatively discussed below.

In both configurations, the incident magnetic energy is partly used to accelerate electrons to keV energies. These electrons then follow magnetic field lines towards the central magnetized body (the planet in both cases), where they generate cyclotron-maser radio emission with a maximum efficiency about 1% of their total energy (Pritchett, 1986; Galopeau et al., 1989). An overall radio-to-magnetic efficiency $\approx 3 - 10\,10^{-3}$ implies that the production of accelerated electrons taps the incident unperturbed magnetic energy flux with an efficiency of ~30 to 100%. But the incident magnetic field is strongly enhanced through pile-up ahead of the obstacle, so that, even taking into account the associated slow-down of the flow, the actual conversion efficiency from magnetic energy into electrons energy should rather be 3–10%.

In the case of Io-Jupiter interaction, the Jovian magnetic field pressure largely dominates the plasma flow pressure and the interaction is sub-Alfvénic, so that the field lines pile-up is moderate. Downstream, 'released' field lines are perturbed by Alfvén waves, whose associated parallel electric field may accelerate electrons to keV energies. The ultimate energy source tapped by the Io-Jupiter electrodynamic interaction, via Jovian magnetic field lines, is actually the planet's rotational energy. Conversely, the solar wind is weakly magnetized, and its interaction with a planetary magnetosphere is strongly super-Alfvénic and causes tight IMF draping around the magnetopause (as an extreme case of Alfvén wings), leading to magnetic field amplitude enhancement (for example up to ×7 at Saturn, i.e. ×50 for the magnetic pressure (Ness *et al.*, 1981)), in this case at the expense of the solar wind flow (kinetic) energy. keV electrons are thought to be accelerated by parallel electric fields associated with Kelvin-Helmholtz waves along the dayside magnetopause (e.g. Galopeau *et al.*, 1995), or via magnetic reconnection at the nose of the magnetopause or in the magnetotail (e.g. Russell, 2000).

Taken as an empirical law, it is interesting to apply the above magnetic radio Bode's law to other solar system objects:

- To estimate the intensity of the Ganymede-Jupiter electrodynamic interaction, we shall simply use equation (4) with B the Jovian field at Ganymede's orbit ($\sim$120 nT), v the Jovian magnetospheric flow velocity relative to the satellite ($\sim$176 km/s), and R_{obs} the radius of Ganymede's magnetosphere ($\sim 2R_g$ with $R_g \approx 2600$ km (Kivelson *et al.*, 1998)). We obtain thus $P_d \approx 9.10^{10}$ W, i.e. one order of magnitude less than the Io-Jupiter interaction. We expect thus a Ganymede-induced radio emission $\sim 10\times$ weaker than the Io-induced one. Such a weak 'Ganymede control' indeed seems to exist in Galileo's long-term observations of Jupiters radio emissions (Menietti *et al.*, 1998).
- The same calculation applied to the satellite Dione, suspected to exert some control on Saturn's radio emissions (Desch and Kaiser, 1981), gives $P_d \approx 2.10^8$ W, and thus a negligible expected effect. The electrodynamic influence of Titan seems even smaller with $P_d \approx 10^7$ W.
- Finally, the case of Mercury in the solar wind, with a magnetosphere radius $\sim 1.5R_m$ and an incident magnetic pressure $\sim 5\times$ that at the Earth, leads to $P_d \approx 4.10^8$ W, and thus to $P_r \approx 10^6$ W. This radio power is probably trapped in Mercury's magnetosphere, because the electron cyclotron frequency at Mercury's magnetosphere is lower than the plasma frequency in the surrounding solar wind.

6. Radio Emissions from Hot Jupiters ?

Hot Jupiters, i.e. giant planets orbiting a few solar radii away from their solar-type star, represent $\sim$50% of the $\simeq$ 40 exoplanets discovered up to now (see www.obspm.fr/planets). Assuming that their parent stars emit a solar-like wind, the extrapolation of equations (4) and (5) to the case of these planets is very interesting, because closer than $\sim$0.2 AU ($\sim 40R_s$) from the central star, the radial component of the IMF ($B_r = B_s R_s^2/r^2$) begins to dominate the azimuthal one. As this radial component increases in r^{-2}, the associated magnetic pressure increases in r^{-4}, and due to the aberration effect caused by the planet's orbital velocity, it still contributes significantly to the perpendicular field component B in (4).

If the hot Jupiter is magnetized, its magnetopause radius will shrink with decreasing distance from the star due to the increase of the stellar wind ram and magnetic pressures, as illustrated on Figure 2a for a Jupiter-like planet. Figure 2b combines this magnetospheric shrinking with the ram and magnetic pressure in-

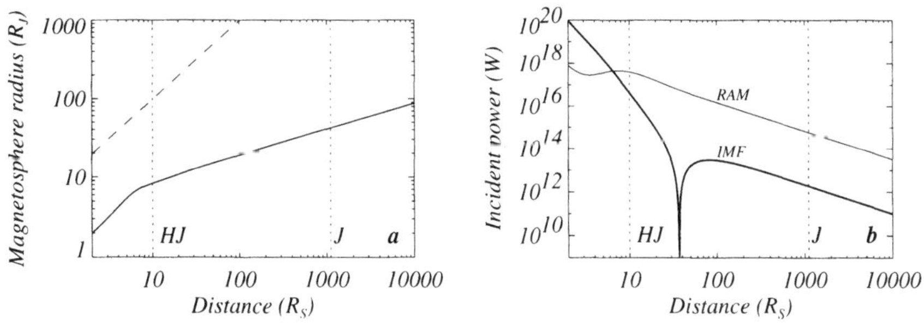

Figure 2. (a) Jovian magnetosphere radius versus distance from the Sun (or solar-type star). The magnetosphere shrinks but remains detached from the star (first bissecting line is dashed). 'J' indicates Jupiter's orbital distance (5.2 AU), and 'HJ' typical hot Jupiters' orbits (at $\sim 10R_S$). (b) Incident flow kinetic (ram) and magnetic (IMF) powers on a Jovian magnetosphere's cross-section, versus distance from the star. The dip in the magnetic power comes from the aberration effect due to the planet's orbital velocity (at the dip, the IMF is parallel to the flow in the planet's frame).

creases to display the incident kinetic (ram) and magnetic powers on the magnetosphere cross-section. It appears that at 10 R_s from the star, (i) both contributions become roughly equivalent, and (ii) the incident magnetic power is $\geq 10^4\times$ higher than at Jupiter's orbit (~5 AU). According to (5), the radiated radio power should be increased by about the same factor, and make the auroral radio emission of magnetized hot Jupiters detectable above galactic background fluctuations from a range of up to 15-20 parsecs, with the largest available radiotelescopes (Zarka *et al.*, 1997). Of course, according to (4), still more intense radio emissions could be produced by hot Jupiters orbiting stars blowing a faster and/or more strongly magnetized wind than the Sun.

If the hot Jupiter is unmagnetized, for example because its magnetic field has decayed due to the tidal lock of the planet's rotation and orbital periods (see e.g. Farrell *et al.*, 1999), then the planet-star electrodynamic interaction may be a giant version of the Io-Jupiter one (the distance of $\sim 10R_S$ also corresponds to the limit between the sub- and super-Alfvénic regimes). Electrons may be accelerated by the Alfvénic disturbances of the stellar magnetic field lines sweeping by the planet's ionosphere, and precipitate towards the star itself. However, due to the presence of the star's corona, much denser than a planetary magnetospheric environment, the condition $f_{pe} \ll f_{ce}$ is generally not fulfilled close to the star, so that the cyclotron-maser will not be able to produce intense radio waves. This process may work occasionally when the planet crosses exceptionally large magnetic loops connecting intense magnetic spots on the star's surface, or on a more regular basis in the case of strongly magnetized stars (typically with a magnetic field $\geq 10-100\times$ that of the Sun). It might then be a good idea to search (e.g. through radial velocity measurements) for exoplanets near magnetic dwarves or other radio flaring stars, but the more fundamental search, which could validate the above magnetic radio

Bode's law, consists in searching for radio emissions from already discovered hot Jupiters. This search is in progress.

Acknowledgements

We thank J. Queinnec and W. Macek for useful discussions.

References

Acuña, M.H., Neubauer, F.M. and Ness, N.F.: 1981, *J. Geophys. Res.* **86**, 8513–8522.
Connerney, J.E.P. *et al.*: 1993, *Science* **262**, 1035–1038.
Crary, F.J.: 1997, *J. Geophys. Res.* **102**, 37–49.
Desch, M.D.: 1983, *J. Geophys. Res.* **88**, 6904–6910.
Desch, M.D. and Kaiser, M.L.: 1981, *Nature* **292**, 739–741.
Desch, M.D. and Kaiser, M.L.: 1984, *Nature* **310**, 755–757.
Farrell, W.M., Desch, M.D. and Zarka, P.: 1999, *J. Geophys. Res.* **104**, 14025–14032.
Galopeau, P., Zarka, P. and Le Quéau, D.: 1989, *J. Geophys. Res.* **94**, 8739–8755.
Galopeau, P., Zarka, P. and Le Quéau, D.: 1995, *J. Geophys. Res.-Planets* **100**, 26397–26410.
Kivelson, M.G. *et al.*: 1998, *J. Geophys. Res.* **103**, 19963–19972.
Louarn, P. *et al.*: 1998, *Geophys. Res. Lett.* **25**, 2905–2908.
Menietti, J.D. *et al.*: 1998 *Geophys. Res. Lett.* **25**, 4281–4284.
Ness, N.F. *et al.*: 1981 *Science* **212**, 211–217.
Neubauer, F.M.: 1980, *J. Geophys. Res.* **85**, 1171–1178.
Prangé, R. *et al.*: 1996, *Nature* **379**, 323–325.
Pritchett, P.L.: 1986, *Phys. Fluids* **29**, 2919.
Reiner, M.J., Kaiser, M.L. and Desch, M.D.: 2000, *Geophys. Res. Lett.* **27**, 297–300.
Russell, C.T.: 2000, *Adv. Space Res.* **26** (3), 393–404.
Zarka, P.: 1992, *Adv. Space Res.* **12** (8), 99–115.
Zarka, P.: 1998, *J. Geophys. Res.* **103**, 20159–20194.
Zarka, P.: 2000, *Geophys. Monogr.* **119**, AGU, 167–178.
Zarka, P. *et al.*: 1997, in: H.O. Rucker *et al.* (eds.), *Planetary Radio Emissions IV*, pp. 101–127, Austrian Acad. Sci. Press, Vienna.
Zarka, P., Treumann, R.A. and Ryabov, B.P.: 2000, *J. Geophys. Res.*, submitted.

V

OBSERVATIONS / NEW DETECTION TECHNIQUES

OBSERVATIONS / NOUVELLES TECHNIQUES DE MESURE

RADIO OBSERVATIONS OF IONS IN COMET HALE-BOPP

FLORENCE HENRY, JACQUES CROVISIER and DOMINIQUE BOCKELÉE-MORVAN
Observatoire de Paris, 92195 Meudon, France

HEIKE RAUER
DLR, Institute of Space Sensor Technology and Planetary Exploration, Rutherfordstrasse, 12489 Berlin, Germany

DAREK LIS
California Institute of Technology, MS 320-47, Pasadena, CA 91125, USA

Abstract. We present radio observations of HCO^+, H_3O^+ and CO^+ in Hale-Bopp.

Radio spectroscopy is a powerful technique for observing molecules in cometary environments and for determining their kinematics. Recently, the exceptionnally bright comets C/1996 B2 (Hyakutake) and C/1995 O1 (Hale-Bopp) offered us the opportunity to make unprecedented observations, resulting in the detection of no less than two dozens of new cometary molecular species (Lis *et al.*, 1999; Bockelée-Morvan *et al.*, 2000). Among them, three molecular ions were discovered at radio wavelengths in comet Hale-Bopp: HCO^+ (Veal *et al.*, 1997), H_3O^+ and CO^+ (Lis *et al.*, 1999) (this latter was already known from its visible spectrum). We observed these ions with the Caltech Submillimetre Observatory (CSO 10 m), the IRAM 30-m telescope and the IRAM interferometer at Plateau de Bure (PdB 15 m) while comet Hale-Bopp was near perihelion ($r_h \sim 1$ AU).

When cometary ices sublimate, gaseous molecules are ejected and expand with small velocities (~ 1 km s^{-1}) from the nucleus. But ionized molecules are accelerated by interaction with the interplanetary magnetic field frozen in the solar wind flow. A way to study the acceleration is to measure the velocities at some distances from the nucleus. Such observations have been performed for HCO^+ on several days (e.g. we observed the comet with shifts from the nucleus of 1' (about 100 000 km), 2', 3', 4' and 6' in the tail direction and 2' in the sun direction on April 18, 1997, cf Figure 1) and CO^+ on April 4, 1997 with the IRAM 30-m.

The evolution of the mean velocity of the HCO^+ ions with distance to the nucleus is plotted in Figure 1. Models assuming a constant acceleration fit well the observations. The acceleration of HCO^+ is between 1 and 5 cm s^{-2}. That of CO^+ (whose data are not shown) seems to be smaller (≤ 1 cm s^{-2}). A modelization with a better account of the solar wind interaction should improve the fit.

Astrophysics and Space Science **277:** 303–304, 2001.

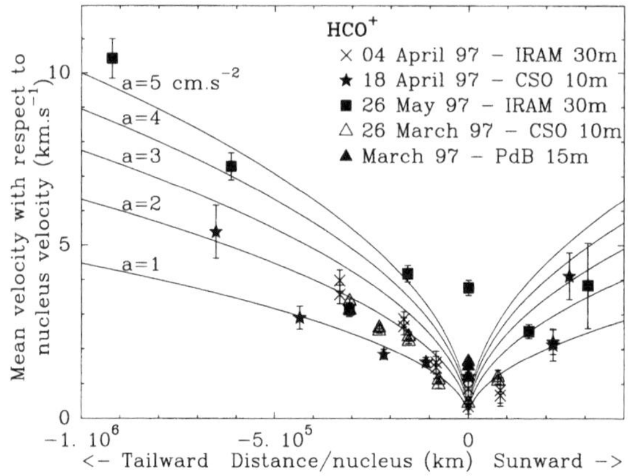

Figure 1. Evolution of the mean velocity of HCO^+ J(3–2) (267.557 GHz) and J(1–0) (89.188 GHz) rotational lines with distance to nucleus for each day of observation. The curves plotted represent the evolution of the velocity with a constant acceleration **a**.

From the observations, the column densities of the ions of interest – H_3O^+, HCO^+ and CO^+ – were evaluated assuming local thermal equilibrium and uniform excitation temperature. In order to interpret these column densities, we have elaborated a simple chemical model of the inner coma. This model starts from the sublimation of parent molecules from the nucleus, which are photodissociated and photoionized into daughter species. We considered reactions that involve parent and daughter molecules, solar photons and electrons, using reaction rates available in the UMIST database (http://saturn.ma.umist.ac.uk:8000 /~tjm/rate/rate.html). These processes lead to a complex composition of the inner coma.

From the results obtained for the initial composition of Bockelée-Morvan *et al.* (2000) and a production rate of H_2O equal to 1.2 10^{31} s^{-1}, we computed the column density of each ion and compared them to the ones derived from the observations. The model results are close to the observations by a factor between 2 and 4. This difference could be explained by the simplistic assumptions that have been made : isotropic distribution of the molecules, the limited number of molecules considered in the nucleus, no extended source, no solar wind interaction and thermal equilibrium. This work is in progress and a more realistic model should improve the results.

References

Bockelée-Morvan, D. *et al.*: 2000, *Astronomy and Astrophysics* **353**, 1101–1114.
Henry, F.: 1999, Rapport de Stage de DEA, Univerité Paris 6.
Lis, D.C. *et al.*: 1999, *Earth, Moon and Planets* **78**, 13–20.
Veal, J.M. *et al.*: 1997, IAU Circular 6575.

RAPID MEASUREMENTS OF SOLAR WIND IONS WITH THE TRIANA PLASMAG FARADAY CUP

MATTHIAS R. AELLIG, ALAN J. LAZARUS and JUSTIN C. KASPER
MIT Center for Space Research, Cambridge, MA, USA
KEITH W. OGILVIE
NASA/GSFC, Greenbelt, MD, USA

Abstract. The Triana PlasMag Faraday Cup (FC) will be able to determine speed, flow angles, temperature, and density of the main solar wind ion species with a time resolution of better than one second. Thus, the Triana PlasMag FC will enable resolution of spatial structures as small as a few hundred kilometers as the structures convect past the spacecraft. Under typical solar wind conditions, that size is comparable to a few proton gyroradii.

1. Instrument

The Triana PlasMag Faraday Cup will be used to determine the distribution and basic flow parameters of the solar wind protons and alpha particles, i.e., speed, flow angles, temperature and density. Particles entering the sensor pass through a set of planar grids and then strike a circular collector plate divided into three 120° sectors. The particle flux in a given energy per charge range is determined by a square-wave voltage applied to one of the grids: the upper and lower voltages of that square wave select the energy per charge of incident particles. The particle flux in that range produces a current on the collector which is modulated at the square-wave frequency. The modulated portion of the current from each collector sector is measured by the electronics. The relative currents on the individual collector sectors measure the flow angle the solar wind makes to the sensor axis. An entire spectrum is obtained by stepping through a set of pairs of voltages. For a detailed description of the principles of a Faraday Cup see Ogilvie *et al.* (1995). Triana will be a 3-axis-stabilized spacecraft and the Faraday Cup will face the sun all the time. The sensor's large angular acceptance allows the determination of off-axis inflow of solar wind at angles as large as 30° relative to the cup axis. A large angular acceptance is required to accommodate variations of the flow direction of the solar wind and the specifics of the Triana orbit and attitude. Since the instrument is facing the general direction of the solar wind all the time, data in a new energy per charge window can be taken as soon as the power supply and the analog electronics allow ($\approx$ 40 ms for the Triana PlasMag Faraday Cup). For the previous experiment on Wind (a spinning spacecraft), roughly 20 pairs of voltages were used, and each energy per charge step was maintained for an entire

Astrophysics and Space Science **277:** 305–307, 2001.

TABLE I

The time t_{spec} to obtain one spectrum for several instruments on various spacecraft (s/c).

Instrument	s/c stabilization	t_{spec} [s]	Launch year
WIND/SWE/FC	spin	42	1994
WIND/3DP	spin	24*	1994
SOHO/CELIAS/PM	3-axis	30	1995
ACE/SWEPAM	spin	64	1997
Triana/PlasMag/FC	3-axis	0.7	2001**

* 3 s moments in burst mode

** Planned

spacecraft rotation (3 seconds) which limited the time resolution of each spectrum to about one minute. Additionally, the solar wind could be viewed only during part of a rotation which could cause aliasing of the measurements especially when the wind changed its characteristics from rotation to rotation. Table 1 demonstrates that by taking advantage of the 3-axis spacecraft stabilization, the FC on Triana will achieve a much better time resolution compared to previous instruments.

2. Scientific Questions

We discuss only a few of the exciting scientific questions that can be addressed with these high time resolution ion measurements.

Plasma fine structure. As the time resolution of ion measurements improves, the internal fine structure of the plasma in transient phenomena such as shock waves, discontinuities, the heliospheric current sheet, and magnetic holes can be determined. The spatial resolution achieved is a few hundred kilometers which is equal to a few proton gyroradii under most solar wind conditions. Magnetic holes have low magnetic field strength and are typically narrow and individual holes are observable from a spacecraft for a few to a few tens of seconds only. So far, only magnetic field measurements could probe this fine structure. The Triana Faraday Cup will enable, for the first time, detailed analysis of the plasma properties at the boundaries of typical magnetic holes and within them. Because of the very low magnetic field strength in the holes, these particle measurements will be performed on a kinetic scale.

Turbulence and waves. The transition between the inertial range and the dissipation range has not been observed in velocity and ion density measurements; magnetic field power spectra suggest that the transition occurs near the Doppler-shifted proton gyrofrequency. Unlike previous measurements of the ion flux variations, the Triana FC data, combined with even faster magnetic field measurements, will enable the determination of the propagation direction of high-frequency waves and whether they are Alfvénic. This is crucial to assess the nature of the dissipation process and the heating and acceleration of particles. Those processes are expected to take place near the proton cyclotron frequency.

3. Conclusion

Taking advantage of its continuous orientation toward the sun, the Triana PlasMag Faraday Cup will determine density, velocity, and thermal speed of solar wind protons and alpha particles with a time resolution of better than once per second. For the first time, the ion parameters can be determined continuously at a high rate allowing previously unanswered questions such as the nature of waves in the dissipation range of solar wind turbulence to be addressed.

References

Ogilvie, K.W. *et al.*: 1995 SWE, a Comprehensive Plasma Instrument for the Wind Spacecraft, *Space Science Reviews* **71**, 55–77.

QUASI-THERMAL NOISE DIAGNOSTICS IN SPACE PLASMAS

K. ISSAUTIER, M. MONCUQUET, N. MEYER-VERNET, S. HOANG
and R. MANNING
Observatoire de Paris, DESPA, CNRS UMR 8632, Meudon, France

Abstract. We present the method of plasma diagnostics by quasi-thermal noise spectroscopy and show examples of application in the solar wind and the Earth's plasmasphere. Using only an electric antenna and a radio receiver, diagnostics of various space environments (magnetized or not) can be obtained *in situ*. Because of its accuracy, this technique can be used to cross-check other plasma sensors.

The quasi-thermal noise diagnostics of space plasmas is based on the generalization of the simple problem of an antenna immersed in black body radiation. Whereas the analysis of these electromagnetic fluctuations yields the corresponding temperature, an antenna in an equilibrium plasma is excited by thermal Langmuir waves; this enables one to measure the density and temperature of the medium. Space plasmas are generally not in equilibrium, but in a stable case, the electrostatic fluctuations measured at the terminals of an antenna are determined by the velocity distributions of the ambient charged particles. The quasi-thermal noise method is based on the interpretation of these fluctuations, allowing the determination of plasma parameters as the density, the temperature, the bulk speed, etc (Meyer-Vernet *et al.*, 2000).

The quasi-thermal noise spectroscopy was successfully applied to weakly magnetized environments, like the tail of a comet (Meyer-Vernet *et al.*, 1986), to the solar wind (Issautier *et al.*, 1998), and to various planetary environments (Io plasma torus, Earth's magnetosphere, Venus' ionosphere). It is presently implemented routinely on the spacecraft Ulysses, Wind and Cassini (Moncuquet *et al.*, 1998), and its measurements serve as a reference for other techniques (see a review in Meyer-Vernet *et al.* 1998).

In weakly magnetized plasmas, the quasi-thermal noise spectra have the typical shape shown in Figure 1 a), with an intense peak just above the local plasma frequency f_p. The spectral analysis yields the total plasma density $n_e \propto f_p^2$, the core electron temperature T_c and the bulk speed V_{SW}(revealed by the Doppler-shifted proton contribution to the wave spectrum below f_p), in addition to suprathermal parameters deduced from the fine structure of the peak (Issautier *et al.*, 1999).

An extension of the method to magnetized plasmas was first used in the Jupiter's magnetosphere when Ulysses crossed the Io plasma torus (Meyer-Vernet *et al.*, 1993). When the electron gyrofrequency f_g is no longer negligible compared to f_p,

Astrophysics and Space Science **277:** 309–311, 2001.

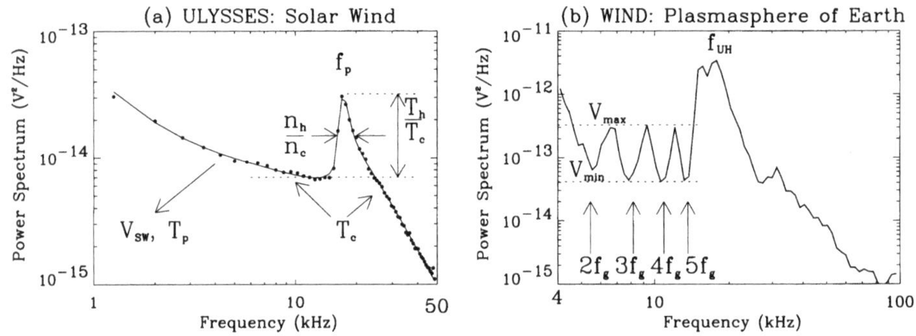

Figure 1. (a) Typical spectrum acquired by Ulysses in the solar wind. The solid line is the theoretical spectrum which best fits the data (dots) within 1.6%. The deduced plasma parameters are $n_e \simeq 3.2$ cm^{-3}, $T_c \simeq 6.6 \times 10^4$ K, $T_h/T_c \simeq 8$, $n_h/n_c \simeq 0.04$, $V_{SW} \simeq 350$ km/s, $T_p \simeq 1.2 \times 10^4$ K.
(b) Voltage power spectrum acquired by Wind in the Earth's plasmasphere (solid line). As explained in the text, the levels V_{min} and V_{max} yield an estimate of the temperatures: in this case, $T_c \simeq 9.5 \times 10^4$ K, and $T_h \simeq 6.6 \times 10^5$ K; the determination of $f_g \simeq 2.7$ kHz and $f_{UH} \simeq 15.6$ kHz yields $n_e \simeq 2.9$ cm^{-3}.

the wave spectrum is modified (Moncuquet *et al.*, 1995), since the electron thermal motion excites Bernstein waves. Figure 1 b) shows a spectrum acquired on WIND in the Earth's plasmasphere. The spectrum peaks at the upper-hybrid frequency f_{UH}, and exhibits weak bands with well-defined minima at gyroharmonics below f_{UH}. Such a spectrum allows one to determine (i) the electron density from the detection of f_{UH} (Moncuquet *et al.*, 1997), (ii) an estimate of the cold (Meyer-Vernet *et al.*, 1993) and hot (Sentman, 1982) electron temperatures, T_c and T_h, from the minimum and maximum levels of the gyroharmonics bands respectively, (iii) the magnetic field magnitude from the detection of the spectral minima at gyroharmonics, which is in agreement with the magnetometer data within a few percent.

This technique will be used on IMAGE to investigate the Earth's magnetosphere. We also plan to measure the electron parameters in the magnetosphere of Saturn, in 2004, with the RPWS experiment on Cassini. In 2000-2001, a second flyby of the solar poles by Ulysses will occur near the solar activity maximum. Accurate electron diagnostics using this method will help to understand the 3-D structure of the solar wind at this stage of the solar cycle.

References

Issautier, K. *et al.*: 1998, *J. Geophys. Res.* **103,** 1969.
Issautier, K. *et al.*: 1999, *J. Geophys. Res.* **104,** 6691.
Meyer-Vernet, N. *et al.*: 1986, *Science,* **232** 370.

Meyer-Vernet, N., Hoang, S. and Moncuquet, M.: 1993, *J. Geophys. Res.* **98,** 21163.
Meyer-Vernet, N. *et al.*: 1998, in: E. Borovsky and R. Pfaff (eds.), *Geophysical Monograph 103: Measurements techniques in Space Plasmas*, **205**.
Meyer-Vernet, N. *et al.*: 2000,in: R.G. Stone et al. (eds), *Geophysical Monograph 119: Space based Radio Observations at Long Wavelengths*, **67**.
Moncuquet, M., Meyer-Vernet, N. and Hoang, S.: 1995, *J. Geophys. Res.* **100,** 21,697.
Moncuquet, M. *et al.*: 1997, *J. Geophys. Res.* **102,** 2373.
Moncuquet, M. *et al.*: 1998, *abstract in Supp. to Eos* **79,** 45, F755.
Sentman, D.D.: 1982, *J. Geophys. Res.* **87,** 1455.

QUASI-THERMAL NOISE SPECTRA MEASURED BY A DIPOLE ANTENNA IN THE UPPER HYBRID FREQUENCY BAND

YURY V. CHUGUNOV, ANNA YU. KAZAROVA and EUGENE A. MAREEV
Institute of Applied Physics RAS, Nizhny Novgorod, Russia

MICHEL MONCUQUET, KARINE ISSAUTIER and NICOLE MEYER-VERNET
DESPA, Observatoire de Paris, Meudon, France

Abstract. Quasi-thermal noise (QTN) spectroscopy is one of the most effective tools for *in situ* diagnostics in space plasmas (Meyer-Vernet *et al.*, 1998; Meyer-Vernet and Perche, 1989; Chugunov and Trakhtengerts, 1978). This method produces routine measurements of the bulk electron density and temperature; recently it has been extended to measure the ion bulk speed. Among the advantages of the method its immunity to spacecraft potential and photoelectron perturbations should be noted. Quasi-thermal noise spectroscopy is used particularly on Ulysses and Wind. However for the interpretation of QTN data the calculation of the noise voltage induced on antennas under different conditions is necessary. This question is especially complicated and so far insufficiently studied in magnetized plasmas. In the present paper we calculate the spectrum of the noise voltage induced on a dipole antenna in the upper hybrid frequency range. The computations are adapted to the interpretation of data acquired on the Ulysses and Wind spacecraft.

1. Model and Results

First we describe briefly the general approach for calculating noise voltage spectra in nonequlibrium plasmas in an external magnetic field (see for details Mareev and Chugunov, 1991). Noise voltage calculations involve essentially the discrete nature of the plasma particles, i.e. require the consideration of the first order terms (over $(N_e r_D^3)^{-1}$, where N_e is the electron density and r_D is the Debye length) in the chain of kinetic equations. The simplest method is based on the test-particle approach. It involves: 1) calculating the signal induced on the antenna terminals by the charge fluctuations in the plasma; 2) averaging over the respective charge distribution functions. In the first stage the most effective (and probably unique) way is to apply the reciprocity theorem which is valid in a wide range of medium conditions, including anisotropic media and media with spatial dispersion. In the quasi-static approach, we write the reciprocity theorem as:

$$\int \rho(\mathbf{r})\varphi_0(\mathbf{r})d\mathbf{r} = \int \rho_0(\mathbf{r})\varphi(\mathbf{r})d\mathbf{r} = \varphi_s \int \rho_0(\mathbf{r})d\mathbf{r} = \varphi_s Q, \tag{1}$$

where $\rho(\mathbf{r})$ is the charge density fluctuation in plasma, which induces the potential perturbation $\varphi(\mathbf{r})$ on the antenna surface. Considering the antenna as a perfect conductor, we have assumed: $\varphi(\mathbf{r}) = \text{const} = \varphi_s$. Generally speaking, the expression

Astrophysics and Space Science **277:** 313–316, 2001.

for the induced potential $\varphi(\mathbf{r}) = \varphi_s$ should be calculated in a plasma where the directions of the external magnetic field $\mathbf{H_0}$, plasma velocity $\mathbf{v_0}$ and wave vector $\mathbf{k}$ are reversed. For quasi-stationary fields, however, (when the plasma motion is described by the Dopler shift $\omega - \mathbf{kv_0}$ only) this does not lead to any changes. $\rho_0(\mathbf{r})$ is the density of the charge Q, which is spread on the antenna surface and creates the potential $\varphi_0(\mathbf{r})$ in the plasma. Note that complex amplitudes are implied. Let us define the power spectrum of the noise voltage as $V_\omega^2 = 2 < \varphi_s\varphi_s^* >$; using the reciprocity theorem, one obtains:

$$V_\omega^2 = \frac{4\pi}{Q^2}\int\int \varphi_0(\mathbf{r}_1)\varphi_0^*(\mathbf{r}_2) < \rho\rho^* >_\omega d\mathbf{r}_1 d\mathbf{r}_2, \tag{2}$$

where $< \rho\rho^* >_\omega$ is the charge correlation function in the plasma. To describe nonequilibrium plasmas, one generally uses a two Maxwellian model – a minor hot electron population of concentration N_h small compared to the concentration N_c of the cold electron population of the ambient plasma, and of temperature T_h great compared to T_c. In the upper hybrid (UH) frequency range $\omega_{pe} < \omega < \sqrt{\omega_{pe}^2 + \omega_{He}^2} = \Omega_{UH}$ the QTN is caused mainly by the hot electron population. The noise level in this band depends substantially on the orientation of the antenna with respect to the resonance cone angle arctg μ around the magnetic fiel direction. As a result, the noise power spectrum V_ω^2 has different asymptotic values in the regions $\operatorname{tg}\alpha > \mu$ and $\operatorname{tg}\alpha < \mu$ (α is the angle between the antenna direction and the magnetic field, $\mu = (|\varepsilon_1/\varepsilon_3|)^{1/2}$, ε_1 and ε_3 are the diagonal components of the dielectric tensor).

It turns out that the resulting expressions are very different depending on the mechanism of absorption of radiation in the cold component. We consider in more detail a case (especially applicable to the Jupiter and Earth's magnetosphere environments) when absorption of radiation in both the hot and cold components is due to the collisionless dissipation. Taking into account the latter circumstance, we follow to the general procedure presented in (Yefimova and Chugunov, 1981; and Mareev and Chugunov, 1991). Then the integral defining the noise power spectrum V_ω^2 takes the form

$$V_{\omega s}^2 = \frac{16T_h}{\omega\pi^2 L\sqrt{|\varepsilon_1\varepsilon_3|}q}\int_0^{2\pi} d\varphi \frac{\sin^4\left[\frac{q}{2y}\sin\alpha(\sin\varphi + \mu\cot\alpha)\right]}{\sin^2\alpha(\sin\varphi + \mu\cot\alpha)^2}\int_0^\infty \frac{dy}{1+\gamma e^{-y^2}} \tag{3}$$

for a wire dipole antenna of half length L. Here $q = \omega L/\mu v_{Tc} \gg 1$, $\gamma = (N_c/N_h)(v_{Th}/v_{Tc})^3$ and $\mu^2 = |\varepsilon_1/\varepsilon_3| = (1 - w)/w$. For the Jupiter environment $w = v/u$, wher e $u = \omega_{He}^2/\omega_{pe}^2 \ll 1$, $v = 2(\omega - \omega_{pe})/\omega_{pe}$.

The integral over φ has been calculated asymptotically when $q/2y \gg 1$; the calculations are substantially different in the regions $\mu^2\cos^2\alpha - \sin^2\alpha > 0$ and $\mu^2\cos^2\alpha - \sin^2\alpha < 0$.

The noise voltage dependencies on the frequency and the angle α in the Jupiter environment have been calculated numerically. We have also calculated the noise

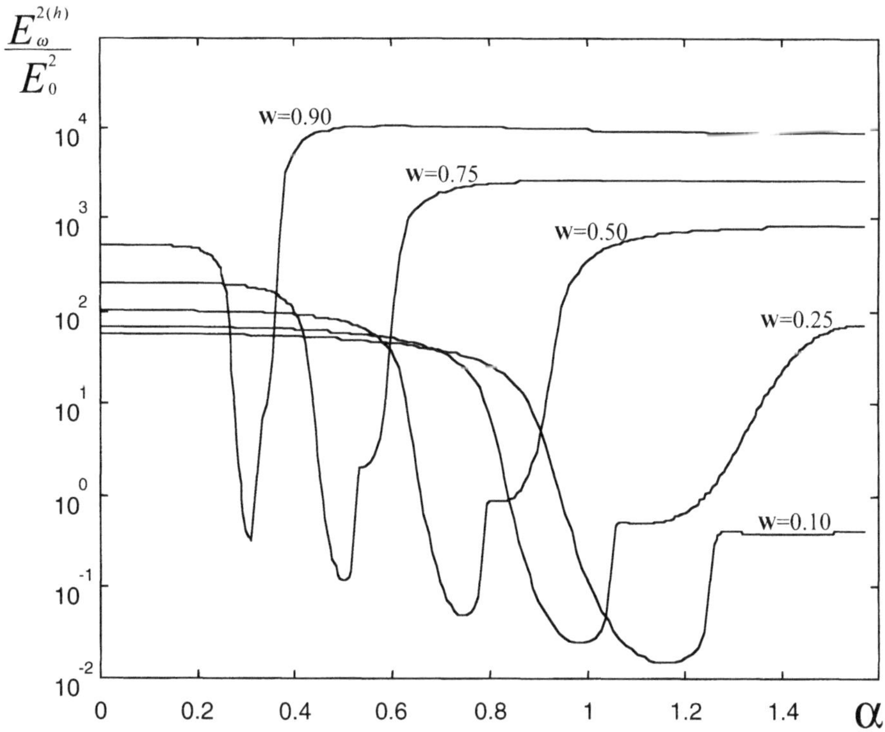

Figure 1. Typical dependencies of noise voltage spectra on the angle α between the antenna and the magnetic field at different frequencies calculated for Ulysses applications, with $w = 2(f - f_{pe})f_{pe}/f_{He}^2$. The noise level is normalized to $E_0^2 = 2T_h/\pi\omega L$. The other parameters are fixed here as: $u = f_{He}^2/f_{pe}^2 = 0.04$, $\gamma = (N_c/N_h)(V_{Th}/V_{Tc})^3 = 10^4$, $q_0 = \omega L/V_{Tc} = 10$.

voltage dependence on the frequency and on the angle under magnetospheric conditions. These results have been applied to the analysis of routine plasma measurements on board the Wind and Ulysses spacecraft. An example of typical curves, showing the noise voltage spectrum as a function of the angle α in UH frequency band, is presented in Figure 1. It is seen from the figure that the noise voltage spectrum has a fine structure with a pronounced decrease just below the resonance angle. Besides, the noise magnitude generally increases with frequency and reaches a maximum near the UH frequency. This property could be helpful for measuring the electron density in space plasmas.

2. Conclusion

In the present paper we calculate the spectrum of the noise voltage induced on a dipole antenna in the upper hybrid frequency range in order to use the results for the interpretation of data from the Ulysses and Wind satellites. The general approach to the calculation of noise voltage spectra under nonequlibrium plasmas, including non-Maxwellian plasma in an external magnetic field, has been developed. It is shown that the noise voltage spectrum has a fine structure in the resonance frequency band. The noise magnitude generally increases with frequency and reaches a maximum at the UH frequency, which can be helpful for measuring the electron density in space plasmas.

Acknowledgements

This work was supported by the INTAS grant N 97-1372, and the Russian Fund of Basic Research, grants N 99-02-17745 and N 00-02-17758.

References

Chugunov, Yu.V. and Trakhtengerts, V.Yu.: 1978, On the low frequency noises, induced on antenna in ionospheric plasma. *Kosm. Issled.* **16** (2), 238.

Mareev, E.A. and Chugunov, Yu.V.: 1991, *Antennas in Plasmas* (in Russian), IAP RAS Edition, Nizhny Novgorod.

Meyer-Vernet N., Hoang S., Issautier K., Maksimovic M., Manning R., Moncuquet M. and Stone R.G.: 1998, Measuring plasma parameters with thermal noise spectroscopy, in: R. Pfaff, J. Borovsky and D. Young (eds.), *Measurement Techniques in Space Plasmas*, pp. 121–136, AGU, Washington, D.C.

Meyer-Vernet, N., Hoang, S., and Moncuquet, M.: 1993, Bernstein waves in the Io torus: a novel kind of electron temperature sensor, *J. Geophys. Res.* **98**, 21163.

Meyer-Vernet, N. and Perche, C.: 1989, Tool kit for antennae and thermal noise near the plasma frequency, *J. Geophys. Res.* **94** (A3), 2405.

Yefimova, T.V. and Chugunov, Yu.V.: 1981, High-frequency noise induced on antenna in the ionospheric plasma, *Geomagn. Aero.* **21** (1), 34.

ROSETTA MISSION MUTUAL IMPEDANCE PROBE MODELLING: THE SHORT AND LONG DEBYE LENGTH PLASMA CASES

J. GEISWILLER, J.G. TROTIGNON and C. BÉGHIN
CNRS, Laboratoire de Physique et Chimie de l'Environnement, 3A, Avenue de la Recherche Scientifique, F-45071 Orléans Cedex 02, France

E. KOLESNIKOVA[2]
Department of Physics and Astronomy, University of Leicester, University Road, Leicester LE1 7RH, United Kingdom

Abstract. The Rosetta spacecraft (S/C), which is planned to meet comet 46P/Wirtanen in 2011, will carry a set of five wave and plasma instruments (i.e. the Rosetta Plasma Consortium). This is to measure the cometary plasma properties from the minimum value of activity of the comet to its maximum value at perihelion. The mutual impedance probe, MIP, is one of those (Trotignon *et al.*, 1999) five.

The MIP sensor is a quasi-electrostatic emitting-receiving device derived from the quadrupolar probes of Storey *et al.* (1969). An alternative current, $I(\omega)$, of variable frequency, ω, lying in the range of frequencies from 28 kHz to 3.5 MHz is injected through the emitting electrode. Simultaneously, the difference in voltage, $V(\omega)$, is measured between two receiving electrodes located at 40 cm and 60 cm from the MIP transmitter. The ratio $V(\omega)/I(\omega)$ is called the mutual impedance, $Z(\omega)$ and depends on the plasma parameters. The electron density is deduced from the modulus of $Z(\omega)$, which peaks at the plasma frequency and is shown in Figure 1. The local Debye length, λ_D, is deduced from the location of antiresonances above the plasma frequency which are formed by the beating of Landau waves. The wave length of these waves is dependent on the electron temperature, the plasma frequency and the emitting frequency. The MIP mode will study plasmas in the diamagnetic cavity of the comet whenever the Debye length will belong to the range from 3 mm to 20 cm.

The LDL mode will be used in long Debye length plasmas, i.e. in the prime instants of the coma formation and the period of spacecraft crossings in the regions between coma and solar wind. In this case, a spherical Langmuir probe located at about 4 m from the MIP sensor is used as the emitting electrode. The difference in voltage will then measured between the two MIP receiving electrodes in the range of frequencies from 7 kHz to 168 kHz.

A computer simulation of the corresponding frequency responses has been made in a collisionless, homogeneous, isotropic and kinetic plasma to precisely evaluate the range of measurements of those modes. Here, we have not choosen to take into account either the presence of the ion-sheath around the sensor and the spacecraft, or problems due to a non-maxwellian distribution of electrons which are

Astrophysics and Space Science **277:** 317–318, 2001.

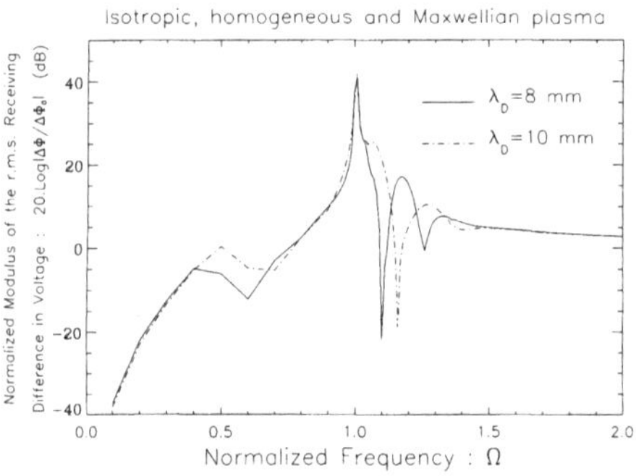

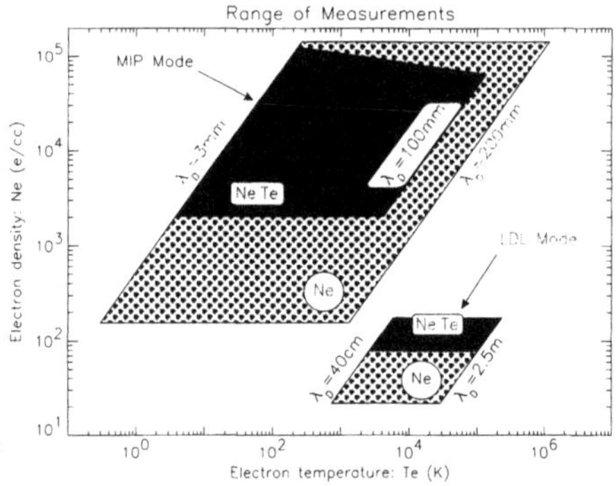

Figure 1. [left] Modelling of the expected normalized modulus of the MIP difference in voltage in a Maxwellian and isotropic plasma. The influence of the Debye length is well illustrated by the variation of the antiresonances location when $\Omega = \omega/\omega_{pe} > 1$. Here, the numerical frequency resolution is of 0.1 ω_{pe} below the plasma frequency and 0.01 ω_{pe} above.

Figure 2. [right] Range of measurements of the electron density and temperature expected in the MIP mode and the LDL mode. These theoritical values are deduced from the DSCD numerical simulation of the instruments in a Maxwellian plasma.

considered to be the object of future papers. We only focus on the influence of the S/C conductive structure in the LDL mode and that of the sensor structure in the MIP mode. In both cases the Discrete Surface Charge Distribution method (Béghin and Kolesnikova, 1998) is used which is basically based on the finite elements method. Regarding the MIP mode, all the conductive structures of the sensor are modelled as a distribution of point sources that are induced on the sensor surface by the emitting monopole and its neighbouring charges. The surface potential of the sensor is calculated using the expression of the electrostatic potential induced by a point source in a maxwellian plasma (Béghin, 1995). Regarding the LDL mode, the S/C presence is taken into account by a wire structure transparent for the plasma. Rigorously, this assumption is valid whenever the plasma Debye length is twice the longest dimension of the S/C body, i.e. $\lambda_D > 1.7$ m.

The necessity to model very carefully the conductive structures which are in the vicinity of the MIP sensor must be pointed out. The LDL mode are indeed found to be strongly dependent on the spacecraft whose dimensions are in the order of the emitting-receiving distance. The range of measurements of the MIP mode and the LDL mode are summarized in Figure 2.

References

Béghin, C.: 1995, *Radio Sci.* **30,** 307.

Béghin, C. and Kolesnikova, E.: 1998, *Radio Sci.* **33,** 503.

Storey, L.R.O., Aubry, M.P. and Meyer, P.: 1969, in: J.O. Thomas and B.J. Landmark (eds.), *Plasmas Waves in space and in Laboratory*, University Press Edinburgh 1, 303.

Trotignon, J.G. *et al.*: 1999, MIP – RPC: the Mutual Impedance Probe in the Rosetta Plasma Consortium. ESA SP-1165.

ANALYTICAL APPROACH TO CALCULATE QUASI-THERMAL NOISE SPECTRUM IN IRREGULAR PLASMA

[1]NIKOLAY A. ZABOTIN, [1]ANDREY G. BRONIN, [2]YURY V. CHUGUNOV and [2]EVGENE A. MAREEV
[1]*Rostov State University, Rostov on-Don, Russia*
[2]*Institute of Applied Physics RAS, Nizhni Novgorod, Russia*

Abstract. Basic properties of quasi-thermal noise spectrum in irregular plasma have been investigated using an analytical point of view. A simple formula for the plasma frequency splitting effect has been obtained for ionospheric conditions. A passive electric antenna, immersed in a stable plasma, detects the fluctuations of the electric potential due to thermal motion of the ambient particles. Properties of this quasi-thermal noise spectrum in homogeneous plasma are relatively well known and are effectively used for diagnostics of space plasma (Aksenov *et al.*, 1978; Trakhtengerts and Chugunov, 1978; Kellog, 1981; Meyer-Vernet and Perche, 1989). Especially, in the Earth's ionosphere or solar wind plasma, random irregularities of electron density are always present. These irregularities may substantially change properties of these media through electromagnetic radiation and may also modify quasi-thermal noise spectrum, which depends on the effective dielectric permittivity tensor. This tensor is defined as the dielectric permittivity tensor of some imaginary 'effective' regular medium in which the field of point source is the same as the mean field in the medium with random irregularities (Ryzhov, Tamoikin and Tatarskii, 1965; Ryzhov 1968). Since the correlation function of electric field fluctuations in random medium may be expressed through the effective dielectric constant tensor (Ryzhov, 1968), it may be used for direct calculation of quasi-thermal noise spectrum. In Zabotin et al. (2000), the Born approximation was used to calculate numerically the effective dielectric permittivity tensor and the modified noise spectrum while we analytically estimate herein the modified noise spectrum.

1. Quasi-Thermal Noise Spectrum and Antenna Impedance

If the plasma is in the thermal equilibrium at temperature T, which will be assumed further, the voltage spectral density may be expressed through antenna impedance by the formula (Meyer-Vernet and Perche, 1989):

$$V^2 = 4k_B T \Re(Z) \tag{1}$$

where k_B is Boltzmann constant, Z is the antenna impedance and $\Re$, $\Im$ denotes the real and imaginary part respectively. For the wire dipole antenna one can get (Meyer-Vernet and Perche, 1989; Kuhl, 1966; Kellog, 1981):

$$\Re(Z) = \frac{4}{\pi^2 \varepsilon_0 \omega} \int_0^\infty dk \frac{\Im \varepsilon_\parallel(\omega, k)}{|\varepsilon(\omega, k)|^2} F(k) \tag{2}$$

Astrophysics and Space Science **277:** 319–322, 2001.

where

$$F(k) = [J_0(ka)]^2 F_1(kL) \tag{3}$$
$$F_1(kL) = \left\{Si(kL) - Si(2kL)/2 - 2\sin^4(kL)/kL\right\}/kL$$

L and a are the length and diameter of antenna wire respectively, J_0 is the Bessel function of the first kind and $Si(x)$ is the integral sinus function.

For frequencies close to plasma frequency, $\left|\Im\varepsilon_\parallel\right| \ll \left|\Re\varepsilon_\parallel\right|$, making possible the passage to the limit

$$\lim_{\Im\varepsilon_\parallel \mapsto 0} \frac{\Im\varepsilon_\parallel}{\Re^2\varepsilon_\parallel + \Im^2\varepsilon_\parallel} = \pi\delta[\Re\varepsilon_\parallel] \tag{4}$$

Substituting (4) into (2) and (1), we approximately obtain

$$V^2 = \frac{16k_B T}{\pi\varepsilon_0\omega} \frac{F(k^*)}{\left|\frac{\partial}{\partial k}\Re\varepsilon_\parallel(\omega, k)\right|_{k=k^*}} \tag{5}$$

where $k^*(\omega)$ is the root of equation $\Re\varepsilon_\parallel(\omega, k) = 0$.

For frequencies close to plasma frequency and for Maxwellian collisionless plasma the longitudinal dielectric permittivity takes the simple expression:

$$\varepsilon_\parallel(\omega, \vec{k}) = 1 - (\omega_p^2/\omega^2)[1 + 3r_D^2 k^2] + i0 \tag{6}$$

where ω_p is the plasma frequency, r_D is the Debye length. From (6) one gets

$$k^{*2} = \frac{\omega^2/\omega_p^2 - 1}{3r_D^2}, \qquad \left|\frac{\partial}{\partial k}\varepsilon_\parallel(\omega, k)\right|_{k=k^*} = 6\frac{\omega_p^2}{\omega^2} r_D^2 k^* \tag{7}$$

and we finally obtain

$$V^2 = \frac{8k_B T\omega^2/\omega_p^2}{\pi\sqrt{3}\varepsilon_0 r_D \omega\sqrt{1 - \omega_p^2/\omega^2}} F\left[\sqrt{\frac{1 - \omega_p^2/\omega^2}{3r_D^2}}\right] \theta(\omega - \omega_p) \tag{8}$$

where $\theta(x)$ is equal to 1 if $x > 0$ and equal to 0 otherwise. The position of the spectrum maximum is determined in this approximation by the function $F_1(kL)$ since $ka \ll 1$ for both ionosphere and solar wind. This function reaches its maximum value at $kL \approx 3.4$ which corresponds to the frequency $\omega^2 \approx \omega_p^2\left[1 + 35(r_D/L)^2\right]$.

For ionospheric plasma, since $L \gg r_D$ (see (Zabotin *et al.*, 2000) for discussion on specific parameters for different kinds of plasma) the spectrum has a sharp maximum at $\omega \approx \omega_p$. In the solar wind typical values are $r_D \sim 10$m and $L \sim 50$m, so the spectrum pike is fairly schifted with respect to plasma frequency. To calculate the quasi-thermal noise spectrum with random irregularities, one should substitute in the above relations the longitudinal dielectric permittivity $\varepsilon_\parallel(\omega, \vec{k})$, for a regular plasma, by the effective longitudinal dielectric permittivity $\varepsilon_\parallel^{eff}(\omega, \vec{k})$ (Ryzhov, 1968).

2. Calculation of Effective Dielectric Permittivity Tensor

To calculate the effective dielectric permittivity tensor, one proceeds from the Dyson equation for the mean field in a random medium (Rytov *et al.*, 1978). It may be easily shown that the effective dielectric permittivity tensor is proportional to the Fourier transform of the mass operator. Taking into account only initial term in the mass operator series expansion (which corresponds to the Born approximation for the mean field) one can obtain a following expression for longitudinal dielectric permittivity of a media with isotropic irregularities:

$$\varepsilon_{\parallel}^{eff}(\omega,\vec{k}) = \varepsilon_{\parallel}(\omega,\vec{k}) - \left[\varepsilon_{\parallel}(\omega,\vec{k}) - 1\right]^2 \times$$
$$\left\{\int \frac{\vec{k}\cdot\vec{p}}{k^2p^2}\frac{\Phi_N(\vec{k}-\vec{p})}{\varepsilon_{\parallel}(\omega,\vec{k})}d^3p + k_0^2\int\left(1-\frac{\vec{k}\cdot\vec{p}}{k^2p^2}\right)\frac{\Phi_N(\vec{k}-\vec{p})}{k_0^2\varepsilon_{\perp}(\omega,\vec{k})-p^2}d^3p\right\} \tag{9}$$

For analytic estimations we choose the model spatial spectrum of irregularities in the following form:

$$\Phi_N(\vec{k}) = \frac{\delta_R^2 l^3}{2\pi^2\left(1-\exp(-R/l)\right)}\left(1+l^2k^2\right)^{-2} \tag{10}$$

where $l = L_m/2\pi$, L_m is the outer scale of spectrum, $\delta_R^2 = D(R)$ is the structure function of irregularities at scale length R. Substituting (10) into (9) and since for both ionospheric and solar wind cases $kL \gg 1$, the real and imaginary parts of the longitudinal dielectric tensor are respectively:

$$\begin{aligned}\Re\varepsilon_{\parallel}^{eff}(\omega,k) &= \Re\varepsilon_{\parallel}(\omega,k)\left[1-\frac{1}{2}\frac{\delta_R^2}{1-\exp(-R/l)}\left(1+\frac{1}{\left|\varepsilon_{\parallel}(\omega,k)\right|^2}\right)\right] \\ &\quad+\delta_R^2/[1-\exp(-R/l)]\end{aligned} \tag{11}$$

$$\Im\varepsilon_{\parallel}^{eff}(\omega,k) = \Im\varepsilon_{\parallel}(\omega,k)\left[1-\frac{1}{2}\frac{\delta_R^2}{1-\exp(-R/l)}\left(1-\frac{1}{\left|\varepsilon_{\parallel}(\omega,k)\right|^2}\right)\right] \tag{12}$$

3. Calculation of Plasma Noise Spectrum with Random Irregularities

To calculate quasi-thermal noise spectrum in plasma with random irregularities, one should substitute (11) and (12) into (2) using the limit condition (4). Since for frequencies close to the plasma frequency, $\Im\varepsilon_{\parallel}$ is very small, one can assume that, (i) $\Im\varepsilon_{\parallel} \ll \Delta$, where $\Delta = \delta_R^2/[1-\exp(-R/l)]$, and (ii) $\left|\varepsilon_{\parallel}(\omega,k)\right| \approx \Re\varepsilon_{\parallel}(\omega,k)$. Introducing new unknown quantity $\psi = \Re\varepsilon_{\parallel}(\omega,k)$ one may reformulate the equation $\Re\varepsilon_{\parallel}^{eff}(\omega,k) = 0$ as a quadratic one:

$$\psi\left[1-(\Delta/2)\left(1+\psi^{-2}\right)\right] + \Delta = 0 \tag{13}$$

So equation (13) splits for couple of equations:

$$\Re\varepsilon_{\parallel}(\omega, k_{\pm}) = (\pm\sqrt{\Delta/2})/(1 \pm \sqrt{\Delta/2}) \tag{14}$$

In hydrodynamic approximation these equations have the following roots:

$$k_{\pm}^2 = (\omega^2/\omega_{\pm}^2 - 1)/(3r_D^2) \tag{15}$$

where $\omega_{\pm}^2 = \omega_p^2(1 \pm \sqrt{\Delta/2})$.

One can find out from (11) that $\left|\partial/\partial k(\Re\varepsilon_{\parallel}^{eff})\right| = \left|\partial/\partial k(\Re\varepsilon_{\parallel})\right|$. The final expression of the quasi-thermal noise with irregularities is

$$V^2 = \frac{4k_BT}{\pi\sqrt{3}\varepsilon_0 r_D\omega}\left\{\frac{\omega^2/\omega_+^2}{\sqrt{1-\omega_+^2/\omega^2}}F\left[\sqrt{\frac{1-\omega_+^2/\omega^2}{3r_D^2}}\right]\theta(\omega-\omega_+) + \frac{\omega^2/\omega_-^2}{\sqrt{1-\omega_-^2/\omega^2}}F\left[\sqrt{\frac{1-\omega_-^2/\omega^2}{3r_D^2}}\right]\theta(\omega-\omega_-)\right\} \tag{16}$$

If $L \gg r_D$ one can show that (16) has two sharp maxima at frequencies $\omega \approx \omega_+$ and $\omega \approx \omega_-$ respectively. The distance between these maxima is then $\Delta\omega \approx \omega_p\sqrt{\Delta/2}$. These analytical results are in agreement with numerical calculations obtained in the ionosphere case by [Zabotin *et al.*, 2000]. The work has been done under support of the INTAS Grant 97-1372.

References

Aksenov, V.I., Efimova, T.V., Komrakov, G.P. *et al.*: 1978, Studies of the Sun and the Earth's Ionosphere Parameters on the Satellite 'Interkosmos-Kopernik-500'. *Kosmicheskiye Issledovaniya* **16** (1), 78.

Kellog, P.J.: 1981, Calculation and observation of thermal electrostatic noise in solar wind plasma, *Plasma Phys.* **23** (8), 735.

Kuhl, H.H.: 1966, Resistance of a short antenna in a warm plasma, *Radio Sci.* **1**, 971.

Meyer-Vernet, N. and Perche, C.: 1989, Tool kit for antennae and thermal noise near the plasma frequency, *J. Geophys. Res.* **94** (A3), 2405.

Rytov, S.M., Kravtsov, Yu.A. and Tatarskii, V.I.: 1978, *Introduction to Statistical Radiophysics*, **2**, Random Fields. Nauka, Moscow, (in Russian).

Ryzhov, Yu.A., Tamoikin, V.V. and Tatarskii, V.I.: 1965, On space dispersion of inhomogeneous media, *JETP* **48** (2), 656 (in Russian).

Ryzhov, Yu.A.: 1968, Analytic properties of the effective dielectric permeability of randomly inhomogeneous media, *JETP* **55** (2:3), 567 (in Russian).

Trakhtengerts, V.Yu. and Chugunov, Yu.V.: 1978, On the HF noise induced in an antenna in the ionospheric plasma, *Kosmicheskiye Issledovaniya* **16** (2), 238.

Zabotin, N.A., Chugunov, Yu.V., Mareev, E.A. and Bronin, A.G.: 2000, Influence of random irregularities on quasi-thermal noise spectrum of plasma, *LANL Archive*, http://arXiv.org/abs/physics/0004021

CORRELATION DEPENDENCES DETERMINED BY SIMULTANEOUS MULTISPACECRAFT OBSERVATIONS OF SOLAR WIND AND IMF STRUCTURES

P.A. DALIN and G.N. ZASTENKER
Space Research Institute, Profsojuznaia 84/32, Moscow, 117810, Russia

K.I. PAULARENA and J.D. RICHARDSON
Center for Space Research, M.I.T., Cambridge, MA, 02139 U.S.A.

Abstract. Solar wind measurements on board several spacecraft were used to study the two-points correlations of the solar wind plasma structures. The factors having the most influence on the correlation level are the density variability and IMF cone angle. The characteristic length of large solar wind structures is estimated at 500–1000 R_E.

1. Available Data and Method

We compared simultaneous ion flux data (1 min. resolution) measured by widely-separated spacecraft (INTERBALL-1, IMP 8 and WIND) in 1995–98. Many cases show similar features at each spacecraft, but in some cases plasma parameters behave very disparately. The global (averaged by the large statistics) value of the correlation coefficient is about 0.73.

Multifactoral analysis was used to obtain the dependencies of the correlations on the spacecraft separation and the different plasma and magnetic field parameters. The parameters available for our study are the solar wind speed, number density (or ion flux) and magnetic field components obtained with 1 or 1.5 min. time resolution. All measurements were divided into 6-hour segments. For each segment the WIND and IMP 8 data were time-shifted to the INTERBALL-1 times, compensating for the solar wind propagation time.

2. Results

The plasma parameters influence on the correlation level was studied. The influence of density (or ion flux) variability (as estimated by the relative standard deviation rsd) is more than one of absolute values of the density, or the solar wind speed. We conclude that the structures with small variations ($rsd < 20\%$) have small spatial scales and these small scale structures are poorly correlated. On the other hand, plasma structures with large variations ($rsd > 20\%$) have a large

Astrophysics and Space Science **277:** 323–324, 2001.

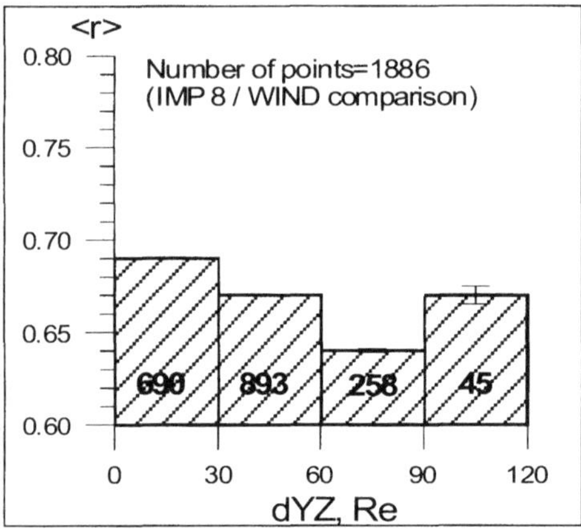

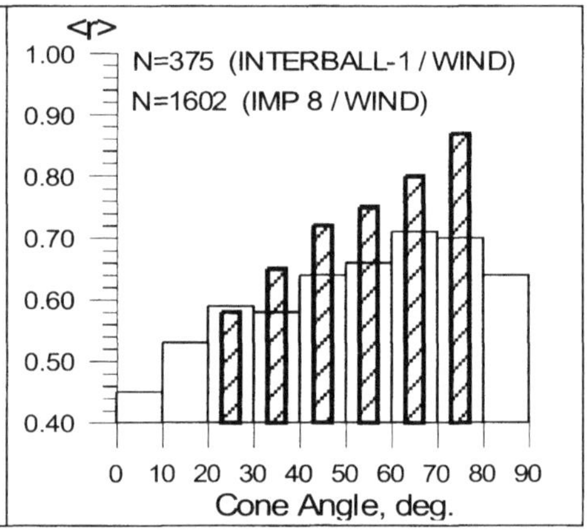

Figure 1. [left] The dependence of the averaged correlations on the distance between IMP 8 and WIND s/c in the plane perpendicular to the Sun-Earth line (numbers of points are shown).

Figure 2. [right] The dependence of the averaged correlations on the IMF cone angle. The empty boxes are for IMP 8/WIND comparison; the shaded ones are for INTERBALL-1/WIND comparison.

spatial scale, about one million km, and thus correlations for these features are high.

We have found a weak but certain dependence of the correlation value on spacecraft separation perpendicular to the Sun-Earth line (Figure 1). It means that the solar wind plasma characteristic length is very large (on average from 500 up to 1000 R_E or $3\text{–}6\cdot10^6$ km) in the plane perpendicular to its propagation.

The role of interplanetary magnetic field was studied also. The IMF cone angle (the angle between the IMF vector and the Sun-Earth line) is an important factor affecting the plasma correlation (Figure 2). As the value of the cone angle increases the correlation of ion flux also increases.

Acknowledgements

This work was partly supported by Russian Foundation for Basic Research (grant 01-02-16182) and NASA contract NAG5-8720 (JURRISS) to MIT and IKI (administered by the CRDF award RP0-847).

References

Paularena, K.I., Zastenker, G.N., Lazarus, A.J. and Dalin, P.A.: 1998, *J. Geophys. Res.* **A7 103**, 14601.

Zastenker, G.N., Dalin, P.A., Petrukovich, A.A, Nozdrachev, M.N., Romanov, S.A., Paularena, K.I., Richardson, J.D., Lazarus, A.J., Lepping, R.P. and Szabo, A.: 2000, *Phys. Chem. Earth* **25** (1-2), 137.

DEVELOPMENTS IN JOVIAN RADIO EMISSIONS TOMOGRAPHY AND OBSERVATIONS TECHNIQUES

H.O. RUCKER, M.Y. BOUDJADA and M. LEITNER
Space Research Institute, Graz, Austria

A. LECACHEUX and M. AUBIER
Observatoire de Paris-Meudon, France

A. KONOVALENKO
Institute of Radioastronomy, Kharkov, Ukraine

P.H.M. GALOPEAU
Centre d'Etude des Environnements Terrestre et Planetaire, Velizy, France

V. SHAPOSHNIKOV
Institute of Applied Physics, Nizhny Novgorod, Russia

Abstract. Jupiter radio emission is known to be the most powerful nonthermal planetary radiation. In recent years specifically space-based observations allow us to permanently cover a large frequency band (from 100 kHz up to 40 MHz combined with ground-based telescopes) of the Jovian spectrum. The Plasma and Wave Science experiment onboard Galileo enables the observation of Jovian kilometric and hectometric emissions; Wind/WAVES and ground-based telescopes (mainly Decametric Array in Nancay, France, and UTR-2 in Kharkov, Ukraine) cover also hectometric and mainly decametric emissions. Specific geometrical configurations between Cassini approaching Jupiter and Wind spacecraft orbiting Earth, with Galileo orbiting Jupiter and Wind, in combination with ground-based observations provide a new approach to perform Jovian radio tomography. The tomography technique is used to analyze ray paths of Jovian radio emission observed in different directions (e.g. solar and anti-solar direction) and for different declination of Earth. The developments of Jovian radio emission tomography in recent years treated refraction effects and its connection to the local magnetic field in the radio source as well as the radio wave propagation through the Io torus and the terrestrial ionosphere. Most recently ground-based multi-site and simultaneous Jupiter decametric radio observations by means of digital spectropolarimeter and waveform receiver provide the basis of a new data analysis treatment. The above addressed topics are without exemption deeply connected to the plasma structures the radio waves are generated in and propagating through.

1. Refraction Effects and Radio Wave Propagation

Wind/WAVES observations combined with simultaneous ground-based observations made by the Nancay Decametric Array (France) enabled the coverage of the entire frequency spectrum of the Jovian decametric emission. Selection of Io-controlled events have been analyzed in terms of models of the Jovian magnetic environment (Lecacheux *et al.*, 1998; Queinnec and Zarka, 1998). Special emphasis was laid on the comparison of the shape of the observed decametric arc

Astrophysics and Space Science **277:** 325–328, 2001.

structure and the calculated change of angle between the radio source local magnetic field direction and direction to observer, with regard to time and frequency. Both studies conclude that refraction effects may play a role at the low frequency ends of radio arcs.

Propagation of the Jovian decametric radio emission from its source to the observer, through the Jovian magnetosphere, Io plasma torus, interplanetary medium and terrestrial ionosphere, results in Faraday rotation of the radio polarization ellipse (Budden, 1985):

$$\Omega = \frac{\pi f}{c} \int_{raypath} (n_o - n_x).ds = C_0 + \frac{C_1}{f^2} \tag{1}$$

where f is the wave frequency, c the speed of light, n_o and n_x the real parts of O- and X-modes refractive indices and C_0 and C_1 are constants. Several studies, from Warwick and Dulk (1964) to Ladreiter *et al.* (1995, and references therein) consistently showed that no more than 10% of the observed Faraday rotation originates from the Jovian magnetosphere.

Investigations analysed the amount of Faraday rotation assuming that the different emission frequencies f are emitted from different regions within the Jovian magnetosphere. Two variants of equation (1) have been analysed, the parabolic approximation depending on f^{-1} and f^{-2} (Shaposhnikov *et al.*, 1999) and the quadratic approximation f^{-2} and f^{-4} (Boudjada and Lecacheux, 1991). The term in f^{-2} is related to the terrestrial ionospheric contribution, whereas terms in f^{-1} and f^{-4} are associated to the plasma inhomogeneities along the ray path in the Io plasma torus. Although the amount of Faraday rotation Ω of non-terrestrial origin is small it may be possible under favorable conditions to monitor in a sort of ground-based remote sensing the latitudinal inhomogeneity of the Io plasma torus, as originally indicated by Dulk *et al.* (1992).

Another most promising approach in Jovian radio emission investigation is the stereoscopic observation by at least two spacecraft, as performed in a recent study by Kaiser *et al.* (2000). The Cassini Radio and Plasma Wave instrument (RPWS) and the Wind/WAVES experiment were able to simultaneously observe Jupiter's decametric (DAM) and hectometric (HOM) radio emissions, leading to the determination of the instantaneous widths of the walls of the hollow conical radiation beams, approximately $1.5° \pm 0.5°$. Future studies should emphasize simultaneously observed common Jupiter HOM features under particular geometric configurations, performed by the Galileo and Wind spacecraft, which could lead to an estimate on specific radiation patterns and their change by propagation effects.

2. New Techniques of Observations

Jupiter decametric radio emissions have recently been observed using the newly developed digital spectropolarimeter DSP (France-Austria) (Kleewein, 1997) and

TABLE I

Quantitative comparison between the digital spectropolarimeter and the waveform receiver

Criteria	*Waveform receiver*	*Digital Spectropolarimeter*
data product	time series data	dynamic spectrum
time resolution	20 ns	1 ms
bandwidth	25 MHz	12.5 MHz
dynamic range	70 dB	70 dB
data loss (sensitivity)	0 %	up to 80 %
data transfer rate	up to 100 MB per sec	up to 4 MB per sec

the waveform receiver (Austria) (Leitner, 1999). Contrary to the realtime spectral analysis the waveform receiver is able to completely store the output signal of the antenna system without any data manipulation. The realisation of this technical challenge was possible due to recent developments concerning high speed data transfer. The waveform of an observed radioburst event can be converted into the common dynamic spectral representation, with free choice of time and frequency resolution. Furthermore alternative data transformation methods like the Wavelet transformation can be applied in order to yield further insight into the physical properties (e.g. very localized and short-periodic behavior of the signal) involved in radio emission processes. Table I compares specific parameters of both the digital spectropolarimeter and the waveform receiver. This newly developed instrument with a bandwidth of 25 MHz enables the recording of a five seconds snapshot of the true received waveform. Following the pioneering work on Jovian S-bursts waveform analysis by Carr and Reyes (1999), we also try to reach the microsecond domain, but with the advantage of broadband observations at the world's largest decametric antenna array at Kharkov (Ukraine).

References

Boudjada, M.Y. and Lecacheux, A.: 1991, Faraday rotation of Jupiter's decametric radiation, *Astron. Astrophys.* **247**, 235.

Budden, K.G.: 1985, The Propagation of Radio Waves in the Ionosphere, *Univ. Press, Cambridge.*

Carr, T.D. and Reyes, F.: 1999, Microstructure of Jovian decametric S bursts, *J. Geophys. Res.*, 25.127.

Dulk, G.A., Lecacheux, A. and Leblanc, Y.: 1992, The complete polarization state of Io-related radio storms from Jupiter: a statistical study, *Astron. Astrophys.* **253**, 292.

Kaiser, M.L., Zarka, P., Kurth, W.S., Hospodarsky, G.B. and Gurnett, D.A.: 2000, Cassini and Wind stereoscopic observations of Jovian nonthermal radio emissions: Measurements of beam widths, *J. Geophys. Res.* **105**, 16053.

Kleewein, P.: 1997, Spectral Analysis of Magnetospheric Radio Emissions, Doctoral Thesis, University of Graz, Austria.

Ladreiter, H.P., Litvinenko, G., Boudjada, M.Y. and Rucker, H.O.: 1995, Faraday rotation in Jupiter's decametric radio emissions used for remote sensing of the terrestrial ionosphere and the emission's source region at Jupiter, *Planet. Space Sci.* **43**, 1595.

Lecacheux, A., Boudjada, M.Y., Rucker, H.O., Bougeret, J.L., Manning, R. and Kaiser, M.L.: 1998, Jovian decametric emissions observed by the Wind/WAVES radioastronomy experiment. *Astron. Astrophys.* **329**, 776.

Leitner, M.: 1999, Waveform Analysis of Radio Emissions in the Decametric Range, Diploma Thesis, University of Graz, Austria.

Queinnec, J. and Zarka, P.: 1998, Io-controlled decametric arcs and Io-Jupiter interaction, *J. Geophys. Res.* **103**, 26.649.

Shaposhnikov, V.E., Boudjada, M.Y., Rucker, H.O., Zaitsev, V.V. and Aubier, M.: 1999, On features of Faraday rotation of the decametric emission in the Jovian magnetosphere, *Astron. Astrophys.* **344**, 709.

Warwick, J.W. and Dulk, G.A.: 1964, Faraday rotation on decametric radio emissions from Jupiter, *Science* **145**, 380.

RADIO OBSERVATIONS OF FILAMENTS AT METRIC AND DECIMETRIC WAVELENGTHS

CH. MARQUÉ, P. LANTOS and J.M. DELOUIS
DASOP–LPSH, Observatoire de Meudon, 92195 Meudon Cedex, France
C.E ALISSANDRAKIS
University of Ioannina, 45110 Ioannina, Greece

Abstract. A filament eruption at decimetric wavelength is illustrated here, involving a quiescent filament seen in absorption. A CME occurs in the vicinity of the event.

1. Introduction

Radio investigations of filaments have been performed at millimetric, and centimetric wavelengths since the beginning of the sixties. Both quiescent and eruptive filaments are seen, essentially in absorption, in this range. At frequencies lower than $\sim$ 15 GHz, radio waves essentially probe the Prominence Corona Transition Region (PCTR) (see Chiuderi Drago, 1990 for a review). Very few observations of filaments were reported in the past at metric and decametric wavelengths; Dulk and Sheridan (1974) described elongated brightness enhancements associated to quiescent filaments, while at 327 MHz (decimetric range) Lang and Willson (1989) saw an emission structure overlaying a quiescent filament. The first example of a quiescent filament seen in depression at decimetric wavelengths was performed with the Nançay Radioheliograph (Delouis, 1999), and other cases were discovered (Marqué et al., 1999) suggesting that it is a common quiet sun feature at these wavelengths. We present here a convincing example of an eruptive filament seen in depression in the decimetric and metric range.

2. The Event

The event described here is very faint; it involves a meridian crossing quiescent filament near an active region. The eruption is seen with SOHO/EIT at 195 Å beginning around 13:00 UT, as a faint outgoing dark feature. A weak post-flare loops system appears after 16:00 UT. The radio event occurs after 14:00 UT, when the filament erupts. The EUV filament is embeded in the radio depression. Since the motion in EUV is slow and less than the beam size of the radioheliograph, a different departure time exists between the radio and EIT. A slow CME ($\sim$ 280

Astrophysics and Space Science **277:** 329–330, 2001.

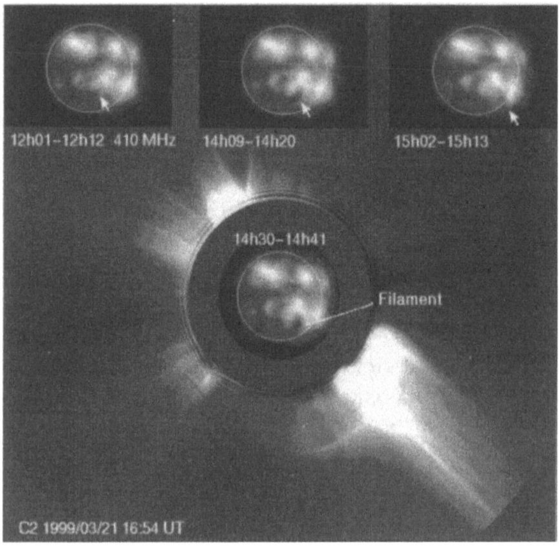

Figure 1. Snapshots of radio eruption at 410 MHz and comparison with the shape of the LASCO-C2 CME. Radio images are synthesis maps integrated over 10 mn for short scales, and the entire day for the sun background.

km/s) occurs in LASCO/C2 image (see figure 1) in the vicinity of the radio event. No significant radio burst is detected. Further investigations will be performed to explain why the filament is seen in absorption (at 410, 327, 236 and 164 MHz), and why it remains in this state during all the eruption.

Acknowledgement

The LASCO image was kindly provided by the SOHO/LASCO consortium.

References

Chiuderi Drago, F.: 1990, IAU Colloq. 117: Dynamics of Quiescent Prominences, 70.
Delouis, J.M.: 1999, PhD Thesis of University Paris VII. *Solar Phys.* **36**, 191.
Dulk, G.A. and Sheridan, K.V.: 1974, *Solar Phys.* **36**, 191.
Lang, K.R. and Willson, R.F.: 1989, *ApJ* **344**, L73.
Marqué, Ch., Lantos, P., Delouis, J.M. and Alissandrakis, C.E.: 1999, *Proc. 8th SOHO Workshop*, ESA SP-446, 483.

DECAMETRIC SOUNDING OF NEAR EARTH PLASMA: CORRELATION AND FRACTAL ANALYSIS OF SCINTILLATION DATA

NATALIA MURAVIEVA and YURY TOKAREV
Radiophysical Research Institute, B. Pecherskaya 25, N. Novgorod 603600, Russia;
E-mail: yt@nirfi.sci-nnov.ru; http://www.nirfi.sci-nnov.ru/plasma

Abstract. Solar wind observations by means of bi-static decametric radar SURA-WIND allow studying a fine structure of solar wind plasma irregularities with resolution up to 40 km. Spectral analysis and Hurst exponents method were applied to analyze the scintillation of radio signal transmitted through turbulent layers of Ionosphere and Solar wind plasma.

1. Observation Technique

In our experiments (Tokarev, 2000) solar wind turbulence was investigated by radio scintillation technique. This method is based on sounding media by monochromatic signal and is broadly used for investigations of nonuniform structure of interplanetary and sun surrounding plasma. The radar we used for observations includes ground based SURA transmitter (Karashtin, 1999) and RAD2 WAVES receiver of WIND spacecraft (Bougeret *et al.*, 1995). Radio signal propagating through the turbulent layer of plasma turbulence is scattered by electron density irregularities. Scattering efficiency grows with diminution of operating frequency. The SURA-WIND transmissions were performed at 5–10 MHz frequency range, which is close to lowest ones for transionospheric traces. Spectral and fractal analysis of time series of received signal allows us to obtain information about characteristic scales, shape of spatial spectrum of solar wind turbulence, scintillation index, power index, Hurst exponent, etc.

2. Experimental Results

We analyzed the data of more than 30 sounding sessions at 9 MHz frequency (Tokarev, 2000). Almost every sample of scintillation spectra calculated for SURA signals has a high frequency singularity located near 0.46 Hz, that might be associated with solar wind scales about 50. High frequency spectral peculiarity exists all the time when spacecraft stays outside of bow shock area and disappears after spacecraft crosses magnetopause toward Earth, see Figure 1. Thus bump on the upper curve reflects radio waves scattering by solar wind clouds. Low frequency part of the plot is associated with influence of ionosphere turbulence which is

Astrophysics and Space Science **277:** 331–334, 2001.

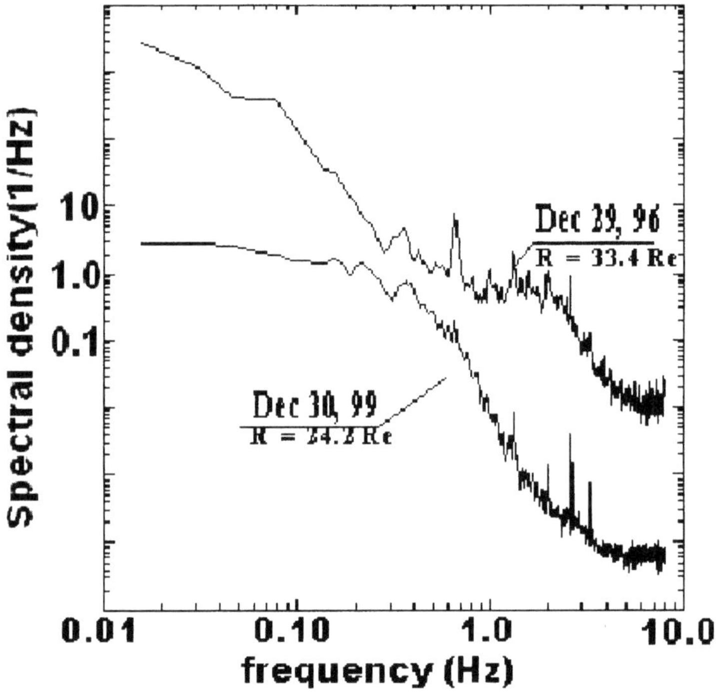

Figure 1. Power spectra of scintillation of 9 MHz signal transmitted through ionosphere (Dec 30, 99) and through both ionosphere and Earth bow shock (Dec 29, 96)

known to have a power law spectra with power law index about $\beta \approx 3$. High frequency plateau results from sampling noises.

Similarity of time records on various time scales, see Figure 2, suggests that the data be of fractal nature and can be analyzed with Hurst method. Calculated Hurst curves (Feder, 1989) are shown in the Figure 3. Data, obtained while spacecraft was located inside magneto-pause, shows persistent character with Hurst exponent $H \approx 0.7$–0.9 , that is in an agreement with value obtained by Hurst for natural phenomena. Data becomes antipersistent with $H \approx 0.4$–0.5 when spacecraft is outside bow shock and radiosignal propagates through both ionosphere and bow shock scattering layers. The value of Hurst exponent varies with averaging time and the scale when behavior of data changes can be estimated as 1000 km. According Gouyet (Gouyet, 1996) spectral index and fractal dimension can be estimated for this process as: $\beta = 2H + 1 \approx 2.5$, $D = 2 - H \approx 1.3$

3. Conclusions

In the experiment with bistatic decameter SURA–WIND radar a double screen scattering by ionosphere and solar wind turbulence has been observed. Rescaled

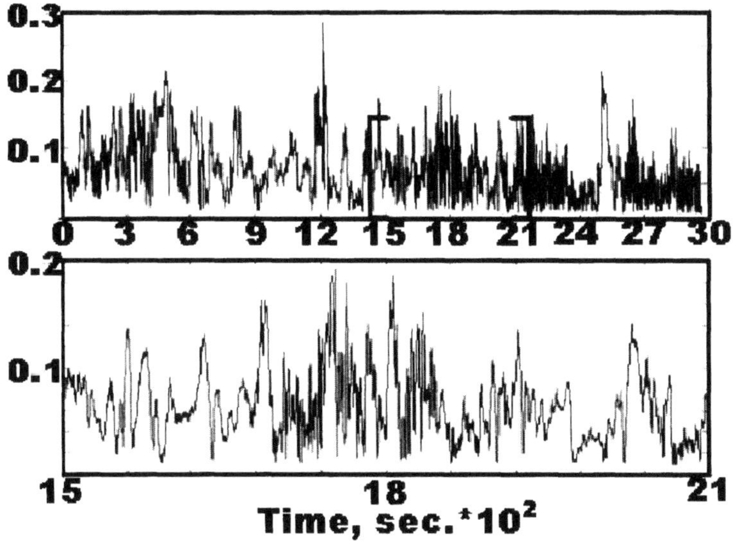

Figure 2. Zoom of time record of signal. January 18, 1999.

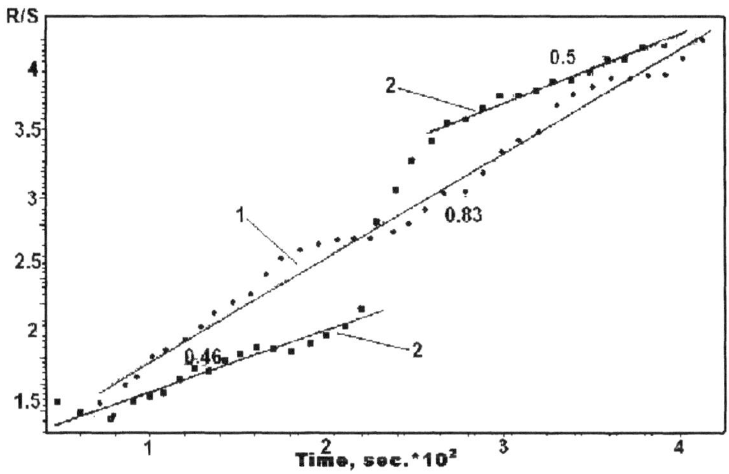

Figure 3. Calculated Hurst curves with linear approximation. 1–spacecraft is inside bow shock H = 0.83; 2–spacecraft is outside bow shock H = 0.46, 0.5.

range analysis shows that scintillation of radio signal transmitted through turbulent media is of fractal nature. For ionospheric layer fractal dimension of scattered signal can be estimated as $D \approx 1.3$ and scintillation spectral index is found as $\beta \approx 2.5$, that is in good agreement with results of spectral analysis. Hurst exponent calculation shows that the influence of solar wind may be recognized through antipersistent behavior of signal scattered by Earth bow shock turbulence. We intend to continue consideration of results of SURA–WIND experiments using this technique of statistical analysis and data processing.

Acknowledgements

This work was supported by INTAS (grant N 97-1964), INTAS – CNES (grant N 97-1450) and RFBR (grant N 99-02-17285).

References

Bougeret, J.-L., Kaiser, M.L., Kellogg, P.J., Manning, R. *et al.*: 1995, WAVES: The radio and plasma wave investigation on the WIND spacecraft, *Space Science Review* **71**, 231.

Feder, J.: 1989, *Fractals*, Masson, Paris.

Gouyet, J.F.: 1996, *Physics and Fractal Structures*, Masson, Paris.

Karashtin, A.N., Komrakov, G.P., Tokarev, Yu.V. and Shlugaev, Yu.V.: 1999, Radar investigation with SURA facility, *Quantum Radiophysics and Electronics* **XLII**(8), 765.

Tokarev, Yu.V., Kaiser, M.L., Belov, Yu.I., Boiko, G.N. and Murav'eva, N.V.: 2000, Small scale turbulence of solar wind around the Earth bow shock, *Astronomical Messenger* **34**(2), 1.

VI

FUTURES

FUTURS

THE HELIOSPHERE AFTER ULYSSES

R.G. MARSDEN
Space Science Department of ESA, ESTEC, Noordwijk, NL

Abstract. Launched in October 1990, the ESA-NASA Ulysses mission has conducted the very first survey of the heliosphere within 5 AU of the Sun over the full range of heliolatitudes. With polar passes taking place in 1994 and 1995, the timing of the mission has enabled Ulysses to characterise the global structure of the heliosphere at solar minimum, when the corona adopts its simplest configuration. The most important findings to date include a confirmation of the uniform nature of the high-speed (~750 km/s) solar wind flow from the polar coronal holes, filling two-thirds of the volume of the inner heliosphere; the sharp boundary, existing from the chromosphere through the corona, between fast and slow solar wind streams; the latitude independence of the radial component of the heliospheric magnetic field; the lower-than-expected latitude gradient of galactic and anomalous cosmic rays; the continued existence of recurrent increases in the flux of low-energy ions and electrons up to the highest latitudes. Without doubt, the Ulysses mission has provided a unique set of observations of the heliosphere at solar minimum, resulting in a good understanding of many aspects of its behaviour. In this review, we will highlight some of the key findings to date, and also look ahead to the challenges that await as Ulysses returns to high latitudes to explore the heliosphere at solar maximum and beyond. Finally, a brief summary is given of the prospects for heliospheric research in the post-Ulysses era.

1. Introduction

It is a great pleasure to deliver a lecture on the Ulysses mission at this meeting held in honour of Prof. Jean-Louis Steinberg. It is particularly appropriate, because Prof. Steinberg was one of the early champions of the Out of Ecliptic Mission (as Ulysses was then known). In the Spring of 1974, ESRO and NASA established a Joint Mission Definition Study Group to 'investigate the scientific significance and technical feasibility' of a combined Out-of-Ecliptic/Solar Stereoscopic mission. The three European scientists who participated were Jean-Louis Steinberg, W. Ian Axford, and D. Edgar Page, together with Gene Parker, Bob Leighton, John Simpson, and Len Fisk. Prof. Steinberg was also one of the Scientific Consultants for the Phase A Study held between September 1975 and April 1976, playing a critical role in obtaining project approval in Europe. Given the extraordinary success of the mission, it is clear that the efforts of Prof. Steinberg and his colleagues were not in vain.

Launched by the space shuttle Discovery in October 1990, Ulysses' primary objective was to characterize, for the first time, the properties of the Sun's environment at all latitudes from the equator to the poles, and at distances ranging from 1 to 5 astronomical units. The scientific investigations address a wide range of

Astrophysics and Space Science **277:** 337–347, 2001.

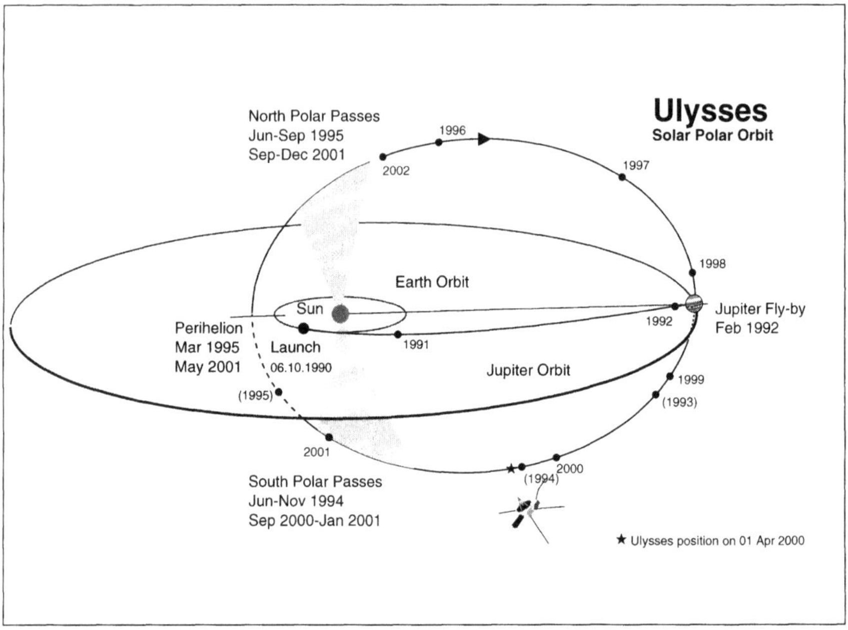

Figure 1. The Ulysses out-of-ecliptic orbit viewed from 15° above the ecliptic plane. Dots mark the start of each year.

heliospheric phenomena, including the solar wind, the heliospheric magnetic field, energetic particles and cosmic rays, natural radio emissions, interstellar gas and dust, and gamma-ray bursts (Wenzel *et al.*, 1992).

A gravity-assist manoeuvre at Jupiter was used to place the spacecraft into its unique, out-of-ecliptic orbit. This orbit is such that Ulysses spends a total of 234 days above 70° solar latitude (typically 132 days at high southern latitudes, and 102 days in the north) and achieves a maximum heliographic latitude of 80.2°. Since there was no compelling scientific argument to explore one polar region before the other, the position of Jupiter (slightly south of the solar equator at the time of the flyby) was exploited to optimize the flight path by sending Ulysses south. Figure 1 shows a perspective view of the trajectory as seen from 15° above the ecliptic plane. The first high-latitude passes in 1994 and 1995 took place under near-quiet solar conditions. Phasing the mission this way was the preferred option scientifically, although, as a result of several lengthy delays in the course of the project, this ultimately occurred by good luck rather than design.

In the ten years since launch, Ulysses has provided an unprecedented perspective of the heliosphere around solar minimum. The observations obtained so far have resolved a broad range of questions in the space sciences, due in large part to the unique orbit of the spacecraft. They have also raised questions unanticipated from our previous knowledge of the heliosphere, and provided a firm base on which to continue our exploration of the heliosphere. The purpose of this article is firstly to highlight the major findings of Ulysses' 'solar minimum mission', and secondly to look ahead to the equally exciting opportunities offered by the current 'solar maximum mission'.

2. Scientific Highlights

The scientific accomplishments of Ulysses to date have been reported in more than 850 publications covering a wide range of solar, heliospheric and astrophysical phenomena (e.g., Marsden, 1995; Marsden *et al.*, 1996; Marsden and Smith, 1997, and references therein). These accomplishments include the following:

- The characterisation of two distinct solar wind states, a fast high-latitude wind that only occasionally extends down to low latitudes, and a slow low-latitude wind centred about the heliospheric current sheet. These are separated by a sharp boundary extending from the corona down to the chromosphere (e.g., Geiss *et al.*, 1995a).
- The discovery that the magnitude of the radial component of the heliospheric magnetic field does not increase towards the poles (Smith and Balogh, 1995). The constancy of the radial field implies that the dipole-like configuration of the Sun's surface field is not maintained, and that as a result, the polar solar wind undergoes significant non-radial expansion.
- The discovery that corotating solar wind stream structures with forward and reverse shock waves, well-studied at low latitudes (e.g., Gosling *et al.*, 1993; Gosling and Pizzo, 1999) and expected to be confined to those regions, produce effects extending to the highest latitudes explored by Ulysses. These effects include the recurrent modulation of galactic cosmic rays (Simpson, 1998) and injection of accelerated lower-energy particles into the polar regions (Sanderson *et al.*, 1994; Simnett *et al.*, 1994), suggesting a revised global structure for the heliospheric magnetic field, or enhanced cross-field diffusion (Kota and Jokipii, 1995).
- The discovery that the influx of cosmic rays at high latitudes is smaller than predicted for this phase of the solar activity cycle (Simpson *et al.*, 1995).
- The determination of the flux and flow direction of interstellar dust grains passing through the solar system (Gruen *et al.*, 1993).
- The measurement of the flow parameters of interstellar helium, leading to an improved description of the motion of the solar system through the local interstellar cloud (Witte *et al.*, 1996).
- The derivation of the density of interstellar atomic hydrogen and helium, leading to improved knowledge of the interaction of the local interstellar cloud with the heliosphere (Gloeckler, 1996).
- The first-ever measurement of the interstellar ^{3}He/^{4}He ratio (Gloeckler and Geiss, 1996), the value of which suggests that the amount of dark matter produced in the Big Bang was greater than previously thought.

- The measurement of individual isotopes of cosmic ray nuclei, showing a source composition that is generally consistent with solar system matter. This suggests that cosmic rays are accelerated in the interstellar medium rather than being the products of explosive nucleosynthesis (Connell and Simpson, 1997).
- The determination of the positions of gamma-ray bursts with unprecedented accuracy (Hurley, *et al.*, 1995), including a contribution to the first plausible identification of an optical counterpart.
- The discovery of a new class of forward-reverse shock pairs associated with coronal mass ejections (CMEs) at high latitudes (Gosling *et al.*, 1994). Overexpansion of the CME caused by internal pressure is the source of these shocks.

In the following sections, we will describe a number of these findings in more detail.

2.1. Global Heliospheric Structure

One of the main goals of the Ulysses mission was to determine the global structure of the heliosphere, in particular with regard to the distribution of solar wind plasma, and its frozen-in magnetic field. The latitude survey carried out by Ulysses, the first ever, has resulted in the following picture of the heliosphere at solar minimum. The three-dimensional structure is characterised by a basic north-south symmetry, and is dominated by the presence of the fast solar wind from polar and high-latitude regions that expands to occupy a large fraction of the heliospheric volume; the slow wind is confined to low latitudes. Observations made by Ulysses during its rapid pole-to-pole transit near perihelion have revealed that the transition from slow to fast wind is surprisingly abrupt. This is graphically illustrated in Figure 2, which shows a polar plot of the solar wind speed and other parameters measured by Ulysses as a function of heliolatitude.

Note that nearly all other space missions have been confined to the narrow region of the heliosphere at low latitudes dominated by slow wind, and that Ulysses has provided the first direct, detailed view of the 'true' solar wind flowing from the polar coronal holes. This fast-flowing solar wind has been found by Ulysses to be relatively constant near solar minimum, with a speed of approximately 750 km/s. On the other hand, while clearly much less variable than the low-latitude slow wind, closer inspection has shown that even the fast wind is far from quiescent. New observations from the SOHO spacecraft present a picture of the solar atmosphere that, even at solar minimum, is highly dynamic, and this is clearly reflected in the Ulysses solar wind data.

The Ulysses magnetic field measurements indicate that the radial field component, and consequently, the magnetic flux, are independent of latitude (Smith and Balogh, 1995). This has lead to the unanticipated conclusion that the magnetic field pressure controls the solar wind flow near the Sun, driving a non-radial expansion by more than a factor of 5 from the polar coronal holes. Another characteristic property of the fast high latitude wind that has been revealed by Ulysses is the continual presence of large amplitude transverse waves in the magnetic field (Balogh *et al.*, 1995). The wave amplitudes typically equal or exceed the magnitude of the background field and, as such, are an example of strong turbulence. The waves are observed over a wide range of periods, the longer period variations probably originating in the random-walk of field lines at the sun (Jokipii and Kota, 1989), which in turn is a signature of the non-quiescent solar surface. Although the spiral magnetic field structure predicted by the earliest models of the solar wind (Parker, 1958) is preserved to the highest latitudes, a significant departure from this so-called Parker field configuration has been discovered. A likely explanation for this is that the observed deviations are the result of the large amplitude waves referred to above.

The north-south symmetry discussed earlier is associated with the heliospheric current sheet (HCS) that separates oppositely directed magnetic fields in the two hemispheres and defines the Sun's magnetic equator. The corresponding magnetic axis passes through the Sun's polar caps, and it is the tilt of this axis relative to the Sun's rotation axis which causes alternating slow (low-latitude) and fast (high-latitude) solar wind streams to sweep over an observer in the ecliptic. Because of the radial outflow of the solar wind, the fast streams eventually run into the slower wind ahead, forming so-called 'corotating interaction regions' (CIRs) (Smith and Wolfe, 1976). These are regions of compressed solar wind plasma, which, as the name suggests, corotate with the Sun. At distances beyond 1 AU, CIRs are often bounded by forward and reverse shock waves (e.g., Gosling and Pizzo, 1999). As will become apparent, the investigation of CIRs themselves, and their influence on the energetic particles and cosmic rays that populate the heliosphere, forms a major theme in the scientific output of Ulysses.

2.2. Energetic Particles and Cosmic Rays

Prior to Ulysses, it was known that the shocks associated with CIRs are able to accelerate low-energy charged particles. Characteristic increases in particle intensity are frequently observed when a CIR sweeps past a spacecraft once per solar rotation. At much higher energies, the same CIR can act as a temporary shield against incoming cosmic rays, causing the observed intensity to decrease rather abruptly, then slowly recover. Surprisingly, Ulysses discovered that the recurrent effects in energetic particle and cosmic ray intensity extend to much higher latitudes than the CIRs themselves (Kunow *et al.*, 1999). Under the influence of the magnetic configuration of the Sun, the angle between the Sun's rotational and magnetic axes changes with the solar cycle. Near solar minimum, the magnetic and rotational axes are nearly aligned, so that the CIRs are restricted to relatively low heliographic latitudes (typically less than 30 degrees). As shown in Figure 2, the series of recurrent increases and decreases starting at low latitudes clearly extends to regions far beyond the latitude range of the CIRs or their associated shocks.

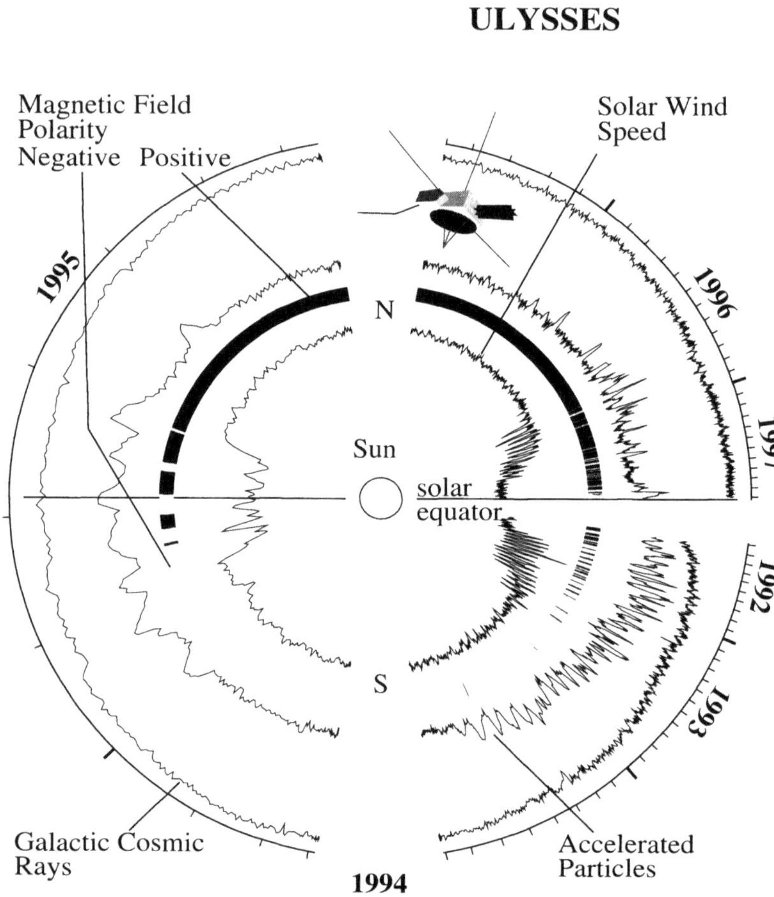

Figure 2. Polar plot of data acquired during the first solar orbit of Ulysses. From the centre outwards: solar wind speed; heliospheric magnetic field polarity (dark is positive, light is negative); intensity of 1.2-MeV protons (log scale); intensity of galactic cosmic ray protons (~1 GeV, log scale). All parameters are plotted as a function of spacecraft latitude. Also shown is the corresponding calendar date.

The Ulysses observations could be explained if the particles are able to move across the mean magnetic field to higher latitudes much more easily than previously thought. An alternative explanation is that the differential rotation of photospheric foot points of heliospheric magnetic field lines interacts with the non-radial near-Sun expansion of the solar wind to produce field lines that deviate drastically from the traditional constant-latitude spirals. The result is to bring field lines from high latitudes near the Sun to low latitudes at 15 AU or more from the Sun, thereby connecting Ulysses to the acceleration region of the energetic particles. This remarkably different magnetic topology is not detectable near the ecliptic plane; it could only be discovered by observations at high heliolatitudes (Fisk, 1996; Zurbuchen *et al.*, 1999). If confirmed by further work, this theory based exclusively on Ulysses data will revolutionize our understanding of the heliospheric magnetic field and cosmic ray transport.

A major goal of Ulysses was to investigate the physics of propagation of energetic particles in the heliosphere. In particular, the question was posed as to whether or not cosmic rays would have easier access to the inner heliosphere along relatively straight magnetic field lines expected over the poles of the Sun. Ulysses showed that the cosmic ray intensity increased by less than a factor of two from the equator to the poles, implying nearly equal difficulty of access (Simpson *et al.*, 1995). A possible explanation is found in the Ulysses discovery that large amplitude waves in the magnetic field, which obstruct cosmic ray propagation, are a characteristic feature of the high latitude fast solar wind.

2.3. Interstellar Gas and Pickup Ions

Ulysses has observed for the first time a variety of atoms entering the heliosphere from interstellar space. With a new technique to directly detect low energy (>30 eV) atomic helium, the local angular distribution of He was measured and, from these observations the velocity vector and kinetic temperature of the interstellar neutral helium at the boundary of the heliosphere have been determined with unprecedented precision (Witte *et al.*, 1993). The velocity vector describes the motion of the solar system through the surrounding Local Interstellar Cloud (LIC). Furthermore, from the neutral and pickup helium (see below) observed locally with Ulysses it was possible to infer for the first time the neutral helium density in the LIC (0.0155 particles cm^{-3}).

Ulysses discovered a vast heliospheric population of ions (so-called 'pickup' ions) produced from interstellar atoms and dust grains by photoionization and charge exchange with the solar wind. Measurements of these pickup ions (H^+, He^{++}, $^3He^+$, $^4He^+$, C^+, N^+, O^+ and Ne^+) by Ulysses have lead to important new results. For example, the abundance of atomic N, O and Ne in the local interstellar cloud was established for the first time (Geiss *et al.*, 1994). The discovery of C^+ (Geiss *et al.*, 1995b) revealed new sources of neutral particles in the heliosphere: 're-cycled' solar wind ions that have been absorbed by interplanetary dust grains and then re-emitted in the inner solar system.

2.4. Cosmic Dust

The dust detector on the Ulysses spacecraft is the first to directly detect dust at high ecliptic latitudes. The experiment has discovered a flux of interstellar grains passing through the solar system (Gruen *et al.*, 1993; Gruen *et al.*, 1994). The detections occur at the rate of about one detection every 3 or 4 days. Streams, or bursts, of dust were also discovered to be emanating from the Jovian system. An analysis of all Ulysses data has identified a total of 11 such streams, and showed that the dust grains are electrically charged and their trajectories are bent by the solar wind magnetic field.

2.5. Astrophysics

The study of the Sun and its environment by Ulysses has an obvious importance to stellar astrophysics, providing the only possible detailed analysis of the interaction of a typical star with its surroundings. Ulysses has also made important contributions to astrophysical studies that reach far beyond the heliosphere. For example, the gamma ray burst (GRB) experiment provides a distant point in space to obtain arc-minute positions for cosmic gamma-ray bursts and observe solar flare X-radiation stereoscopically. It has become the flagship in the 3rd Interplanetary Network of gamma-ray burst detectors. This network includes, or has included, Pioneer Venus Orbiter, the Russian GRANAT spacecraft, WIND, Yohkoh, the Compton Gamma-Ray Observatory, and the Italian SAX mission, among others. Important contributions from the Ulysses experiment include the detection of a giant periodic flare from a soft gamma-ray repeater (Hurley *et al.*, 1999).

Ulysses has also carried out cosmic ray studies with significance for astrophysics, providing isotopic abundance determinations of unequaled precision for many of the more abundant elements in the cosmic radiation. The measurements of the isotopes of iron and nickel have already eliminated some models for heavy element nucleosynthesis and for the origin of cosmic rays (Connell and Simpson, 1997). Isotopes of less abundant elements such as Cl and Ar, which have important implications for nucleosynthesis and propagation theory, have also been measured (Connell *et al.*, 2000).

The Ulysses measurements of the ratio of helium isotopes (^{3}He/^{4}He) in the local interstellar gas (Gloeckler and Geiss, 1996), the first of their kind, have made it possible to compare present-day light element abundances in the Local Interstellar Cloud with their values at the time of the formation of the Solar System. The abundances were found to have remained essentially unchanged, placing new constraints on models of Galactic chemical evolution. These data have also lead to important refinements in another fundamental cosmological parameter, providing information on conditions that obtained in the Big Bang.

3. Ulysses at Solar Maximum

The main goal of the Ulysses mission is to explore the heliosphere in the three spatial dimensions. Without doubt, this goal has been successfully achieved for the heliosphere close to solar minimum. In reality, however, there is also a fourth dimension to be considered: time. The heliosphere is a dynamic structure that changes quite dramatically over the period of a solar cycle (approximately 11 years). In order to understand fully the way the heliosphere evolves with time, measurements are also needed at solar maximum. Fortunately, the spacecraft and its scientific instruments were still in excellent condition at the end of the first

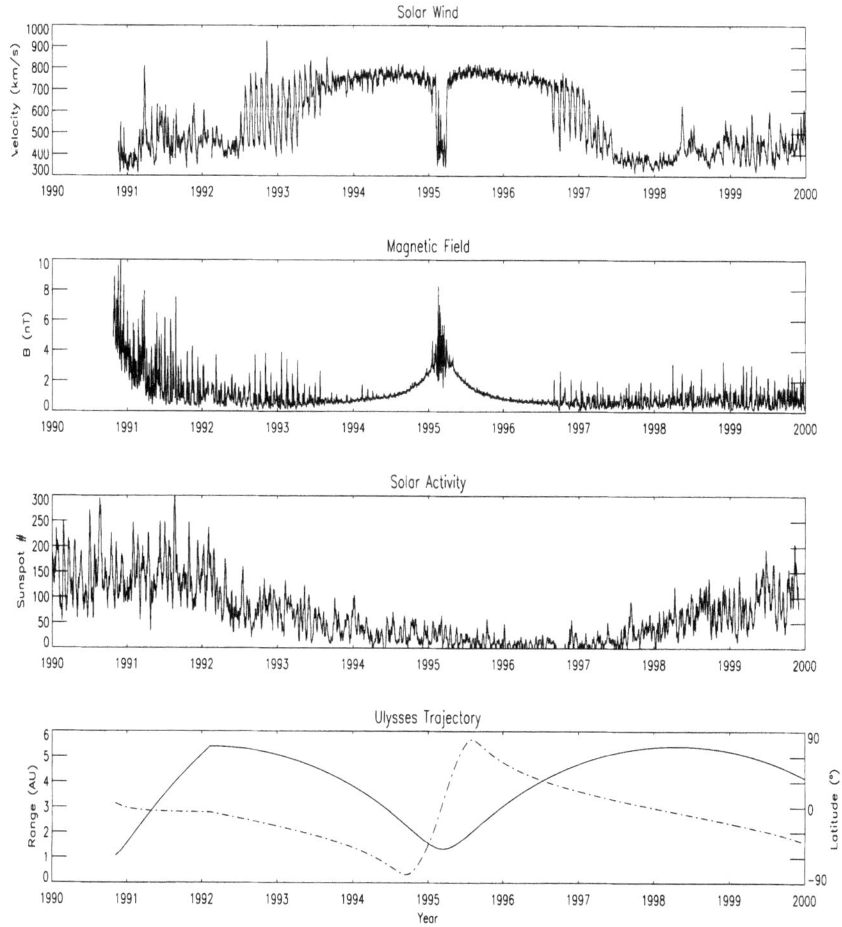

Figure 3. A mission overview of key parameters measured by Ulysses: (top panel) Solar wind speed (SWOOPS expt.); (second panel) magnetic field magnitude (VHM/FGM expt.); (third panel) sunspot number; (bottom panel) heliocentric range (solid line) and heliolatitude of Ulysses.

phase of the mission in October 1995. It was therefore decided to continue Ulysses' voyage of discovery for a second 6-year orbit around the Sun. A mission overview of solar wind speed and heliospheric magnetic field magnitude, plotted together with sunspot number and trajectory information, is shown in Figure 3.

Now, five years later, Ulysses is all set to visit the Sun's southern polar regions for a second time. The actual pass over the south pole will occur in November, with the northern polar pass one year later. The main goals for this phase of the mission will be to map the large-scale source regions of the slow and fast solar wind at solar maximum, and to observe the switch in polarity of the Sun's magnetic field. About once every 11 years, the Sun's polar caps change their polarity from positive to negative (or vice versa). These changes have an effect on the global structure of the heliosphere, and Ulysses will be ideally placed to investigate this. The additional observation time will also be beneficial to the experiments measuring rare cosmic ray isotopes and interstellar pickup ions, and for continued detection of gamma ray bursts.

Ulysses is not the only space mission currently studying the Sun. In fact, it is a member of an impressive 'fleet' of spacecraft that includes the ESA-NASA Solar and Heliospheric Observatory (SOHO), and NASA's ACE and TRACE. These will soon be joined by ESA's four Cluster II satellites. Together, these missions form an impressive tool to study the active Sun and its influence on our environment, in space and on the ground, helping us better understand 'space weather'.

What about future plans to study the Sun and the heliosphere? NASA intends to send a space probe very close to the Sun in 2007, and to put a pair of satellites into orbit around the Sun that will make stereoscopic images of the corona. ESA is also studying a Solar Orbiter that will go inside the orbit of Mercury to make high-resolution images of the solar surface. But for now, Ulysses, travelling on its unique out-of-ecliptic orbit, represents the only opportunity to explore the heliosphere from equator to poles, and over a complete cycle of solar activity. Its achievements to date have been remarkable, and there is every reason to expect that the findings at solar maximum and beyond will be equally exciting.

Acknowledgements

The author gratefully acknowledges the contributions of the Ulysses Principal Investigators and their teams (on whose collective work this review is based), and the JPL Project Scientist for Ulysses, Dr Edward J. Smith. The referee is thanked for suggesting improvements to the original manuscript. Ulysses is a programme of international co-operation between ESA and NASA.

References

Balogh, A., *et al.*: 1995, *Science* **268**, 1007.
Connell, J.J. and Simpson, J.A.: 1997, *Ap. J. Lett.* **475**, L61.
Connell, J.J. *et al.*: 2000, *Ap. J.* in press.
Fisk, L.A.: 1996, *J. Geophys. Res.* **101**, 15547.
Geiss, J. *et al.*: 1994, *Astron. Astrophys.* **282**, 924.
Geiss, J. *et al.*: 1995a, *Science* **268**, 1033.
Geiss, J. *et al.*: 1995b, *J. Geophys. Res.* **100**, 23373.
Gloeckler, G.: 1996, in: R. v. Steiger, R. Lallement and M.A. Lee (eds.), *The Heliosphere in the Local Interstellar Medium*, **335**.
Gloeckler, G. and Geiss, J.: 1996, *Nature* **381**, 210.
Gosling, J.T. *et al.*: 1993, *Geophys. Res. Lett.* **20**, 2789.
Gosling, J.T. *et al.*: 1994, *Geophys. Res. Lett.* **21**, 2271.
Gosling, J.T. and Pizzo, V.: 1999, *Space Sci. Rev.* **89**, 21.

Gruen, E. *et al.*: 1993, *Nature* **362**, 428.
Gruen, E. *et al.*: 1994, *Astron. Astrophys.* **286**, 915.
Hurley, K. *et al.*: 1995, *Astrophys. Space Sci.* **231**, 227.
Hurley, K. *et al.*: 1999, *Nature* **397**, 41.
Jokipii, J.R. and Kota, J.: 1989, *Geophys. Res. Lett.* **16**, 1.
Kota, J. and Jokipii, J.R.: 1995, 268, 1024.
Kunow, H. *et al.*: 1999, *Space Sci. Rev.* **89**, 221.
Marsden, R.G. (ed.): 1995, The High Latitude Heliosphere, *Space Sci. Rev.* **72**(1/2).
Marsden R.G. *et al.*: 1996, *Astron. Astrophys.* **316**, 279.
Marsden, R.G. and Smith, E.J.: 1997, *Adv. Space Res.* **19**, (6)825.
Parker, E.N.: 1958, *Ap. J.* **128**, 664.
Sanderson, T.R. *et al.*: 1994, *Geophys. Res. Lett.* **21**, 1113.
Simnett, G.M. *et al.*: 1994, *Geophys. Res. Lett.* **21**, 1561.
Simpson, J.A. *et al.*: 1995, *Science* **268**, 1019.
Simpson, J.A.: 1998, *Space Sci. Rev.* **83**, 7.
Smith, E.J. and Wolfe, J.H.: 1976, *J. Geophys. Res.* **3**, 137.
Smith, E.J. and Balogh, A.: 1995, *Geophys. Res. Lett.* **22**, 3317.
Wenzel, K.-P. *et al.*: 1992, *Astron. Astrophys. Suppl.* **92**(2), 207.
Witte, M. *et al.*: 1993, *Adv. Space Res.* **13**, (6)121.
Witte, M. *et al.*: 1996, in: R. v. Steiger, R. Lallement and M.A. Lee (eds.), *The Heliosphere in the Local Interstellar Medium*, **289**.
Zurbuchen, T.H. *et al.*: 1999, in: S.R. Habbal, R. Esser, J.V. Hollweg and P.A. Isenberg (eds.), *Solar Wind Nine*, AIP Conf. Proc. **471**, 87.

MAJOR UNSOLVED PROBLEMS IN SPACE PLASMA PHYSICS

MELVYN L. GOLDSTEIN *
NASA Goddard Space Flight Center

Abstract. There are many space plasma physics problems that are both *major* and *unsolved*, there are other problems for which the categorization of solved or unsolved depends on one's point of view, and there are still other problems that are well understood but unsolved in the sense that quantitative predictions cannot be made although the basic physics is known. The following discussion will, of necessity, be limited and selective. The nature of the Alfvénic turbulence in the solar wind remains a major unsolved mystery: Why is the power spectrum of this anisotropic, compressible, magnetofluid often Kolmogoroff-like, with a power spectral index close to the $-5/3$ value characteristic of normal fluids? What is the three-dimensional symmetry of the turbulence? Are the magnetic fields quasi-two-dimensional and stochastic, or have they been highly refracted by small velocity shears? What is the origin of the -1 slope of the energy containing scales? What is the relationship between the turbulent fields and the diffusion coefficients for energetic particle transport parallel and perpendicular to the ambient magnetic field? A general problem in turbulence research is the relationship between the fluid approximation and the kinetic physics that describes the dissipation and damping of fluctuations. There is still much to learn about solar flares, coronal mass ejections and magnetospheric substorms. Another major puzzle is how to quantitatively describe the interaction of the solar wind with the interstellar medium; a problem probably not amenable to solution using fluid equations.

1. Introduction: What Constitutes a 'Major Unsolved Problem'?

It is nearly impossible to provide a credible summary of the major unsolved problems in space plasma physics, not the least because most attendees at the Paris 2000 conference have their own opinion as to what is significant and unsolved. Below I try to address some issues surrounding this topic and hope to raise a few important issues.

1.1. Possible definitions of a 'major' (unsolved) problem

What is a *major* unsolved problem? One possibility is that it is what the person being addressed is working on. Sometimes, however, the definition is more focused; for example, a problem that is being proposed in a grant application is by definition both major and unsolved. For a graduate student, the thesis is a *major* problem. For the postdoctoral associate, when the goal it so obtain permanent employment, the next project must also be a *major* problem.

Astrophysics and Space Science **277:** 349–369, 2001.

1.2. Possible definitions of a (major) 'unsolved' problem

Once we have consensus on what constitutes a *major* problem, we must define what we mean by an *unsolved* problem. Perhaps this conundrum comes into clearest focus when research is submitted for publication. At that time, what is a *major* problem often depends on whether the person being asked is the author or the referee. Often an *unsolved* problem is one that an individual has worked on, but hasn't yet solved. A different point of view on how to approach the question of whether or not a problem has been solved was a comment reportedly made by a graduate student at a conference on magnetospheric substorms: 'If all researchers in this field over the age of 40 would retire, the substorm problem would be solved.'

1.3. What constitutes a 'solution'?

After agreeing on a list of major unsolved problems, it remains to determine when a problem is *solved*. (The list of major *solved* problems is impressive; see the accompanying article by Axford.) There are several issues to consider, and although some of these may appear obvious, they have sometimes been ignored. The first is the need to incorporate into theories constraints imposed by observations. This is often not trivial. For example, researchers expert in solar physics, but not particularly conversant with the solar wind literature have sometimes proposed theories of heating and acceleration of the solar wind plasma that are not viable when the implications of the theory are extended into the solar wind and compared with *in situ* data.

Space plasma physics is a messy discipline; the experiments and phenomena we conduct and observe are complex and difficult, if not impossible, to control. We are continually trying to make sense out of phenomena that have many degrees of freedom and for which the plasma parameters are continually changing. This complexity makes it difficult to discern what parameters and what variations in those parameters are the controlling influences. Occasionally, however, problems can be defined in such a way as to make possible definitive experimental tests. In those relatively rare cases, it becomes possible to test theories in the classic way, by trying to find counterexamples. Even in our messy discipline, there are theories that can be refuted by finding a single counterexample.

2. A Personal View

In this section, I discuss a few problems in space plasma physics that, in my opinion, are both *major* and *unsolved*. This personal list is not meant to be comprehensive; several of these examples are research areas I've worked in (see, Section 1) and therefore, have some familiarity with. I believe that these examples illustrate general areas of complexity and difficulty that apply to many other unsolved problems.

2.1. Interplanetary Turbulence

The solar wind, in addition to its intrinsic interest as an astrophysical phenomenon, is a wonderful laboratory for studying both kinetic plasma phenomena and the properties and puzzles of magnetofluid turbulence. Many plasma instabilities have been studied in detail in either the solar wind or the Earth's magnetosphere. For example, the identification and verification of the ion-cyclotron beam instability was first described in detail in the solar wind and later in the magnetosphere (Viñas *et al.*, 1984; Tsurutani *et al.*, 1985). One of the richest sources of plasma phenomena in the solar wind is found in the vicinity of comets, where the interaction of the solar wind with both the neutral and charged atoms and molecules excite a wonderful zoo of waves and instabilities. A good review of this phenomenology can be found in Tsurutani (1991).

In addition to kinetic phenomena, the macroscopic solar wind behaves in many respects as if it were a fluid containing fully-developed turbulence. Thus, in addition to its uniqueness as a laboratory for the study of kinetic physics, the solar wind is also a unique laboratory for the study of magnetohydrodynamic (MHD) turbulence.

That there is no generally accepted theory of MHD turbulence is not surprising given the fact that after more than a century of effort, there is no theory of fluid turbulence. Turbulence in a magnetized fluid is intrinsically more complex than is fluid turbulence, primarily because the magnetic field introduces additional degrees of freedom and additional symmetries (and invariants) not present in fluids. The asymmetries introduced by a d.c. magnetic field have no counterpart in fluid turbulence, because of the pseudo-vector nature of the magnetic field. Whereas, a d.c. velocity can be removed by a Galilean transformation, that is not true of a magnetic field. A magnetofluid also supports a greater variety of wave modes than does an ordinary fluid.

2.1.1. *Alfvén Waves*

The fundamental fluctuation whose evolution and properties describe many phenomena in magnetofluids is the Alfvén wave (Alfvén, 1950), defined by $\delta\mathbf{v} = \pm\delta\mathbf{b}$, where $\delta\mathbf{v}$ and $\delta\mathbf{b}$ are the fluctuating velocity and magnetic fields, respectively, and $\delta\mathbf{b}$ is normalized to Alfvén speed units, i.e., $\delta\mathbf{b} = \delta\mathbf{B}/\sqrt{4\pi\rho_0}$, where ρ_0 is the average plasma mass density and $\delta\mathbf{B}$ is the fluctuating magnetic field (in this case, in cgs units). The sign of the correlation indicates whether the waves are propagating parallel to $(-)$, or anti-parallel to $(+)$, the direction of the background magnetic field $\mathbf{B}_0$.

The earliest observations of plasma and magnetic fields in the solar wind revealed that fluctuations resembling Alfvén waves were common (Unti and Neugebauer, 1968; Coleman, 1966; Belcher and Davis, 1971). An example of interplanetary Alfvénic fluctuations is shown in Figure 1. This data interval was chosen nearly at random and consisted of two days of 8-minute-averaged magnetic field

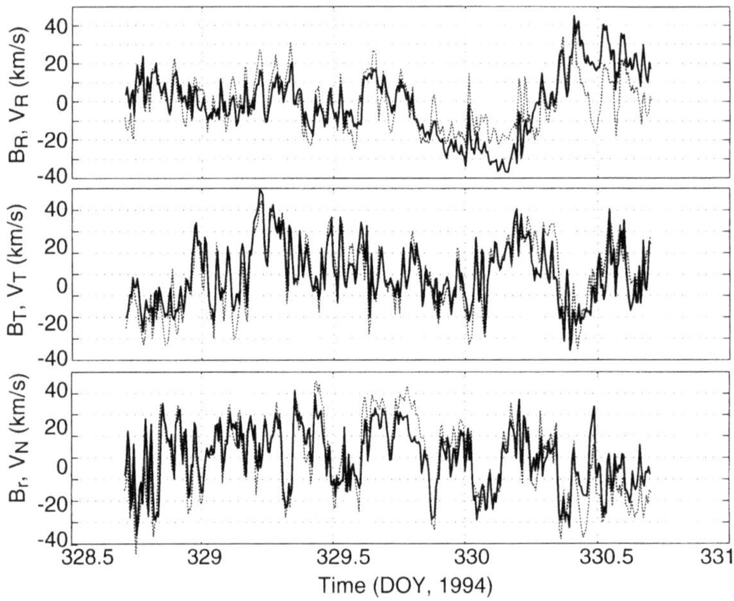

Figure 1. Ulysses data from December 1994. Plotted are the three vector components (in the standard *RTN* heliospheric coordinate system) of 8-minute averages of the fluctuating magnetic field (thin lines) and velocity field (thick lines) illustrating the presence of Alfvén waves.

data from the Ulysses spacecraft. At the time Ulysses was well above the heliospheric plasma sheet in fast solar wind. A high degree of correlation between $\delta\mathbf{b}$ and $\delta\mathbf{v}$ is evident, and the sign of the correlation indicates that the fluctuations are propagating outward from the Sun.

Pure Alfvén waves are exact solutions of the ideal (i.e., dissipationless) incompressible equations of MHD. Thus, to the extent that solar wind fluctuations are pure Alfvén waves and to the extent that the solar wind is an incompressible magnetofluid, nonlinear interactions should be suppressed and there should be no actively evolving turbulence. In such a situation, the outward propagation of the fluctuations would indicate that one was observing the remnants of turbulence from the corona. (At the super-Alfvénic point, only outward propagating fluctuations will be convected into the solar wind.) Observations similar to these raised the question of just how ubiquitous were Alfvén waves with such high degrees of correlation, and how close to being incompressible was the solar wind.

A convenient diagnostic, helpful in addressing the first of those questions, is the cross helicity, which is one of the global invariants of three-dimensional incompressible ideal MHD and is defined by $H_c = ½ \int dx^3 \delta\mathbf{v} \cdot \delta\mathbf{b}$. More useful than H_c itself is the dimensionless normalized cross helicity defined by $\sigma_c(k) \equiv$

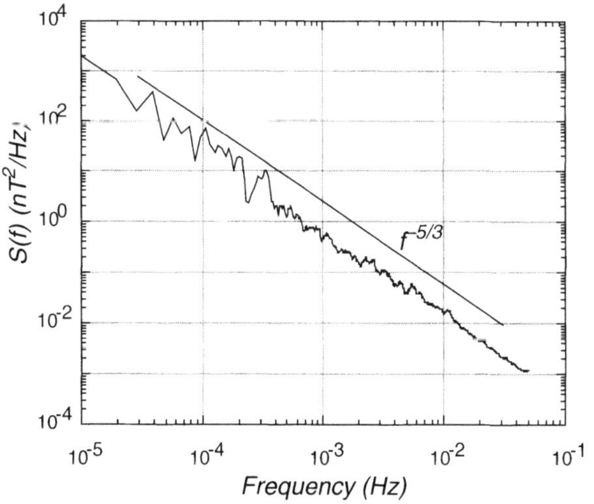

Figure 2. A power spectrum of Voyager 2 data from 10 AU. A line with slope of $-5/3$ is shown for comparison.

$2H_c^r(k)/E^r(k)$, where the numerator contains the 'reduced' cross helicity formed by integrating the Fourier transform of the power spectral tensor over the two directions orthogonal to the direction of the solar wind velocity. Similarly, $E^r(k)$ is the reduced spectral total energy. The reduced cross helicity is constructed from the reduced power spectral tensor (Matthaeus and Goldstein, 1982). The resulting one-dimensional spectrum is a function only of wave number $\mathbf{k}$ parallel to $\mathbf{V}_{\mathbf{sw}}$. In configuration space, an essentially equivalent expression for $\sigma_c(k)$ is

$$\sigma_c \equiv 2 < \delta\mathbf{b} \cdot \delta\mathbf{v} > / < \delta b^2 + \delta v^2 >, \tag{1}$$

where $<>$ indicates averaging over some time or length scale. Here we choose the sign of σ_c so that positive values indicate Alfvén propagation outward from the Sun.

2.1.2. *Characteristics of Solar Wind Turbulence*

The strongest indication that the Alfvénic fluctuations may be indicative of evolving turbulence comes from the power spectrum of both the magnetic and velocity fluctuations. In Figure 2 we show a power spectrum of magnetic field fluctuations from data collected by Voyager 2 when it was near 10 AU. The spectrum has a power-law shape with a spectral index of $-5/3$. As first noted by Coleman (1968), a $-5/3$ slope is characteristic of fluid turbulence (Kolmogorov, 1941); the expectation for magnetofluid turbulence is a slope of $-3/2$. Coleman (1968) argued that the $-5/3$ slope of the power spectra and strong velocity shears separating fast and slow

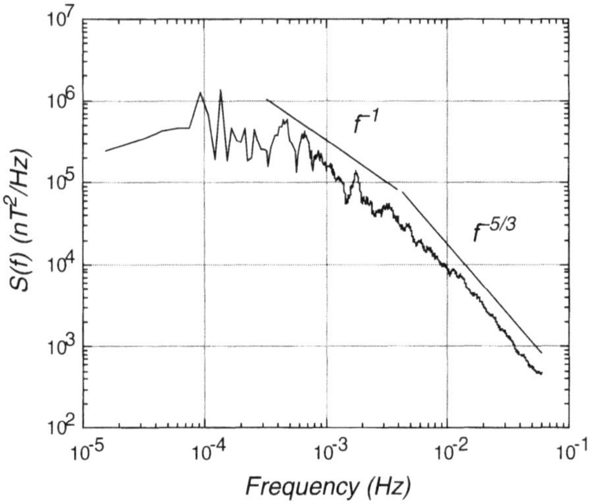

Figure 3. Trace of the power spectral matrix constructed from 8-s averaged Helios 2 data obtained near 0.3 AU. Lines with slopes of $-5/3$ and -1 are shown for comparison.

solar wind streams indicated that the solar wind was an actively evolving turbulent magnetofluid. Alternatively, Belcher and Davis (1971) argued that the very high values of σ_c indicated that the Alfvénic fluctuations were fossil remnants of coronal processes.

A resolution of these two viewpoints was offered by Roberts *et al.* (1992) (see also the review by Goldstein *et al.* (1995)) who argued that the nearly pure outward propagating Alfvénic fluctuations did indeed reflect coronal processes, but that these fluctuations were stirred by *in situ* velocity shears leading to an evolution of the spectrum with heliocentric distance. This stirring by velocity shears also produced inward-propagating fluctuations. Their arguments were bolstered by numerical solutions of the incompressible and compressible MHD equations together with analyses of data from the Helios, IMP, ISEE, and Voyager spacecraft.

At low frequencies (long wavelengths) the spectrum is flatter, with a spectral index of -1. This is most apparent in data obtained in the inner heliosphere, as shown in Figure 3 where a power spectrum computed from the Helios 2 magnetometer experiment is plotted. The data were obtained near 0.3 AU and the -1 portion of the spectrum may reflect the source spectrum in the solar atmosphere.

Many unsolved issues remain. Why typical spectra of magnetic (or velocity) fluctuations (Matthaeus and Goldstein, 1982) have spectral slopes of $-5/3$ is a mystery. The dimensional arguments for a $-5/3$ slope made by Kolmogoroff for ordinary fluids assume that the fluid is both incompressible and isotropic. While it could be argued that the solar wind is approximately incompressible, at least in

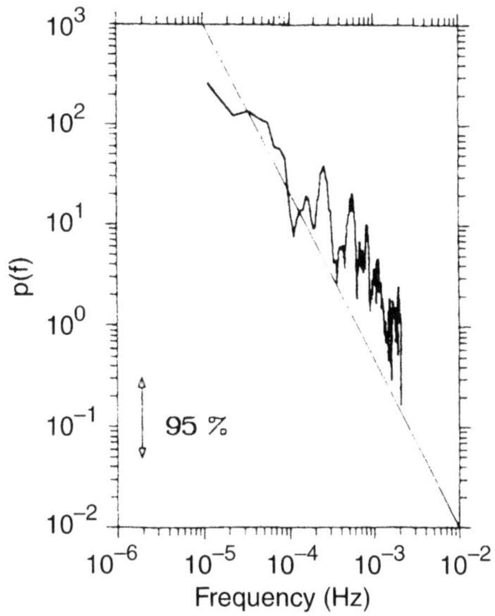

Figure 4. A power spectrum of density fluctuations Helios 2 obtained when the spacecraft was at 0.3 AU. A line with slope of $-5/3$ is shown for comparison. (Adapted from Marsch and Tu (1990).)

regions devoid of shock waves and co-rotating interaction regions, the presence of a relatively strong DC magnetic field indicates that isotropy is not a good assumption, and one might therefore expect that Kraichnan's prediction of a $-3/2$ for Alfvénic turbulence would be observed. Also puzzling is the fact that spectra of the density fluctuations also have spectral slopes of $-5/3$. In Figure 4 we show as an example, a power spectrum constructed from proton density data obtained from Helios 2 when the spacecraft was close to 0.3 AU. Spectra of fluctuations in the magnitude of the magnetic field, B, also typically have slopes of $-5/3$ (Montgomery *et al.*, 1987; Matthaeus and Brown, 1988; Matthaeus *et al.*, 1991). While spectra of the components of the magnetic field and/or velocity steepen at high wave numbers, reflecting a dissipation range that begins near the ion cyclotron scale, the density spectra and the spectra of the fluctuating power in B often show a flattening at high frequencies and that, too, has not been explained.

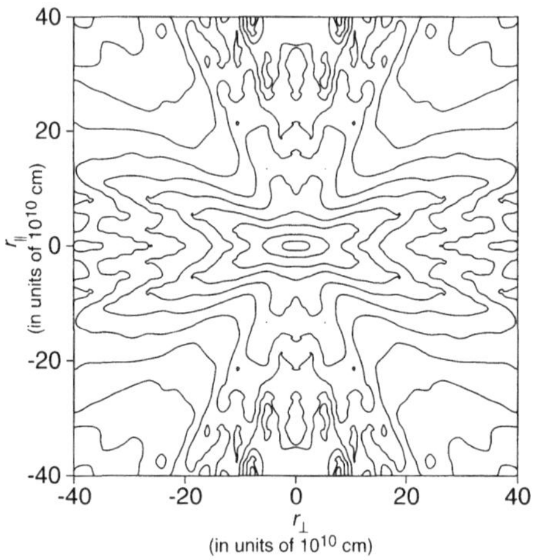

Figure 5. A two-dimensional correlation function constructed from nearly two years of magnetometer data obtained by ISEE-3. Pure parallel-propagating planar Alfvén waves would result in contours parallel to the $r_\perp$-axis while quasi-two-dimensional turbulence would result in contour lines parallel to the $r_\parallel$-axis. (Adapted from Matthaeus *et al.* (1990).)

2.2. Three-Dimensional Symmetries of Interplanetary Fluctuations

2.2.1. *The Correlation Function*

Most of the information we have about solar wind fluctuations comes from measurements of reduced one-dimensional power spectra. Although the concept of a reduced spectrum implies integration over the directions orthogonal to the flow velocity (Batchelor, 1970); in fact, the spectra we construct are by their very nature 'reduced' because the solar wind is both supersonic and superAlfvénic (Matthaeus and Goldstein, 1982). It is of fundamental importance to learn the nature of the three-dimensional symmetries of this turbulence because it is the three-dimensional properties that determine how both solar energetic particles and galactic cosmic rays propagate throughout the heliosphere; determining, e.g., whether or not solar energetic particles from flares and Coronal Mass Ejections will impact the Earth and how solar magnetic fields produce the 22-year modulation of galactic cosmic rays.

The simplest interpretation of the high Alfvénicity of solar wind fluctuations is that they are both planar and parallel propagating. Such waves pitch-angle scatter charged particles efficiently as they propagate along the background magnetic field, but allow for relatively little transverse diffusion across the magnetic field. The first indication that this simple point of view was inadequate was an analysis by Matthaeus *et al.* (1990) of nearly two years of magnetometer data from the ISEE-3

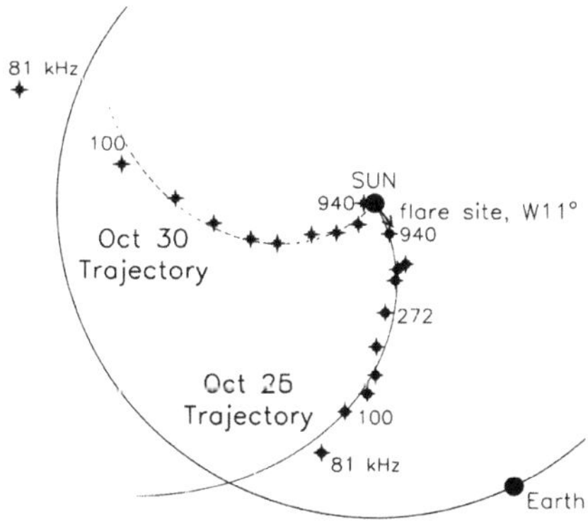

Figure 6. The signature of solar type III radio burst exciters as detected by the URAP experiment on Ulysses. (Adapted from Reiner *et al.* (1995).)

spacecraft which accumulated solar wind magnetic field data nearly continuously at the Earth's libration point.

By organizing the magnetometer data into statistically stationary subsets of what they assumed to be a single ensemble of the interplanetary magnetic field, Matthaeus et al. succeeded in constructing a two-dimensional correlation function of the interplanetary fluctuations (Figure 5). The correlation function revealed the existence of a second component, in addition to the expected Alfvénic fluctuations, that had the symmetry of quasi-two-dimensional structures. This second component could also result from either pressure-balanced structures or waves propagating highly obliquely to the background magnetic field.

Since that first attempt to determine the symmetry of interplanetary fluctuations, researchers have used a variety of methods in an attempt to extract two- and three-dimensional information from one-dimensional power spectra (Bieber *et al.*, 1996; Carbone *et al.*, 1995). Bieber and Matthaeus in particular have concluded that as much as 80% of the power in interplanetary consists of quasi-two-dimensional fluctuations. Whether the fluctuations are truly quasi-two-dimensional in the sense that $\delta\mathbf{k}$ and $\delta\mathbf{b}$ are both perpendicular to $\mathbf{B}$ (as opposed to, say, being highly refracted waves or pressure-balanced structures) has yet to be determined.

2.2.2. *Consequences for Charged Particle Propagation*

Planar, parallel-propagating Alfvén waves are efficient pitch-angle scatterers (Jokipii, 1971), but only inefficiently transport particles across field lines. In contrast, quasi-

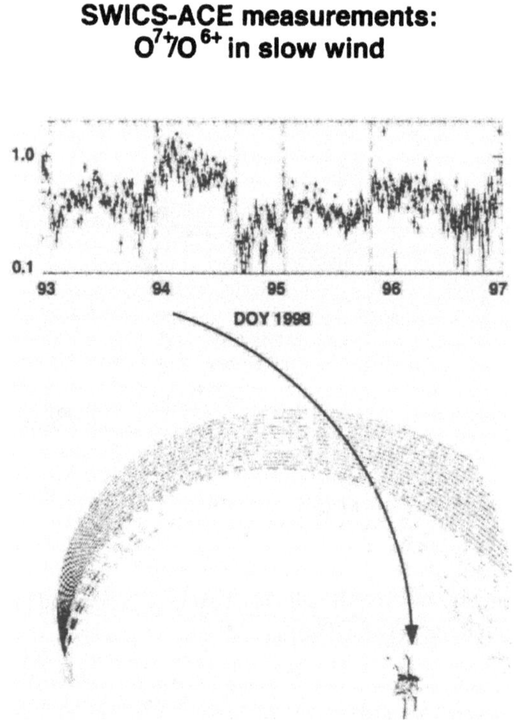

Figure 7. The sharp transitions observed in the charge states of ionized Oxygen during April 3–7, 1998 suggest that there is little random motion of the magnetic field. (Zurbuchen, private communication (2000).)

two-dimensional fluctuations do not resonantly scatter particles, however, as shown by Matthaeus *et al.* (1995), quasi-two-dimensional turbulence causes field lines to diffuse stochastically, which should lead to significant perpendicular diffusion.

To date there has been no direct observational confirmation that 80% of interplanetary magnetic fluctuations are quasi-two-dimensional, although considerable evidence suggests that the wave vectors of much of the fluctuating power lies in directions highly oblique to the background magnetic field. For example, the energetic electrons that are the exciters of solar type III radio bursts appear to move along the background mean magnetic field and show no indication stochastic diffusion. Figure 6 shows an example constructed from Ulysses radio data by Reiner *et al.* (1995). [The Ulysses Unified Radio And Plasma wave (URAP) experiment owes its existence, in very large part, to the leadership to Jean-Louis Steinberg.]

The observation by the Advanced Composition Explorer (ACE) of sharp transitions in the fluxes of O^{7+}/O^{6+} across magnetic flux tubes (Figure 7) also indicates that interplanetary field lines are not stochastic, at least on large scales. The observations suggest that solar wind ions remain attached to the magnetic field, making them tracers of field lines, especially in slow wind. The sharp transitions indicate little random motion of the field, at least across flux tubes (Zurbuchen *et al.*, 2000). Understanding the symmetries of the interplanetary turbulence and how and why the observed fluctuation spectra have Kolmogoroff-like power-laws remains a major challenge.

2.3. Subgridscale Modeling of the Dissipation Range

In addition to the general lack of understanding of how and why the inertial range of interplanetary turbulence forms and evolves, the dissipation range of the spectrum is poorly studied and poorly understood. The fluid description used to study inertial range turbulence breaks down as one approaches spatial scales comparable to the ion Larmor radius or ion skin depth and kinetic effects need to be included. Accurate descriptions of magnetic reconnections require, at least approximate, inclusion of election physics (Biskamp *et al.*, 1997; Shay *et al.*, 1998; Shay *et al.*, 1999). Even at larger scales, fluid descriptions become questionable in regions of space where the plasma becomes collisionless, such as in the outer heliosphere near the termination shock. Even closer in, but still beyond $\approx$ 5 AU, the addition of interstellar pickup ions to the ambient solar wind introduces an energetically important non-Maxwellian component that cannot be described well within a fluid paradigm.

2.3.1. *Hyperdissipation, Nonlinear Operators and Hall MHD*

There are several possible approaches as to how one addresses this situation. One could, of course, just wait for bigger and faster computers to enable kinetic simulations to be carried out over macroscopic spatial scales. More practical approaches currently being used include modifying the fluid equation using various forms of sub-grid-scale modeling, such as adding hyperdissipation and hyperresistivity (Biskamp, 1982; Biskamp and Welter, 1989; Passot and Pouquet, 1988) to the MHD equations (i.e., by including terms proportional to $\nu\nabla^n$ in addition to the classical $\nu\nabla$ term, where ν represents either viscosity or resistivity). A somewhat more physically motivated approach is to concentrate the dissipation in spatial regions of strong gradients, where, in fact, most turbulent dissipation occurs. This approach was pioneered by Passot and Pouquet (1987) and Passot and Pouquet (1988) for turbulence in ordinary fluids and generalized to MHD by Siregar *et al.* (1995).

Although neither hyperdissipation nor nonlinear operators incorporate kinetic phenomena, finite Larmor radius and Hall terms can be added to the equations and, because those terms are not intrinsically dissipative, they can be included while still retaining either the nonlinear or hyperdissipation terms (Huba, 1994; Huba, 1995; Huba, 1996a; Huba, 1996b; Ghosh *et al.*, 1996; Ghosh and Goldstein, 1997). As one moves below the ion cyclotron scale, one can use the two-fluid equations to include the electron fluid explicitly (Shay *et al.*, 1998; Biskamp *et al.*, 1995;

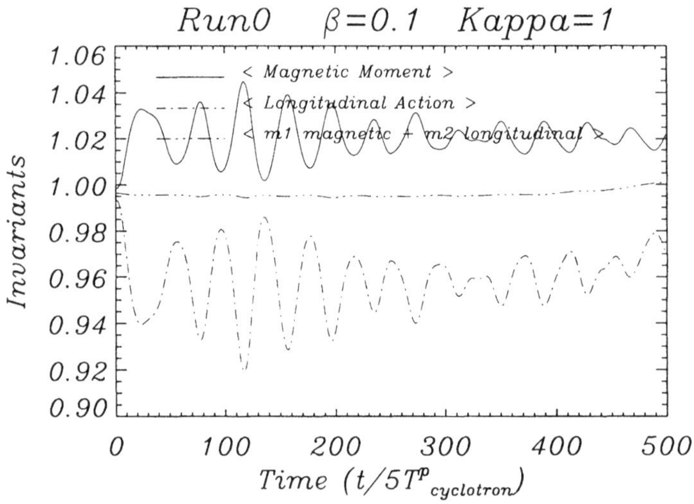

Figure 8. Illustration of the quasi-invariant that was derived from the coarse-grained fluid treatment of the resonant interaction of a spectrum of proton cyclotron waves in a low beta plasma. (From Siregar *et al.* (1998b).)

Biskamp *et al.*, 1995; Biskamp *et al.*, 1997; Birn *et al.*, 1998; Hesse and Winske, 1998; Hesse *et al.*, 1998).

A Coarse-Grained Fluid Approximation. Until kinetic simulations on global scales become practical, one can try to incorporate kinetic theory into modeling the dissipation range. A first step in this direction was taken by modelling the damping of parallel propagating proton cyclotron waves. Advances in this area have required parallel use of kinetic codes to guide the analytical developments (Siregar *et al.*, 1998a; Siregar *et al.*, 1998b). Even the simple case of parallel propagating waves, low plasma beta, and a single fluid treatment shows interesting new physics, *viz.* the presence of a quasi-invariant in a coarse-grained magnetofluid solution (Figure 8). Much still needs to be done in this area, but even the preliminary results suggest that application of linear theory to the damping of ion cyclotron waves in the corona should be reevaluated because the quasi-invariant indicates that ion cyclotron waves may not damp as rapidly as predicted by linear theory.

2.4. Heating and Accelerating the Solar Corona and Solar Wind

The temperature of the solar atmosphere rises sharply between the chromosphere, where it is thousands of degrees Kelvin, and the solar corona, where it exceeds a million degrees. Exactly how the corona is heated is not understood, but the heating mechanism determines both the speed and temperature of the solar wind, especially the fast wind above coronal holes. Fluid models of the solar atmosphere can account for neither the temperature nor the speed of fast solar wind without postulating additional heat sources (see, e.g., Parker (1958)).

2.4.1. *Heating by Alfvén Waves*

One of the most attractive scenarios for heating the corona involves damping Alfvén waves (Wentzel, 1974; Wentzel, 1976; Stein, 1981; Heyvaerts and Priest, 1983; Marsch *et al.*, 1982; Kuperus *et al.*, 1981; Golub *et al.*, 1980; Ofman and Davila, 1995). Alfvén waves are easily generated and are ubiquitous. Dissipating them can both heat the corona and accelerate the wind.

In spite of the attractiveness of the wave heating scenario, the theory has had difficulty because of the strong observational constraints imposed by interplanetary observations of Alfvén wave energy. In particular, extrapolations of Helios data by Roberts (1989) and of Ulysses data by Smith *et al.* (1995) indicate that the observed fluxes of interplanetary Alfvénic fluctuations are inadequate to account quantitatively for the observed temperature and speed of fast solar wind. One solution to this conundrum is to 'hide' the needed Alfvén waves in high frequency fluctuations that will undergo linear ion cyclotron damping as the atmosphere convects outward into the interplanetary medium (Marsch and Tu, 1997; McKenzie *et al.*, 1995; McKenzie *et al.*, 1997; Roberts, 1996). It is not known if the nonlinear analyses of ion cyclotron damping described above (Siregar *et al.*, 1998a; Siregar *et al.*, 1998b) will modify significantly the linear scenarios. Such waves could be generated by microflares at the boundaries of supergranules, as suggested from SOHO and TRACE observations. A related scenario has been proposed recently by Viñas *et al.* (2000), who suggest that kinetic Alfvén waves will undergo Landau damping, thereby heating the corona.

A Three-Dimensional Model with Wave Acceleration If one merely assumes that the needed Alfvén wave flux is present and that it can be damped in such a way as to avoid the limits set by the Helios and Ulysses observations, then it is possible to construct fluid models of the solar wind that agree quantitatively with many aspects of the Ulysses measurements of solar wind plasma parameters made as the spacecraft passed through the polar regions of the heliosphere. In Figure 9 we show one such example taken from Usmanov *et al.* (2000).

2.4.2. *Reconnection and Other Scenarios*

One alternative to the wave heating scenario is to have the corona heated by small reconnection events (Parker, 1983a; Parker, 1983b; Parker, 1983c; Ryutova *et al.*, 1997; Tarbell *et al.*, 1999). Although reconnection avoids the observational constraint noted above, it has other difficulties, not the least of which is the challenge of observing such highly localized small energy releases (Feffer *et al.*, 1997; Krucker *et al.*, 1999; Lin *et al.*, 1991) and the theoretical challenge of proving that the mechanism can, in fact, occur (van Ballegooijen, 1985; van Ballegooijen, 1986). Recently, Nordlund and Galsgaard (1997), Galsgaard and Nordlund (1997a), and

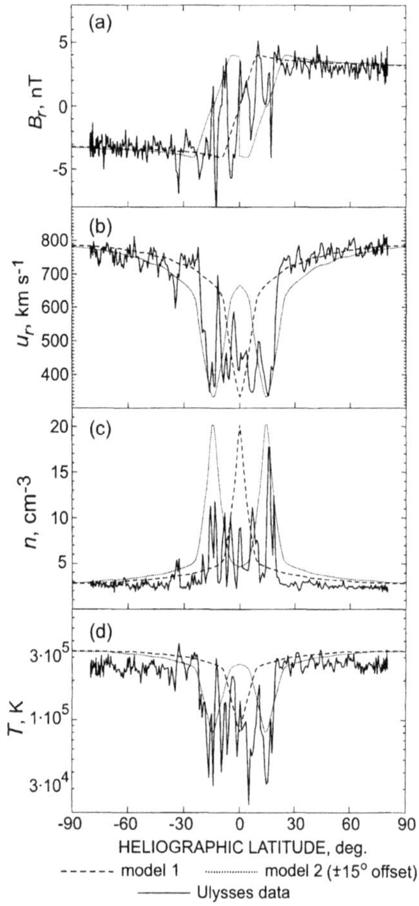

Figure 9. Daily averages of (a) radial magnetic field, (b) radial velocity, (c) number density, and (d) plasma temperature observed by Ulysses (solid lines). Plots are versus heliographic latitude and compared to latitude profiles computed in the present paper. The model curves are shown with dashed and dotted lines. The radial magnetic field, number density, and plasma temperature measured by Ulysses are normalized to 1 AU. (From Usmanov *et al.* (2000).)

Galsgaard and Nordlund (1997b), using high resolution three dimensional simulations, showed that 'topological' dissipation as proposed by Parker (1972) can occur in the corona. This process is probably confined to closed magnetic structures because it is difficult to envision how significant twist could buildup in open topologies.

Other mechanisms have been proposed that may also be important. The velocity filtration proposal (Scudder, 1994; Scudder, 1996) relies on the existence of non-Maxwellian distribution functions in the corona – as one moves higher in the corona, the hot non-Maxwellian tails of the distribution function encounter fewer Coulomb collisions than do the particles in the cool, dense, Maxwellian core of the

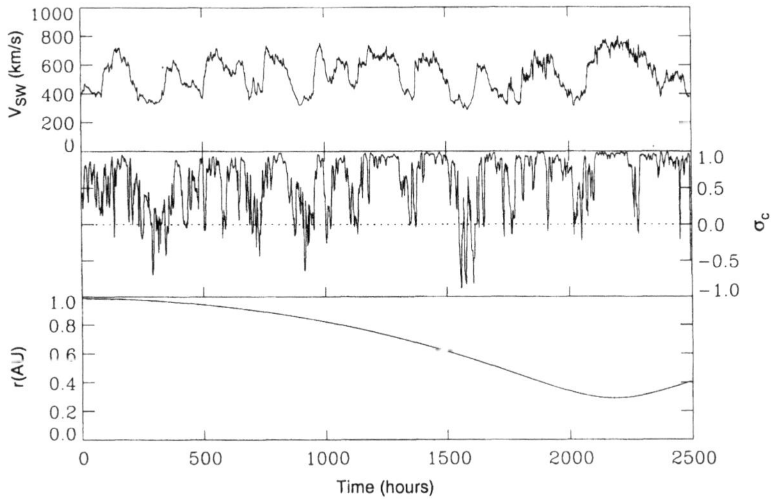

Figure 10. Large-scale features of about 100 days of hour-averaged data from the primary mission of Helios 2 (early 1976). From top to bottom the panels are: (1) the speed of the solar wind, showing recurrent high-speed streams; (2) σ_c; (3) the distance of the spacecraft from the Sun. (Adapted from Goldstein *et al.* (1995).)

distribution, so that distributions become systematically 'hotter.' Determining how the corona and solar wind are heated and accelerated continues to be a challenging unsolved problem.

3. Issues Surrounding the Nature of Solutions

3.1. Are there complex problems that have been solved?

The previous sections I included discussion of a few major unsolved problems in space plasma physics. There are, of course, many others, but there is one issue that eventually needs to be considered as solutions to problems are proposed, *viz.*, how do we decided when a problem is solved, and are there instances when a single observation or set of observations can be used as a counterexample to disprove conclusively a theory? This section illustrates the issue with one complex problem that does appear to have been solved. At least one observation, in my opinion, disproves several alternative theories that have been published.

3.1.1. *Why Alfvén Waves Disappear in the Outer Heliosphere*

As discussed in Subsection 2.1.2, when the solar wind and interplanetary Alfvén wave were first discovered, two competing hypotheses arose as to their origin and role. During the past decade it has become clear that both of those early suggestions were, in part, correct: outward propagating Alfvén waves are indicative of

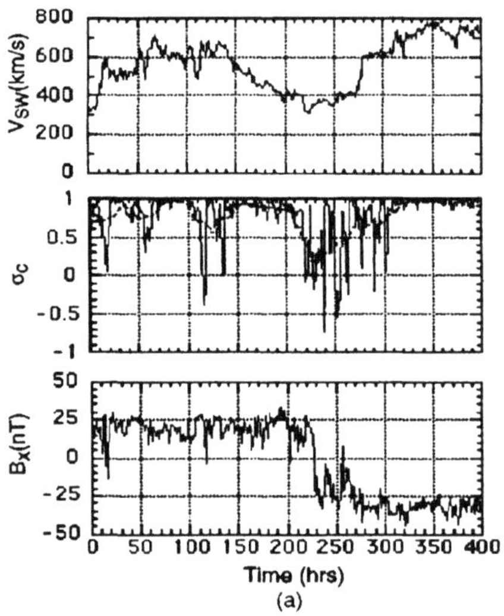

Figure 11. Hourly averaged Helios 2 data obtained near 0.3 AU (days 93-110 of 1976). Plotted are V_{SW}, σ_c at the scale of about 3 hours (solid line) and a 25-hour running mean (dashed line), together with the radial component ('B_x') of the magnetic field. (Adapted from Roberts *et al.* (1992).)

turbulence in the solar corona and the interaction of fast and slow wind does stir the solar wind and cause the turbulence to evolve. The role of velocity shear is illustrated in Figure 10. During periods of strong variation in solar wind speed, σ_c, is small, while near 0.3 AU and in regions of relatively constant wind speed, σ_c is close to $+1$, indicating that nearly pure Alfvénic fluctuations are propagating outward from the Sun until and unless stream shear stirs the medium.

The roles of velocity shear and the heliospheric current sheet are illustrated in Figure 11, which shows seven days of one-hour-averaged Helios data obtained near 0.3 AU. A decrease in σ_c as the wind speed decreases and as the spacecraft approaches the heliospheric current is apparent. This behavior has been modeled in both two- and three-dimensional simulations of the compressible and incompressible MHD equations (Roberts *et al.*, 1992; Roberts and Ghosh, 1999).

Perhaps even more telling are observations made by Voyager 1 near 8 AU (Figure **??**), which show that in the absence of velocity shear Alfvénic fluctuations evolve very slowly and remain outward propagating with high cross helicity ($\approx$ 0.8) out to large heliospheric distances. This time interval was modeled by Whang and Burlaga (1985) who tracked this parcel of plasma from its initial observation by Helios 1 at 0.3 AU early in 1980, to its observation at 8 AU by Voyager 1 in the middle of that year. Whang and Burlaga determined that the region labeled days 23 – 30 in Figure 12 had not interacted with any neighboring material, at least along the radial direction. Thus, velocity shear and proximity to the heliospheric current sheet are critical in the evolution of Alfvénic fluctuations, yet papers continue to get published that use linear theory and ignore stream shear to

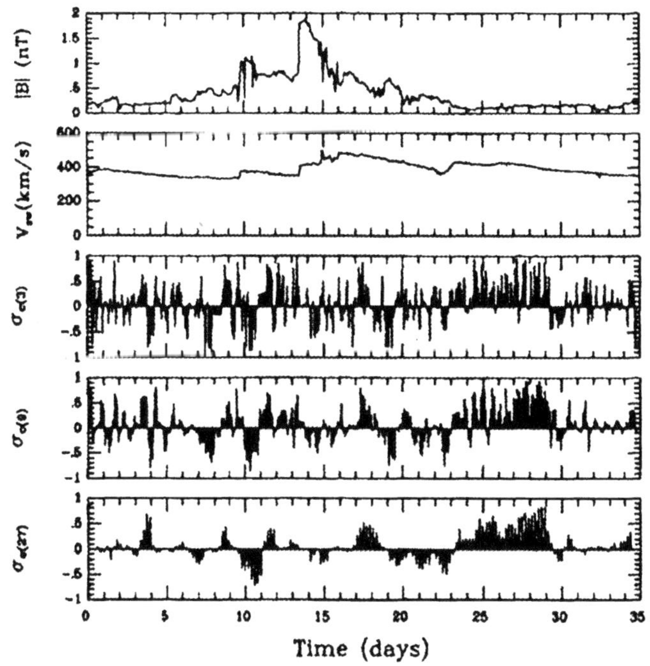

Figure 12. Hour-averaged Voyager 1 data gathered near 8 AU in mid-1980, along with various derived quantities. From top to bottom: (1) $|B|$; (2) V_{SW}; (3) σ_c at the 3-hour scale; (4) σ_c at the 9-hour scale; (5) σ_c at the 27-hour scale. (Adapted from Roberts *et al.* (1987)).

calculate significant evolution of Alfvén waves. such analyses ignore observations showing that without velocity shear Alfvén waves evolve only slowly out to 8 AU. Consequently, such theories are *a priori* wrong.

4. Other Major Unsolved Problems

This short review does not permit an exhaustive discussion of all major unsolved problems in space plasma physics. Because most readers will have their own favorites, I have consciously selected only a few examples to discuss in some depth. These are admittedly areas in which I have worked, but hopefully that has added some depth, and maybe controversy to the discussion. I have not mentioned, except in passing, a few of the most significant unsolved problems in this field. That is not because I think that they are any less important than the ones mentioned above, but limited space and expertise constrain the discussion to a few examples.

One problem at the top of most lists is solar flares and coronal mass ejections (CMEs). Although we may understand how most of the radiation is generated and how particles are energized, fundamental questions remain. With the launch of SOHO, CMEs are certainly better understood but exactly what their geometry is and how they exit from the corona and evolve into the solar wind are still subjects of intense research. Many classes of solar radio emission are well understood, but fundamental questions remain about type II bursts: Where in relation to the collisionless shock wave are the electrons accelerated? Where are the fundamental and second harmonic radiation generated? Certainly no list of unsolved problems would be complete, notwithstanding the graduate student quoted in subsection 1.2, without including magnetospheric substorms.

Acknowledgements

I would like to acknowledge the hospitality of the Institute for Theoretical Physics at the University of California, Santa Barbara, where a portion of this manuscript was written. The visit to the ITP was supported, in part, by the U. S. National Science Foundation under Grant No. PHY94-07194, and, in part, by the NASA Goddard Space Flight Center. I would like to thank the organizers for their support in making this paper possible and the reviewer for the constructive suggestions.

References

Alfvén, H.: 1950, *Cosmical Electrodynamics*. Oxford: Clarendon Press.

Batchelor, G.K.: 1970, *Theory of Homogeneous Turbulence*. New York: Cambridge University Press.

Belcher, J.W. and Davis, L.: 1971, Large-amplitude Alfvén waves in the interplanetary medium, 2. *J. Geophys. Res.* **76**, 3534.

Bieber, J.W., Wanner, W. and Matthaeus, W.H.: 1996, Dominant two-dimensional solar wind turbulence with implications for cosmic ray transport, *J. Geophys. Res.* **101**(A2), 2511–2522.

Birn, J., Hesse, M. and Schindler, K.: 1998, Formation of thin current sheets in space plasmas, *J. Geophys. Res.* **103**(A4), 6843–6852.

Biskamp, D.: 1982, Effect of secondary tearing instability on the coalescence of magnetic islands, *Phys. Lett. A* **87**, 357.

Biskamp, D., Schwarz, E. and Drake, J.F.: 1995, Ion-controlled collisionless magnetic reconnection, *Phys. Rev. Lett.* **75**(21), 3850–3853.

Biskamp, D., Schwarz, E. and Drake, J.F.: 1997, Two-fluid theory of collisionless magnetic reconnection, *Phys. Plasmas* **4**(4), 1002–1009.

Biskamp, D. and H. Welter: 1989, Dynamics of decaying two-dimensional magnetohydrodynamic turbulence, *Phys. Fluids B* **1**, 1964.

Carbone, V., Malara, F. and Veltri, P.: 1995, A model for the 3-D magnetic field correlation spectra of low frequency solar wind fluctuations during Alfvénic periods, *J. Geophys. Res.* **100**, 1763.

Coleman, P.J.: 1966, Hydromagnetic waves in the interplanetary medium, *Phys. Rev. Lett.* **17**, 207.

Coleman, P.J.: 1968, Turbulence, viscosity, and dissipation in the solar wind plasma, *Astrophys. J.* **153**, 371.

Feffer, P.T., Lin, R.P., SlassiSennou, S., McBride, S., Primbsch, J.H., Zimmer, G., Pelling, R.M., Pehl, R., Madden, N., Malone, D., Cork, C., Luke, P., Vedrenne, G. and Cotin, F.: 1997, Solar energetic ion and electron limits from HIREGS observations, *Solar Physics* **171**(2), 419–445.

Galsgaard, K. and Nordlund, A.: 1997a, Heating and activity of the solar corona .2. Kink instability in a flux tube, *J. Geophys. Res.* **102**(A1), 219–230.

Galsgaard, K. and Nordlund, A.: 1997b, Heating and activity of the solar corona .3. Dynamics of a low beta plasma with three-dimensional null points, *J. Geophys. Res.* **102**(A1), 231–248.
Ghosh, S. and Goldstein, M.L.: 1997, Anisotropy in Hall MHD turbulence due to a mean magnetic field, *J. Plasma Phys.* **57**, 129–154.
Ghosh, S., Siregar, E., Roberts, D.A. and Goldstein, M.L.: 1996, Simulation of high-frequency solar wind power spectra using Hall magnetohydrodynamics, *J. Geophys. Res.* **101**(A2), 2493–2504.
Goldstein, M.L., Roberts, D.A. and Matthaeus, W.H.: 1995, Magnetohydrodynamic turbulence in the solar wind, *Ann. Rev. Astron. and Astrophys.* **33**, 283.
Golub, L., Maxson, C., Rosner, R., Serio, S. and Vaiana, G.S.: 1980, Magnetic-fields and coronal heating, *Astrophys. J.* **238**(1), 343–348.
Hesse, M. and Winske, D.: 1998, Electron dissipation in collisionless magnetic reconnection, *J. Geophys. Res.* **103**(A11), 26479–26486.
Hesse, M., Winske, D. and Birn, J.: 1998, On the ion-scale structure of thin current sheets in the magnetotail, *Physica Scripta* **T74**, 63–66.
Heyvaerts, J. and Priest, E.R.: 1983, Coronal heating by phase-mixed shear Alfvén waves, *Astron. Astrophys.* **117**, 220.
Huba, J.D.: 1994, Hall dynamics of the Kelvin-Helmholtz instability, *Phys. Rev. Lett.* **72**(13), 2033–2036.
Huba, J.D.: 1995, Hall magnetohydrodynamics in space and laboratory plasmas, *Phys. Plasmas* **2**(6), 2504–2513.
Huba, J.D.: 1996a, Finite Larmor radius magnetohydrodynamics of the Rayleigh-Taylor instability, *Phys. Plasmas* **3**(7), 2523–2532.
Huba, J.D.: 1996b, The Kelvin-Helmholtz instability: Finite Larmor radius magnetohydrodynamics, *Geophys. Res. Lett.* **23**(21), 2907–2910.
Jokipii, J.R.: 1971, Propagation of cosmic rays in the solar wind, *Rev. Geophys. Space Phys.* **9**, 27.
Kolmogorov, A.N.: 1941, The local structure of turbulence in incompressible viscous fluid for very large Reynolds numbers, *Dokl. Akad. Nauk SSSR* **30**, 299.
Krucker, S., Larson, D.E., Lin, R.P. and Thompson, B.J.: 1999, On the origin of impulsive electron events observed at 1 AU, *Astrophys. J.* **519**(2), 864–875.
Kuperus, M., Ionson, J.A. and Spicer, D.S.: 1981, On the theory of coronal heating mechanisms, *Ann. Rev. of Astron. Astrophys.* **19**, 7–40.
Lin, R.P., Hurley, K.C., Smith, D.M. and Pelling, R.M.: 1991, A search for hard X-Ray microflares near solar minimum, *Solar Physics* **135**(1), 57–64.
Marsch, E., Goertz, C.K. and Richter, K.: 1982, Wave heating and acceleration of solar-wind ions by cyclotron-resonance, *J. Geophys. Res.* **87**(A7), 5030–5044.
Marsch, E. and Tu, C.-Y.: 1990, Spectral and spatial evolution of compressive turbulence in the inner solar wind, *J. Geophys. Res.* **95**, 11,945.
Marsch, E. and Tu, C.-Y.: 1997, The effects of high-frequency Alfvén waves on coronal heating and solar wind acceleration, *Astron. and Astrophys.* **319**(3), L17–L20.
Matthaeus, W.H. and Brown, M.R.: 1988, Nearly incompressible magnetohydrodynamics at low Mach number, *Phys. Fluids* **31**, 3634.
Matthaeus, W. and Goldstein, M.: 1982, Measurement of the rugged invariants of magnetohydrodynamic turbulence, *J. Geophys. Res.* **87**, 6011.
Matthaeus, W.H., Goldstein, M.L. and Roberts, D.A.: 1990, Evidence for the presence of quasi-two-dimensional, nearly incompressible fluctuations in the solar wind, *J. Geophys. Res.* **95**, 20,673.
Matthaeus, W.H., Gray, P.C., Pontius, J.D.H. and Bieber, J.W.: 1995, Spatial structure and field-line diffusion in transverse magnetic turbulence, *Phys. Rev. Lett.* **75**, 2136.
Matthaeus, W.H., Klein, L.W., Ghosh, S. and Brown, M.R.: 1991, Nearly incompressible magnetohydrodynamics, pseudosound, and solar wind fluctuations, *J. Geophys. Res.* **96**, 5421.

McKenzie, J.F., Axford, W.I. and B.M.: 1997, The fast solar wind, *Geophys. Res. Lett.* **24**(22), 2877–2880.

McKenzie, J.F., Banaszkiewicz, M. and Axford, W.I.: 1995, Acceleration of the high speed solar-wind, *Astron. and Astrophys.* **303**, L45.

Montgomery, D., Brown, M. and Matthaeus, W.H.: 1987, Density fluctuation spectra in magneto-hydrodynamic turbulence, *J. Geophys. Res.* **92**, 282.

Nordlund, K. and Galsgaard, K.: 1997, Topologically forced reconnection, in: G. Simnett, C.E. Alissandrakis and L. Vlahos (eds.), *Proc. 8th European Solar Physics Meeting*, Vol. 489 of *Lecture Notes in Physics*. Heidelberg: Springer, pp. 179–200.

Ofman, L. and Davila, J.M.: 1995, Alfvén wave heating of coronal holes and the relation to the high-speed solar wind, *J. Geophys. Res.* **100**(A12), 23,413–23,426.

Parker, E.N.: 1958, Dynamics of the interplanetary gas and magnetic fields, *Astrophys. J.* **128**, 664.

Parker, E.N.: 1972, Topological dissipation and the small-scale fields in turbulent gases, *The Astrophys. J.* **174**, 499.

Parker, E.N.: 1983a, Absence of equilibrium among close-packed twisted flux tubes, *Geophys. and Astrophys. Fluid Dynamics* **23**(2), 85–102.

Parker, E.N.: 1983b, Magnetic neutral sheets in evolving fields .1. General-theory, *Astrophys. J.* **264**(2), 635–641.

Parker, E.N.: 1983c, Magnetic neutral sheets in evolving fields .2. Formation of the solar corona, *Astrophys. J.* **264**(2), 642–647.

Passot, T. and Pouquet, A.: 1987, Numerical simulation of compressible homogeneous flows in the turbulent regime, *J. Fluid. Mech* **181**, 441.

Passot, T. and Pouquet, A.: 1988, Hyperviscosity for compressible flows using spectral methods, *J. Comp. Phys.* **75**, 300.

Reiner, M.J., Fainberg, J. and Stone, R.: 1995, Large-scale interplanetary magnetic field configuration revealed by solar radio bursts, *Science* **270**, 461–464.

Roberts, D.A.: 1989, Interplanetary observational constraints on Alfvén wave acceleration of the solar wind, *J. Geophys. Res.* **94**, 6,899.

Roberts, D.A.: 1996, The role of waves, turbulence, and structures in heating and accelerating the solar wind, invited paper, in: S. Habbal (ed.), *Scientific Basis for Robotic Exploration Close to the Sun*. Marlboro, MA, p. 185.

Roberts, D.A. and Ghosh, S.: 1999, A kinematic analysis of the role of velocity shear in expanding plasmas, *J. Geophys. Res.* **104**(A10), 22395–22399.

Roberts, D.A., Goldstein, M.L., Klein, L.W. and Matthaeus, W.H.: 1987, Origin and evolution of fluctuations in the solar wind: Helios observations and Helios-Voyager comparisons, *J. Geophys. Res.* **92**, 12,023.

Roberts, D.A., Goldstein, M.L., Matthaeus, W.H. and Ghosh, S.: 1992, Velocity shear generation of solar wind turbulence, *J. Geophys. Res.* **97**, 17,115.

Ryutova, M., Shine, R., Title, A. and Sakai, J.I.: 1997, A possible mechanism for the origin of emerging flux in the sunspot moat, *Astrophys. and Space Sci.* **492**, 402–414.

Scudder, J.D.: 1994, Ion and electron suprathermal tail strengths in the transition region: support for the velocity filtration model of the corona, *Astrophys. J.* **427**, 446.

Scudder, J.D.: 1996, Electron and ion temperature gradients and suprathermal tail strengths at Parker's solar wind sonic critical point, *J. Geophys. Res.* **101**(A5), 11,039–11,054.

Shay, M.A., Drake, J.F., Denton, R.E. and Biskamp, D.: 1998, Structure of the dissipation region during collisionless magnetic reconnection, *J. Geophys. Res.* **103**(A5), 9165–9176.

Shay, M.A., Drake, J.F., Rogers, B.N. and Denton, R.E.: 1999, The scaling of collisionless, magnetic reconnection for large systems, *Geophys. Res. Lett.* **26**(14), 2163–2166.

Siregar, E., Ghosh, S. and Goldstein, M.L.: 1995, Nonlinear entropy production operators for magnetohydrodynamic plasmas, *Phys. Plasmas* **2(5)**, 1480.

Siregar, E., Viñas, A.F. and Goldstein, M.L.: 1998a, Coarse graining and nonlocal processes in proton cyclotron resonant interactions, *Phys. Plasmas* **5**, 333.

Siregar, E., Viñas, A.F. and Goldstein, M.L.: 1998b, Topological invariants in cyclotron turbulence, in: T. Chang and J.R. Jasperse (eds.), *The 1998 Conference on Multi-Scale Phenomena in Space Plasmas II*, **15**. Cascais, Portugal, p. 307.

Smith, E.J., Balogh, A., Neugebauer, M. and McComas, D.: 1995, Ulysses observations of Alfvén waves in the southern and northern solar hemispheres, *Geophys. Res. Lett.* **22**(23), 3381–3384.

Stein, R.F.: 1981, Stellar chromospheric and coronal heating by magneto-hydrodynamic waves, *Astrophys. J.* **246**(3), 966–971.

Tarbell, T., Ryutova, M. and Covington, J.: 1999, Heating and jet formation by hydrodynamic cumulation in the solar atmosphere, *Astrophys. J.* **514**(March 20), L47–L51.

Tsurutani, B.: 1991, Comets: A laboratory for plasma waves and instabilities, in: A.D. Johnstone (ed.), *Cometary Plasma Processes*, Geophys. Monogr. Ser. Washington, DC: AGU, p. 364,

Tsurutani, B.T., Richardson, I.G., Thorne, R.M., Butler, W., Smith, E.J., Cowley, S.W.H., Gary, S.P., Akasofu, S.I. and Zwickl, R.D.: 1985, Observations of the right-hand resonant ion-beam instability in the distant plasma sheet boundary-layer, *J. Geophys. Res.* **90**(A12), 2159–2172.

Unti, T.W. and Neugebauer, M.: 1968, Alfvén waves in the solar wind, *Phys. Fluids* **11**, 563.

Usmanov, A.V., Goldstein, M.L., Besser, B.P. and Fritzer, J.M.: 2000, A global model of the solar wind with WKB Alfvén Waves: Comparison with Ulysses data, *J. Geophysical Res.* **105**, 12,675.

van Ballegooijen, A.A.: 1985, Electric currents in the solar corona and the existence of magnetostatic equilibrium, *Astrophys. J.* **298**(1), 421–430.

van Ballegooijen, A.A.: 1986, Cascade of magnetic energy as a mechanism of coronal heating, *Astrophys. J.* **311**(2), 1001–1014.

Viñas, A.F., Goldstein, M.L. and Acuña, M.H.: 1984, Spectral analysis of magnetohydrodynamic fluctuations near interplanetary shocks, *J. Geophys. Res.* **89**(A6), 3762.

Viñas, A.F., Wong, H.K. and Klimas, A.J.: 2000, Generation of electron suprathermal tails in the upper solar atmosphere: Implications for coronal heating, *Astrophys. J.* **528**, 528–523.

Wentzel, D.G.: 1974, Coronal heating by Alfvén waves, *Solar Physics* **39**(1), 129–140.

Wentzel, D.G.: 1976, Coronal heating by Alfvén waves .2, *Solar Physics* **50**(2), 343–360.

Whang, Y.C. and Burlaga, L.F.: 1985, Evolution and interaction of interplanetary shocks, *J. Geophys. Res.* **90**, 10765.

Zurbuchen, T.H., Hefti, S., Fisk, L.A., Gloeckler, G. and Schwadron, N.A.: 2000, Magnetic structure of the slow solar wind: Constraints from composition data, *J. Geophys. Res.* **105**, 18,327.

THE FUTURE OF SPACE SCIENCE IN THE 21ST CENTURY

R.M. BONNET
European Space Agency, 8–10 Rue Mario Nikis, 75738 Paris Cedex 15

Abstract. Space Science helped the start of the open space race after the launch of Sputnik-1 in 1957. Conversely, the use of space vehicles during the cold war allowed the scientists to conduct many observations and make discoveries which have dramatically changed our views of our own Solar System and of the Universe. What will be the future of this activity in the next century, with the disappearance of the cold war justification and in the context of shrinking budgets? Is there a future for space exploration? For what benefit and how will space science programmes be conducted? Who will be the main players? Are there limits to our ability to explore? The pioneers of space research in the post- Sputnik-1 era, like J-L. Steinberg, had both an easier and a more difficult time than space scientists of today. Nevertheless, space science will only survive in the next century if it succeeds in reaching the deep interest and motivation of society at large.

1. Introduction

Space Science in its experimental phase is big science like particle physics. Hence it requires substantial amounts of money and large organisations (big institutions – agencies – governments). Because it is big, it is also relatively costly and bound by big infrastructure elements. Hence, its future is relatively easy to predict since, at least over a period of 20 years, the necessary large investments induce stability.

Space Science was born during the cold war, simply because the cold war created the means which made it possible. But, today, the cold war is over and as a consequence, budgets are more difficult to negotiate. The engine which made it possible came to rest, the motivation, a result of both competition and scientific challenge, has partly faded away from the deciding bodies. Today, Space Science must be justified on its own merits.

What are they? I can quote at least three of them:

- Fundamental advances in science and knowledge
- Technological push and industrial competitiveness
- Prospects for education and formation of new generations of scientists, engineers and managers.

Are they different, however, from other areas of science or big science, such as particle physics, ground-based astronomy, biology or information technologies? Probably not!

Because of their cost, their visibility, especially in the case of failure, Space Science missions will only survive in the next century if a new social contract is established through which society in general sees a benefit in its existence/permanence.

Astrophysics and Space Science **277:** 371–378, 2001.

Although we may regret it, it is a real fact of life, and one which will probably dominate our ways of working in the next century, that science can no longer work autonomously. Interaction with the public is more and more required. 'Science must leave its ivory tower and enter the agora' (Gibbons, 1999).

Nevertheless, space missions do have a unique asset: they have the ability to physically explore new regions of space and also of time. Spacecraft, or spaceships, are enabling human beings to physically cross new territories or volumes of space and to constantly push back the distances from Earth, beyond which no human built hardware has ever travelled. This exploratory characteristic plays on people's imagination and carries with it a strong element of dream and intellectual evasion to which no-one can be insensitive. This asset should be fully exploited.

Indeed, without changing at all its scientific content, the ESA Science Programme had to be directed more towards stimulating, if that is possible, the imagination of politicians and budget officials! I will now describe this new way of presenting that programme.

2. The New Horizons 2000 Programme

ESA's Horizons 2000 programme contains a set of ongoing and planned missions which span a period of more than 15 years. Until recently, the programme was presented discipline by discipline and mission by mission, objective by objective. This worked well, was effective in getting more money for the programme because there was a long-term plan. When money started to shrink however, the allocated budget became incompatible with the schedules, and flexibility started to disappear. The objectives of the programme were still valid but the concept of missions had to be changed in order to become cheaper if not necessarily smaller.

In the new approach, the Cornerstones of the programme are no longer presented as single projects but rather like science themes which should avoid the potential criticism of 'social isolation', in other words, they must become more 'sellable'. The four main themes of the ESA Science Programme address four broad questions of important scientific significance and scope while also responding to the interrogations of the public at large as they are perceived today.

(i) What is the origin of the Universe? How did it evolve and what are the fundamental physical laws and principles which govern it?

(ii) What is the origin and evolution of stars and planetary systems and how many stellar systems have developed or will develop detectable terrestrial planets?

(iii) Can we understand better the formation and future evolution of planets in our own Solar System and how life developed in it?

(iv) How does our own star work so that we can forecast its behaviour and how does it influence our planet and our living conditions?

An implicit advantage of this presentation is that these themes are sufficiently broad to accommodate a large number of missions developed not only by ESA but also by other agencies in Europe or abroad. This should, therefore, make it easier to trigger international cooperation and coordination between several programmes. This is very important because, in the 21st century, unless the same conditions which characterised the cold war prevail again, international cooperation and sharing of the main science objectives between various agencies' programmes will become and should become more and more the rule. In fact, the themes of Horizons 2000 as presented here are not very far from those of NASA's Origins Program and coordination between ESA and NASA should become easier in the future.

3. Physics of Space

Let us now concentrate more specifically on the area of interest to this audience and to the scientific topic which Jean-Louis Steinberg has pursued so remarkably during his career, and which is called here the Physics of Space.

Indeed, advertising this area of science to the public is more arduous and requires more imagination than, for example the search for life on Mars or Europa. Nevertheless, I claim that in spite of this difficulty the same effort should be made to 'socialise' this field.

Whenever you travel in space, performing *in situ* measurements means that you perform space physics. Magnetic fields with intensities varying by more than 15 orders of magnitude pervade stellar coronae and winds, galaxies, the Universe, and the surroundings of neutron stars and black holes. Studying the physics of space is like working in a laboratory where the laboratory is the Universe itself. Certainly, one can predict that this field of research will develop but the directions in which these developments will materialise are extremely hazardous to predict at this time.

What are the missions relevant to this area of Science already in the Horizons 2000 Programme?

3.1. The Near-Term Programme

The near-term programme involves two missions already in orbit and a third one ready for launch:

- Ulysses (web address: http://helio.estec.esa.nl/ulysses), launched in 1990, was discussed in the course of this symposium, and a very clearly oriented heliospheric mission, unique in that it explores the third dimension of the Solar System up to a distance of 5 AU;
- SOHO (web address: http://sohowww.estec.esa.nl/), launched in 1995, is in full operation after its heroic recovery from a spectacular loss in 1998;
- CLUSTER (web address: http://sci.esa.int/cluster), ready for another launch next summer.

SOHO, launched a little more than four years ago, is gradually responding to all the questions it was supposed to answer and clearly deals with the way our star works and how its light and its magnetic field are generated, how and why it becomes magnetically active, with a rough periodicity of 11 years, why and how its corona is so hot and how predictable are the effects of solar eruptions and CMEs and of the Solar Wind. These questions are of fundamental interest to us human beings who live on a planet which is constantly under the influence of the Sun. This explains why the SOHO results are regularly reported in the press as they progressively push back the limits of our knowledge and open the way to forecasting the bursts of activity of our star.

CLUSTER constitutes with SOHO the first Cornerstone of Horizons 2000: the Solar Terrestrial Science Programme (STSP). SOHO and CLUSTER were supposed to fly at the same time because they were essentially studying the Sun-Earth connection, while addressing the same fundamental plasma physics and magnetic phenomena, both in the solar atmosphere and in the Earth's magnetosphere.

CLUSTER was launched in June 1996. Unfortunately, the mission was lost due to the accident of the Ariane 5 launcher during the maiden flight of that new rocket. A tremendous effort was made by both ESA and the scientists involved in the mission to redevelop the four Cluster satellites, which are again ready for launch in June and July of this year from Baïkonour on board two commercially purchased Soyuz launchers. Since SOHO is performing so well, assuming that the four CLUSTER satellites operate according to expectations, the objectives of the STSP programme should be fully restored and implemented during solar maximum.

Of course all deep space missions such as Ulysses, Galileo and Cassini naturally analyse the physics of space. But defining today the future missions dedicated to that area of science after CLUSTER is not totally without risk: the risk that the objectives defined today might be obsolete when the results of CLUSTER arrive.

This is to a large extent the reason why the Survey Committee which advised ESA back in 1994 in defining its future programme faced with difficulties when formulating precisely what follow-on missions should be considered in this field of science on the eve of the SOHO and CLUSTER launches.

3.2. A MISSION TO MERCURY

Nevertheless, they managed to select a Cornerstone mission dedicated both to the study of Mercury, the planet closest to the Sun, and to plasma physics.

The Mercury mission (web address: http://sci.esa.int/home/bepicolombo/index.cfm), now renamed BepiColombo for obvious reasons, is presently under study at Alenia and will be the subject of a public presentation next April at the EGS meeting in Nice, in view of its confirmation as Cornerstone 5 of the Horizons 2000 Programme in the fall of 2000. It consists of an orbiter, a subsatellite and a small lander. The subsatellite is especially dedicated to the study of particles and fields in the vicinity of Mercury.

The modularity of the mission makes it more feasible to implement and to cope in particular with the existence of other Mercury missions planned by other agencies. BepiColombo when confirmed by ESA's Space Science Advisory Committee in September 2000 as Cornerstone 5 will fly in 2009, using Solar Electric Propulsion.

3.3. Space physics missions

At the end of September 1999, ESA released a Call for new mission proposals in the Flexible Mission category, defined as costing ESA's science budget no more than 175 Meuro. Space physics missions are well suited to small or medium size spacecraft. But it is hard to know what these missions will be.

Certainly on the eve of the launch of NASA's IMAGE mission and of the Thermosphere Ionosphere Mesosphere Energetic and Dynamics mission in May, it is difficult to predict at this stage, and clearly too early to give any indications here.

3.4. Solar physics missions

Although not materialised by a Cornerstone mission in Horizons 2000, Solar Physics is considered as a scientific priority in ESA's Space Science Programme. In response to the Call for new proposals, flexible missions in this field might be proposed. They could fly as early as 2010.

However, similar difficulties (abundance of missions and forthcoming new results) confront the Solar Physics community, with SOHO still performing quasi nominally as well as TRACE (NASA) and Yohkoh (Japan).

For nearly 15 years the community has requested ESA to study two new original ways of looking at the Sun.

The first one consists in observing the Sun with two satellites simultaneously. Indeed, long ago, stereoscopic observations were pioneered by Jean-Louis Steinberg. Ironically, as is often the case, the scarcity of budgets in Europe would not permit such a mission realistically before the end of the next decade. Luckily, NASA has decided to include a STEREO mission in its programme and I am very satisfied that European scientists have been accepted by NASA to fly on that spacecraft. If the habit of giving the names of famous scientists to missions successfully operating in orbit is maintained, the present name of STEREO could easily be changed into the name of the 'Steinberg Observatory'.

The second way of studying the Sun is by sending a probe into its outer layers. The Solar Probe has also long been studied by NASA and by ESA, and I remember very well the active involvement of Jean-Louis Steinberg already in the seventies, as he was campaigning, enthusiastically as usual, for the Probe, and how deeply involved he was in its definition.

The European Solar Physics community has also been involved more recently, following the 1994 Survey Committee recommendations, in the study of a mission orbiting close to the Sun, if possible at very high latitudes. SOLO, as it is called, has been studied by ESA at phase A level. This study identifies what are the technical and conceptual limits of the mission, which has to remain within a fixed-capped and realistic budget. In spite of these constraints, the mission could carry a scientific payload including *in situ* fields and particles measuring devices, in addition to spectroscopic and imaging instruments, as well as fundamental physics experiments.

3.5. The Case of Cheap Missions

One thing can be taken for granted: the cheaper the missions, the easier it will be to consider them in future plans. But can all missions be cheap?

Space Physics missions certainly do possess a strong advantage as compared with astronomy missions for example: because they focus on very sharply defined scientific objectives, they fit small and cheap satellites more easily. However, as recently illustrated by the relatively large number of failures in the 'Faster Better and Cheaper' category of NASA missions, the overall efficiency of this approach must be questioned.

Obviously, the laws of physics make it impossible to foresee that all missions in the next century will be small and cheap. One thing is clear however, future missions should always, without exception, be better, because there is no interest in undertaking second class science or in rediscovering the wheel and their managers should take no risk in compromising on the quality of the hardware and the control of procedures.

So far, my prospective journey into the next century has not gone very deep in time, but it is not easy to outline a very long perspective, especially in this domain whose evolution is to a large extent determined by the forthcoming results of the missions already in orbit or in the planning stage.

4. International Cooperation

As for all domains of big science, Space Science is an area where international cooperation might bring important benefits. International cooperation works with maximum efficiency when there is mutual interest, complementarity of objectives and mutual respect among the different partners. However, in the 21st Century, when the money available to space science might, on the best assumptions, be capped at a constant level, international cooperation must be planned sufficiently ahead of time. I raise the concern here whether the Faster-Better-Cheaper approach is compatible with the concept of international cooperation, only because it takes time to plan ahead jointly and also to properly implement sub-parts of small missions between various partners. Nevertheless, the coordination of future plans of the different agencies early enough, as well as the definition of roadmaps among the different space partners in respect of the strategic objectives of each, is a first step and a prerequisite.

Only in this way will the overall amount of space science money available around the world be efficiently used, allowing future missions to be undertaken successfully, at a lower cost for each partner.

As an illustration, ESA and NASA have recently decided to join forces in the study of three very challenging and exciting missions:

- The Next Generation Space Telescope (NGST) the successor of Hubble;
- LISA, a laser interferometer to detect low frequency gravitational waves;
- IRSI(ESA)-TPF(NASA) a near infrared interferometer to detect and make pictures of terrestrial planets around other stars.

This is clearly a way forward to the future and a means to embrace the fantastic challenge which we face in pursuing our quest of the unknown and in exploring the hidden and yet invisible Universe.

5. Jean-Louis STEINBERG

International cooperation is an area which Jean-Louis Steinberg pioneered. Indeed, the missions which he undertook in the sixties and seventies, in particular with the Soviets (at a time when such cooperation was not so easy), were relatively cheap but there was a strong desire to cooperate and to share the scientific goals. At the same time, the missions envisaged by Jean-Louis Steinberg were very innovative. In addition, they have also been successful. They have been successful because J-L. Steinberg never abandoned and never betrayed the fundamental principles which govern the management of space missions, relying strictly and seriously on the respect of quality assurance principles.

When he started his activities in space science in the early sixties, Jean-Louis Steinberg, modest as usual, claimed that he knew nothing of the management of space experiments. In this way he went to visit the various space groups already in existence in France at that time, just to learn. And he learnt well, as can be judged from his record of success.

One of the ways to success for those who may wish to undertake a career in the next century in Space Science is to follow the lesson of Jean-Louis Steinberg: pioneering new territories of science, while at the same time refusing to ignore the lessons of the past. These are the best ingredients for success.

6. Conclusion

It is time to conclude.

I have touched upon several generalities and described some of ESA's near term plans as well. In no way can I envisage the space science missions of the next century without the respect of the scientific social-managerial and international cooperation aspects which I mentioned.

Throughout my presentation I have tried to illustrate the key and exemplary role and behaviour of Jean-Louis Steinberg in his career. He was a pioneer of Space Science and he maintained intact that pioneer spirit throughout his life. He is an example to follow.

I am convinced that there will be no future for Space Science in the next century without a pioneering spirit and a pioneering approach whereby failure is not an option, where the aim is to push back the limits of discovery for the benefit not only of scientists but also of the general public knowledge.

References

Gibbons, M.: 1999, *Nature* **402** (6761), C81.

Some participants of the ‘Rencontres de l’Observatoire’.

LIST OF PARTICIPANTS

AELLIG, Matthias R., MIT, Cambridge, USA
mra@space.mit.edu

ALISSANDRAKIS, Costas, University of Ioannina, Greece
calissan@cc.uoi.gr

ANDERSON, Roger R., The University of Iowa, Iowa City, USA
roger-r-anderson@uiowa.edu

ARANEDA, Jaime A., Universidad de Concepción, Chile
jaraneda@udec.cl

AUBIER, Monique, Observatoire de Paris, Meudon, France
Monique.Aubier@obspm.fr

AXFORD, Ian, Max-Planck-Institut für Aeronomie, Katlenburg-Lindau, Germany
axford@linmpi.mpg.de

BAGENAL, Fran, LASP-University of Colorado, Boulder, USA
bagenal@colorado.edu

BASTIAN, Timothy S., Nat'l Radio Astron. Obs., Charlottesville, USA
tbastian@nrao.edu

BATANOV, Oleg, IKI RAN, Moscow, Russia
obat@romance.iki.rssi.ru

BELENKAYA, Elena S., Institute of Nuclear Physics, Moscow, Russia
elena@dec1.npi.msu.su

BERTHELIER, J.J., CETP, Saint-Maur, France
Jean-Jacques.Berthelier@cetp.ipsl.fr

BESKIN, Vassili, Lebedev Physical Institute, Moscow, Russia
beskin@lpi.ru

BOCKELÉE-MORVAN, Dominique, Observatoire de Paris, Meudon, France
domi@obspm.fr

BONNET, Roger-Maurice, European Space Agency (ESA), Paris, France
c/o: dbauer@hq.esa.fr

BOSQUED, Jean-M., CESR/CNRS, Toulouse, France
bosqued@cesr.fr

BOUGERET, Jean-Louis, Observatoire de Paris, Meudon, France
bougeret@obspm.fr

BOUQUET, Serge, CEA/DIF, Bruyères-le-Châtel, France
bouquets@bruyeres.cea.fr

Astrophysics and Space Science **277:** 381–387, 2001.

CAROUBALOS, Constantin, Université d'Athènes, Grèce
kkarou@di.uoa.gr

CASSE, Fabien, Observatoire de Grenoble, France
Fabien.Casse@obs.ujf-grenoble.fr

CERISIER, Jean-Claude, CETP, Saint-Maur, France
cerisier@cetp.ipsl.fr

CHERNOV, Guennadi, IZMIRAN, Troitsk, Russia
gchernov@izmiran.troitsk.ru

CHIUDERI, Claudio, Universitá di Firenze, Florence, Italy
chiuderi@arcetri.astro.it

CHIUDERI DRAGO, Franca, Universitá di Firenze, Florence, Italy
fdrago@arcetri.astro.it

COWLEY, Stanley W.H., University of Leicester, United Kingdom
swhc1@ion.le.ac.uk

CORNILLEAU-WEHRLIN, Nicole, CETP/UVSQ/CNRS, Vélizy, France
nicole.cornilleau@cetp.ipsl.fr

CRANMER, Steven R., Smithsonian Astrophysical Observatory, Cambridge, USA
scranmer@cfa.harvard.edu

CROVISIER, Jacques, Observatoire de Paris, Meudon, France
crovisie@obspm.fr

DALIN, Peter A., Space Research Institute, Moscow, Russia
pdalin@iki.rssi.ru

DAVYDENKO, Stanislav S., Institute of Applied Physics, Nizhny Novgorod, Russia
davyd@appl.sci-nnov.ru

DELCOURT, Dominique, CETP-CNRS, Saint-Maur des Fossés, France
delcourt@cetp.ipsl.fr

DENISSE, Jean-François, Observatoire de Paris, France

DUBIN, Daniel H.E., University of California, San Diego, USA
dhdubin@ucsd.edu

DUBOULOZ, Nicolas, CETP/CNRS/UVSQ, Saint-Maur, France
Nicolas.DUBOULOZ@cetp.ipsl.fr

DULK, George A., Observatoire de Paris, Meudon, France
DULK@obspm.fr

DUMIN, Yurii, IZMIRAN, Troitsk, Russia
dumin@cityline.ru

ENCRENAZ, Thérèse, Observatoire de Paris, Meudon, France
therese.encrenaz@obspm.fr

FEAUTRIER, Nicole, Observatoire de Paris, Meudon, France
Nicole.Feautrier@obspm.fr

FEDOU, Pierre, Observatoire de Meudon, France
Pierre.Fedou@obspm.fr

FERRIÈRE, Katia M., Observatoire Midi-Pyrénées, Toulouse, France
ferriere@obs-mip.fr

FIALA, Vladimir, Institute of Atmospheric Physics, Prague, Czech Republic
fiala@ufa.cas.cz

FONTAINE, Dominique, CETP – CNRS, Vélizy, France
fontaine@cetp.ipsl.fr

FORME, François, CETP, Vélizy, France
francois.forme@cetp.ipsl.fr

GALOPEAU, Patrick H.M., CETP, Vélizy, France
Patrick.Galopeau@cetp.ipsl.fr

GEISWILLER, Johann, L.P.C.E. – C.N.R.S., Orléans, France
geiswill@cnrs-orleans.fr

GOLDSTEIN, Melvyn L., NASA/Goddard Space Flight Center, Greenbelt, USA
melvyn.goldstein@gsfc.nasa.gov

GOMBEROFF, Luis, Universidad de Chile, Santiago, Chile
lgombero@abello.seci.uchile.cl

GRAPPIN, Roland, Observatoire de Paris, Meudon, France
Roland.Grappin@obspm.fr

GRECO, Antonella, Universitá della Calabria, Arcavacata di Rende, Italia
greco@fis.unical.it

HAMELIN, Michel, CETP-IPSL, Saint-Maur, France
hamelin@cetp.ipsl.fr

HENRY, Florence, Observatoire de Paris, Meudon, France
Florence.Henry@obspm.fr

HOANG, Sang, Observatoire de Paris, Meudon, France
Sang.Hoang@obspm.fr

HUBERT, Daniel, Observatoire de Paris, Meudon, France
daniel.hubert@obspm.fr

ILLIANO, Jean-Marie, CETP/CNRS, Saint-Maur, France
Jean-Marie.ILLIANO@cetp.ipsl.fr

ISSAUTIER, Karine, Observatoire de Paris, Meudon, France
karine.issautier@obspm.fr

JASENSKY, Vaclav, Institut of Atmospheric Physics, Prague, Czech Republic
vjas@alenka.ufa.cas.cz

JOHNSON, Robert, University of Virginia, Charlottesville, USA
rej@virginia.edu

JOKIPII, J.R., University of Arizona, Tucson, USA
jokipii@lpl.arizona.edu

KELLOGG, Paul J., University of Minnesota, Minneapolis, USA
Kellogg@waves.space.umn.edu

KIVELSON, Margaret G., IGPP-UCLA, Los Angeles, USA
mkivelson@igpp.ucla.edu

KLEIN, Karl-Ludwig, Observatoire de Paris, Meudon, France
Ludwig.klein@obspm.fr

KRASNOPOLSKY, Vladimir A., The Catholic University of America, Washington D.C., USA
vladimir.krasnopolsky@gte.net

LACOMBE, Catherine, Observatoire de Paris, Meudon, France
catherine.lacombe@obspm.fr

LALLEMENT, Rosine, CNRS Service d'Aéronomie, Verrières-le-Buisson, France
rosine.lallement@aerov.jussieu.fr

LANDI, Simone, Observatoire de Paris, Meudon, France
landi@despa.obspm.fr

LANTOS, Pierre, Observatoire de Paris, Meudon, France
pierre.lantos@obspm.fr

LAVAL, Guy, CNRS CPHT Ecole Polytechnique, Palaiseau, France
laval@cpht.polytechnique.fr

LE QUÉAU, Dominique, Centre d'Etude Spatiale des Rayonnements (CESR), Toulouse, France
lequeau@cesr.fr

LEBLANC, François, University of Virginia, Charlottesville, USA
fl3x@virginia.edu

LEBLANC, Yolande, Observatoire de Paris, Meudon, France
leblanc@obspm.fr

LECACHEUX, Alain, Observatoire de Paris, Meudon, France
alx@obspm.fr

LELLOUCH, Emmanuel, Observatoire de Paris, Meudon, France
lellouch@obspm.fr

LEMAIRE, Joseph, Institut d'Aéronomie Spatiale de Belgique (IASB/BIRA), Bruxelles, Belgium
jl@oma.be

MAKSIMOVIC, Milan, Observatoire de Paris, Meudon, France
maksimovic@obspm.fr

MANGENEY, André, Observatoire de Paris, Meudon, France
mangeney@megasv.obspm.fr

MAREEV, Eugene A., Institute of Applied Physics RAS, Nizhny Novgorod, Russia
mareev@appl.sci-nnov.ru

MARQUÉ, Christophe, Observatoire de Paris, Meudon, France
christophe.marque@obspm.fr

MARSDEN, Richard G., ESA/ESTEC, Noordwijk, The Netherlands
rmarsden@estec.esa.nl

MASNOU, Jean-Louis, Observatoire de Bordeaux, Floirac, France
masnou@observ.u-bordeaux.fr masnou@obspm.fr

MEYER, Nicole, Observatoire de Paris, Meudon, France
nicole.meyer@obspm.fr

MONCUQUET, Michel, Observatoire de Paris, Meudon, France
moncuquet@obspm.fr

MOTTEZ, Fabrice, CETP/CNRS, Vélizy, France
fabrice.mottez@cetp.ipsl.fr

NAZAROV, Vladimir, IKI RAN, Moscow, Russia
vnazarov@romance.iki.rssi.ru

NEUBAUER, Fritz M., Universität Köln, Kóln, Germany
neubauer@geo.Uni-Koeln.DE

OMURA, Yoshiharu, RASC, Kyoto University, Kyoto, Japan
omura@kurasc.kyoto-u.ac.jp

PANTELLINI, Filippo, Observatoire de Paris, Meudon, France
Filippo.Pantellini@obspm.fr

PARKER, Eugene N., University of Chicago, USA
parker@odysseus.uchicago.edu

PELLETIER, Guy, Observatoire de Grenoble, France
Guy.Pelletier@obs.ujf-grenoble.fr

PERCHE, Claude, Observatoire de Paris, Meudon, France
Claude.Perche@obspm.fr

PICK, Monique, Observatoire de Paris, Meudon, France
pick@obspm.fr

PIERRARD, Viviane, Institut d'Aéronomie Spatiale de Belgique (IASB/BIRA), Bruxelles, Belgique
viviane@oma.be

PINÇON, Jean-Louis, LPCE/CNRS, Orléans, France
jlpincon@cnrs-orleans.fr

PODESTA, John J., New York University, Glen Ridge, USA
j.podesta@ieee.org

POEDTS, Stefaan, Centre for Plasma Astrophysics, K.U.Leuven, Heverlee, Belgium
Stefaan.Poedts@wis.kuleuven.ac.be

PONOMARJOV, Maxim, State Academy of Aviation Technology, Tutaev, Russia
pmg@univ.uniyar.ac.ru

POTTELETTE, Raymond, CETP, Saint-Maur des Fossés, France
Raymond.POTTELETTE@cetp.ipsl.fr

PRANGÉ, Renée, CNRS Institut d'Astrophysique Spatiale, Orsay, France
prange@ias.fr

PULINETS, Sergey A., IZMIRAN, Troitsk, Russia
pulse@izmiran.rssi.ru

RAUER, Heike, DLR, Berlin, Germany
heike.rauer@dlr.de

REME, Henri, Centre d'Etude Spatiale des Rayonnements (CESR), Toulouse, France
henri.reme@cesr.fr

ROSENBERG, Marlene, University of California, USA
rosenber@ece.ucsd.edu

ROSTAS, François, Observatoire de Paris, Meudon, France
Francois.Rostas@obspm.fr

RUCKER, Helmut O., Space Research Institute, Graz, Austria
helmut.rucker@oeaw.ac.at

RYUTOVA, Margarita P., Lawrence Livermore National Laboratory/IGPP, Livermore, USA
ryutova@beowulf.llnl.gov

SAHAL-BRÈCHOT, Sylvie, Observatoire de Paris, Meudon, France
Sylvie.Sahal-Brechot@obspm.fr

SALEM, Chadi, Observatoire de Paris, Meudon, France
Chadi.Salem@obspm.fr

SAMSONOV, Andrei, St. Petersburg State University, Russia
samsonov@geo.phys.spbu.ru

SCHMIEDER, Brigitte, Observatoire de Paris, Meudon, France
schmieder@obspm.fr

SCHWENN, Rainer, Max-Planck-Institut für Aeronomie, Katlenburg-Lindau, Germany
schwenn@linmpi.mpg.de

SMETS, Roch, CETP / UVSQ / IPSL, Vélizy, France
rsm@cetp.ipsl.fr

STEINBERG, Jean-Louis, Observatoire de Paris, Meudon, France
Steinberg@obspm.fr

STROBEL, Darrell F., Johns Hopkins University, Baltimore, USA
strobel@jhu.edu

TAKTAKISHVILI, Alexander, Abastumani Observatory, Tbilisi, Georgia
staktak@caucasus.net

TIMOFEEV, Vladislav, Institute of Cosmophysical Research & Aeronomy, Yakutsk, Russia
v.e.timofeev@sci.yakutia.ru

TOKAREV, Yuri M., RRI (NIRFI), Nizhny Novgorod, Russia
yt@nirfi.sci-nnov.ru

TREUMANN, Rudolf A., Max-Planck-Institute for extraterrestrial Physics, Garching, Germany
tre@mpe.mpg.de

TRICOTTET, Matthieu, CEA, France

TROTIGNON, Jean-Gabriel, LPCE/CNRS, Orléans, France
jgtrotig@cnrs-orleans.fr

VELLI, Marco, Universitá di Firenze, Firenze, Italia
velli@arcetri.astro.it

VESELOVSKY, Igor, Institute of Nuclear Physics, Moscow, Russia
veselov@dec1.npi.msu.su

VILMER, Nicole, Observatoire de Paris, Meudon, France
vilmer@obspm.fr

VONTRAT, Aurélie, CETP, Vélizy, France
aurelie.vontrat@cetp.ipsl.fr

WAITE, J. Hunter, Southwest Research Institute, San Antonio, France
hunter@swri.edu

WLERICK, Gérard, Observatoire de Paris, Meudon, France
Gerard.Wlerick@obspm.fr

ZABOTIN, Nikolay A., Rostov State University, Rostov-on-Don, Russia
zabotin@ip.rsu.ru

ZARKA, Philippe, Observatoire de Paris / CNRS, Meudon, France
Philippe.Zarka@obspm.fr

ZIMBARDO, Gaetano, Universitá della Calabria, Arcavacata di Rende, Italia
zimbardo@fis.unical.it

MIX
Papier aus verantwortungsvollen Quellen
Paper from responsible sources
FSC® C105338

If you have any concerns about our products,
you can contact us on
ProductSafety@springernature.com

In case Publisher is established outside the EU,
the EU authorized representative is:
Springer Nature Customer Service Center GmbH
Europaplatz 3, 69115 Heidelberg, Germany

Printed by Libri Plureos GmbH
in Hamburg, Germany